PHYSICS WITH AN ELECTRON POLARIZED LIGHT-ION COLLIDER

Related Titles from AIP Conference Proceedings

581 Physics of, and Science with, the X-Ray Free-Electron Laser: 19th Advanced ICFA Beam Dynamics Workshop
Edited by C. Pellegini, S. Chattopadhyay, M. Cornacchia, and I. Lindau, August 2001, 0-7354-0022-9

578 Physics and Experiments with Future Linear e^+e^- Colliders: LCWS 2000
Edited by Adam Para and H. Eugene Fisk, July 2001, 0-7354-17-2

576 Application of Accelerators in Research and Industry: Sixteenth International Conf.
Edited by J. L. Duggan and I. L. Morgan, July 2001, 0-7354-0015-6

572 Electron Beam Ion Sources and Traps and Their Applications: 8th International Symp.
Edited by Krsto Prelec, June 2001, 0-7354-0011-3

561 Tours Symposium on Nuclear Physics IV: Tours 2000
Edited by M. Arnould, M. Lewitowicz, Yu. Ts. Oganessian, H. Akimune, M. Ohta, H. Utsunomiya, T. Wada, and T. Yamagata, April 2001, 1-56396-996-3

549 Intersections of Particle and Nuclear Physics: 7th Conference, CIPANP2000
Edited by Zohreh Parsa and William J. Marciano, December 2000, 1-56396-978-5

546 Beam Instrumentation Workshop 2000: Ninth Workshop
Edited by Kenneth D. Jacobs and R. Coles Sibley III, December 2000, 1-56396-975-0

542 Physics Potential and Development of Muon Colliders and Neutrino Factories: Fifth International Conference
Edited by David B. Cline, November 2000, 1-56396-970-X

518 Proton Emitting Nuclei: PROCON'99—First International Symposium
Edited by Jon C. Batchelder, May 2000, 1-56396-937-8

508 Hadron Physics: Effective Theories of Low Energy QCD
Edited by A. H. Blin, B. Hiller, M. C. Ruivo, C. A. Sousa, and E. van Beveren, March 2000, 1-56396-927-0

495 Experimental Nuclear Physics in Europe: ENPE 99, Facing the Next Millennium
Edited by Berta Rubio, Manuel Lozano, and William Gelletly, November 1999, 1-56396-907-6

481 Nuclear Structure 98
Edited by C. Baktash, September 1999, 1-56396-858-4

459 Heavy Quarks at Fixed Target
Edited by Harry W. K. Cheung and Joel N. Butler, February 1999, 1-56396-864-9

To learn more about these titles, or the AIP Conference Proceedings Series, please visit the webpage **http://www.aip.org/catalog/aboutconf.html**

PHYSICS WITH AN ELECTRON POLARIZED LIGHT-ION COLLIDER

Second Workshop
EPIC 2000

Cambridge, Massachusetts   14–15 September 2000

EDITOR
Richard G. Milner
MIT-Bates Linear Accelerator Center
Middleton, Massachusetts

Melville, New York, 2001
AIP CONFERENCE PROCEEDINGS ■ VOLUME 588

Editor:

Richard G. Milner
MIT-Bates Linear Accelerator Center
21 Manning Avenue
Middleton, MA 01949
USA

E-mail: milner@mitlns.mit.edu

The article on pp. 234-239 was authored by U.S. Government employees and is not covered by the below mentioned copyright.

Authorization to photocopy items for internal or personal use, beyond the free copying permitted under the 1978 U.S. Copyright Law (see statement below), is granted by the American Institute of Physics for users registered with the Copyright Clearance Center (CCC) Transactional Reporting Service, provided that the base fee of $18.00 per copy is paid directly to CCC, 222 Rosewood Drive, Danvers, MA 01923. For those organizations that have been granted a photocopy license by CCC, a separate system of payment has been arranged. The fee code for users of the Transactional Reporting Service is: 0-7354-0028-8/01/$18.00.

© 2001 American Institute of Physics

Individual readers of this volume and nonprofit libraries, acting for them, are permitted to make fair use of the material in it, such as copying an article for use in teaching or research. Permission is granted to quote from this volume in scientific work with the customary acknowledgment of the source. To reprint a figure, table, or other excerpt requires the consent of one of the original authors and notification to AIP. Republication or systematic or multiple reproduction of any material in this volume is permitted only under license from AIP. Address inquiries to Office of Rights and Permissions, Suite 1NO1, 2 Huntington Quadrangle, Melville, N.Y. 11747-4502; phone: 516-576-2268; fax: 516-576-2450; e-mail: rights@aip.org.

L.C. Catalog Card No. 2001094615
ISBN 0-7354-0028-8
ISSN 0094-243X
Printed in the United States of America

Contents

Preface ix
Sponsors and Scientific Organizing Committee xi

Hadronic Physics with a Polarized Electron-Ion Collider 1
    J. M. Cameron, J. T. Londergan, and R. G. Milner
Learning from QCD 13
    F. Wilczek
Nucleon Spin Physics: Experimental Status and Potential of
HERMES, COMPASS, and RHIC 34
    A. Miller
Overview of Skewed Parton Distributions 49
    A. Schäfer
The Spin Structure of the Nucleon: Theoretical Overview 54
    R. L. Jaffe
Current Fragmentation in the Semi-inclusive Leptoproduction 75
    P. J. Mulders
Probing "Generalized Parton Distributions" with JLab at 12 GeV 89
    S. Stepanyan
e-A Physics at a Collider 104
    G. T. Garvey
Deeply Inelastic Scattering off Nuclei at RHIC 121
    R. Venugopalan
Luminosity Limitations for Electron-Ion Collider 142
    V. A. Lebedev
Overview of European Plans for Future Lepton/Photon
Scattering Facilities 155
    D. von Harrach

SPIN AND FLAVOR STRUCTURE OF THE NUCLEON

Semi-inclusive Studies of the Quark Spin and Flavor Structure
of the Proton 171
    U. Stösslein and E. R. Kinney
Establishing Evidences of Factorization in Semi-inclusive
Electron Scattering 176
    X. Jiang
(Vector) Meson Production and Duality 182
    R. Ent
Generalized Parton Distributions and Deep Virtual
Compton Scattering 187
    D. Hasell, R. Milner, and K. Takase

SEMI-EXCLUSIVE PROCESSES

Generalized Parton Distributions and the Dependence of Parton
Distributions on the Impact Parameter........................199
    M. Burkardt
An Energy Recovery Electron Linac-on-Ring Collider..................204
    L. Merminga, G. A. Krafft, V. A. Lebedev, and I. Ben-Zvi
Wide Angle Compton Scattering 229
    R. Jakob
Structure of the Goldstone Bosons...........................234
    R. J. Holt and P. E. Reimer
Calculation of Fragmentation Functions in Two-Hadron
Semi-inclusive Processes...................................240
    A. Bianconi, S. Boffi, D. Boer, R. Jakob, and M. Radici
TESLA-N: Electron Scattering with Polarized Targets at TESLA 245
    V. Korotkov (for the TESLA-N Study Group)
Hadronic Wave Functions and Exclusive Processes..................250
    E. Stein
Λ and $\bar{\Lambda}$ Polarization as a Measurement of Distribution and
Fragmentation Functions..................................255
    M. Boglione, M. Anselmino, and F. Murgia
Single Spin Asymmetries in Semi-inclusive Electroproduction:
Access to Transversity 260
    K. A. Oganessyan, N. Bianchi, E. De Sanctis, and W.-D. Nowak

HEAVY QUARKS/TARGET FRAGMENTATION

Quark-Hadron Duality in Electron Scattering......................267
    W. Melnitchouk
Λ Polarization in the Target Fragmentation Region of SIDIS...............272
    A. Kotzinian
Polarized Heavy Flavor Production in Next-to-Leading Order QCD..........277
    M. Stratmann
Soft Diffraction at EPIC Scales 282
    M. A. Pichowsky
The Strange Sea inside the Nucleon............................286
    V. Barone, C. Pascaud, and F. Zomer

e-A PHYSICS

QCD in eRHIC and EPIC...................................293
    J. Jalilian-Marian
QCD at High Parton Density 297
    Y. V. Kovchegov
QCD Instantons in the Soft Pomeron 302
    Y. V. Kovchegov

Diffractive Electroproduction of Vector Mesons from Light Nuclei 307
    L. L. Frankfurt, M. M. Sargsian, and M. I. Strikman
Physics of p-A Collisions and Its Relation to e-A Collisions 312
    J.-C. Peng

MACHINE

Conceptual Design Study of the Electron-Proton Storage Ring
Collider with Polarized Beams 319
    I. A. Koop, M. S. Lorestelev, I. N. Nesterenko, A. V. Otboev,
    V. V. Parkhomchuk, E. A. Perevedentsev, V. B. Reva, V. G. Sahmovsky,
    D. N. Satilov, P. Y. Shatunov, Y. M. Shatunov, A. N. Skrinsky, R. Milner,
    and C. Tschalaer
Electron and Positron Polarisation at HERA: Past and Future 338
    D. P. Barber (for the HERA Polarisation Group)
Polarized Electrons at Bates: Source to Storage Ring 343
    T. Zwart, E. Booth, F. Casagrande, K. Dow, M. Farkhondeh, W. Franklin,
    E. Ihloff, K. Jacobs, J. Matthews, R. Milner, T. Smith, C. Tschalaer,
    E. Tsentalovich, W. Turchinetz, and F. Wang
Proton and Electron Polarisation in Storage Rings:
Some Basic Concepts 350
    D. P. Barber, G. H. Hoffstätter, and M. Vogt
Siberian Snakes and Spin-Flipping in Storage Rings 355
    B. B. Blinov

Agenda 361
Schedule 363
List of Participants 367
Author Index 379

Preface

This volume contains the proceedings of the Second Workshop on Physics with an Electron Polarized light-Ion Collider (EPIC-00). It was held at the Cambridge campus of the Massachusetts Institute of Technology on September 14 – 16, 2000. The EPIC-00 Workshop was sponsored by Brookhaven National Laboratory, Indiana University Cyclotron Facility, and by the MIT-Bates Laboratory.

Over the last several years, physicists interested in understanding the structure of matter at the fundamental partonic (quark and gluon) level have come to realize that an electron-ion collider can address many of the outstanding questions in hadronic physics. This has led to meetings at GSI (1997), IUCF (EPIC-99, April 1999), Brookhaven (December 1999) and Yale (May 2000) which preceded this workshop. In Summer 2000, a new Long Range Planning Exercise was announced for Nuclear Physics in the United States and the proponents of an electron-ion collider came together to make the scientific case for this machine. Thus, the MIT meeting, which had originally been announced as a workshop focussed on the nucleon and light nuclei, was broadened to include presentations on the exciting possibilities with heavy nuclear beams. Further, at the MIT meeting the new machine was renamed the Electron-Ion Collider (EIC) and a steering committee was formed to guide the initiative.

I wish to thank the International Organizing Committee for their advise in establishing the program for this meeting. I also acknowledge the untiring efforts of Virginia Bullard (Bates) in both organizing the workshop and preparing this volume for publication. Finally, I thank the speakers for their excellent presentations and for submitting the written contributions which are collected in this proceedings.

Richard G. Milner
Middleton, MA
March, 2001

Second Workshop on Physics with an Electron Polarized light-Ion Collider

Sponsored by Brookhaven National Laboratory, Indiana University Cyclotron Facility and MIT-Bates Laboratory.

Scientific Organizing Committee
A. Bruell (MIT)
L. Bland (IUCF)
J. Cameron (IUCF) (Co-chair)
L. Cardman (Jefferson Lab)
A. Deshpande (Yale)
R. Holt (Illinois)
N. Isgur (Jefferson Lab)
R. Jaffe (MIT)
K. Jacobs (MIT-Bates)
E. Kinney (Colorado)
S.Y. Lee (IUCF)
T. Londergan (IUCF)
W. Lorenzon (Michigan)
R. McKeown (Caltech)
R. Milner (MIT-Bates) (Co-chair)
R. Redwine (MIT)
A. Schaefer (Regensburg)
C. Tschalaer (MIT-Bates)
M. Vetterli (TRIUMF)
S. Vigdor (IUCF)
W. Vogelsang (Stony Brook)
F. Wilczek (Institute of Advanced Study, Princeton)
S. Wissink (IUCF)

Hadronic Physics with a Polarized Electron-Ion Collider

J.M. Cameron[*], J.T. Londergan[*], and R.G. Milner[†]

[*]*Indiana University Nuclear Theory Center, 2401 Sampson Road, Bloomington, IN 47408, USA*
[†] *MIT-Bates Linear Accelerator Center, 21 Manning Road, P.O. Box 846, Middleton, MA 01949, USA*

Abstract. A high luminosity polarized Electron-Ion Collider (EIC) can provide precise and complete data essential to the ultimate understanding of the microscopic structure of matter. With a luminosity in excess of $10^{33} \cdot \frac{1}{A}$ cm^{-2} s^{-1} and a variable center-of-mass energy in the range of 30 to 100 GeV, EIC would be a powerful new microscope to probe the partonic structure of matter. The scientific highlights motivating the machine are summarized.

INTRODUCTION

It has now been over twenty-five years since the formulation of Quantum Chromodynamics (QCD), the theory which identifies colored quarks and gluons as the basic constituents of matter, and which provides an understanding of the strong interactions in terms of the basic forces between these nucleonic constituents. In the intervening decades, we have learned a great amount of information about the partonic (quark and gluon) structure of hadronic matter. However, our knowledge is still far from complete. Some crucial questions in this field remain open:

- What is the structure of hadrons in terms of their quark and gluon constituents?

- How do quarks and gluons evolve into hadrons?

- What is the role of quarks and gluons in the structure of atomic nuclei?

The answer to these questions is the missing key to our ultimate understanding of the microscopic structure of matter. A high luminosity electron-ion collider turning on at the end of this decade would be the ideal machine to address the above questions. The collider should have a high luminosity, greater than $10^{33} \cdot \frac{1}{A}$ cm^{-2} s^{-1}, and have a variable center-of-mass energy in the range of 30 to 100 GeV.

A "NEXT-GENERATION" FACILITY FOR PARTONIC PHYSICS

Measurements of deep inelastic scattering (DIS) carried out over the past 25 years have determined single-quark probability densities with great precision [1]. However, in inclusive DIS with charged lepton beams, one is unable to distinguish between quark and antiquark contributions to structure functions, and it is difficult to separate contributions from different quark flavors. Recently, semi-inclusive deep inelastic scattering has been used to differentiate between contributions from quarks and antiquarks, and to identify the quark flavors participating in partonic reactions. In the past year or two, there has been great interest in the physics insights which can be obtained from hard exclusive reactions; the proof of factorization theorems, and studies of generalized parton distributions, have made this an exciting area of research.

As a result, there is great interest in using semi-inclusive and exclusive processes to probe new features of the quark-parton structure of matter. The HERMES experiment at HERA used semi-inclusive measurements to determine the contributions of different quark flavors, and quark vs. antiquark effects, in the spin of the proton [2]. They have also obtained first measurements of exclusive (technically, "semi-exclusive") processes. Measurement of such processes is also a significant aspect of the proposed 12 GeV upgrade at Jefferson Laboratory [3].

These facilities, together with the upcoming COMPASS experiment at CERN, will provide a tantalizing first look at these reactions. However, experimental conditions at these facilities (see Fig. 1) are not optimally suited to studies of partonic physics. They typically operate at rather low Q^2 values, where higher-twist contributions could be substantial. They also have a limited range in both x and Q^2, which makes it difficult to observe the evolution of these distributions. As a consequence, it would appear that a dedicated facility, designed from the outset to probe the essential $x - Q^2$ region, and with the detection capabilities required to access this physics, should be the highest priority for a "next-generation" facility in the field of electromagnetic and hadronic physics.

What are the features of a facility necessary to address these issues? First, it must have sufficiently high energy that the cross-sections are dominated by the leading amplitudes. Next, it should have optimal control of spin-flavor degrees of freedom. It should allow full kinematic coverage; in particular, to observe both the fast current jets, and the slow target fragments. Furthermore, it should provide full azimuthal coverage, since azimuthal asymmetries highlight key aspects of quark/parton structure. In addition to continued inclusive measurements, such a facility will concentrate on both semi-inclusive and hard exclusive processes, which require both high luminosity and effective coincidence detection.

To address these issues, we propose an "*E*lectron-*I*on *C*ollider", or *EIC* facility. This would be an asymmetric collider, with an electron beam colliding with a beam of protons or ions. For the lightest ions (p, D and ^{3}He), both beams would

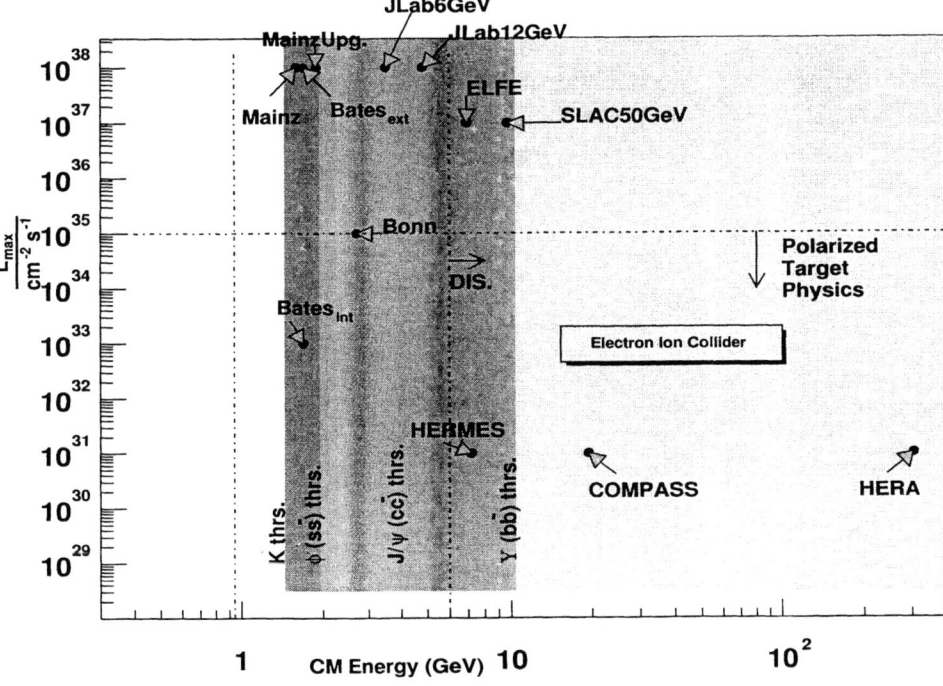

FIGURE 1. A plot of maximum luminosity vs. CM energy for a selection of existing and proposed facilities worldwide using lepton scattering. The shaded area denotes the proposed EIC parameters.

be polarized, in order to provide maximal control of the spin and flavor degrees of freedom. The collider CM energies would be substantially greater than at fixed target facilities such as Jefferson Lab or SLAC, and well below the high energy range of HERA. Such a facility would provide high luminosity (an $e-p$ luminosity of at least 10^{33} cm^{-2}s^{-1}) to enable the study of semi-inclusive and exclusive reactions. Variable CM energy will allow the study of hard exclusive processes, where detection of all final-state particles is easier at lower energies. For inclusive measurements and heavy-nucleus collisions higher energies, up to $\sqrt{s} \sim 100$ GeV, are desirable. The asymmetric collider geometry would allow both current fragment and target fragmentation events to be measured and kinematically separated. Fig. 1 shows the luminosity vs. CM energy for a series of existing and proposed electron facilities. To date, facilities cluster into high luminosity but relatively low CM energy machines, and high energy but low luminosity facilities. Clearly, the EIC collider would occupy a unique regime.

For partonic physics at moderate x, in semi-inclusive and exclusive measurements, a large solid-angle detector is required with excellent particle identification and tracking capabilities. The luminosity and particle ID requirements necessitate that the accelerator and detector design be closely correlated. We request an aggressive R&D effort over the next few years, in order to evaluate competing accelerator designs, and to demonstrate that the luminosity and detector requirements can be simultaneously satisfied.

The physics potential of such a collider has been reviewed in a series of conferences and workshops. The first of these were held in Germany beginning in 1997 [4]. These were followed by a series of workshops on electron colliders [5–7] held in the US. Obviously it is impossible in such a short document to review all physics topics; we refer to the workshop proceedings for a more complete list.

SEMI-INCLUSIVE MEASUREMENTS OF PARTON STRUCTURE

In semi-inclusive measurements induced by leptons, one measures both the scattered lepton, and a fast outgoing final hadron. The resulting cross section is given by

$$\sigma \sim \sum_q e_q^2 \, q(x) \, D_q^h(z)$$

In addition to the quark probability $q(x)$, the cross section contains the "fragmentation function" $D_q^h(z)$, which describes the probability that quark q will fragment to final hadron h with fractional energy z.

The EIC facility could probe parton spin distributions at an order of magnitude smaller x than can be measured at HERMES. Figure 2 shows the precision which could be obtained for spin probabilities for various quark flavors, with one month's running at the EIC projected luminosity [8]. One can obtain impressive statistical measurements, in a kinematic region clearly complementary to that probed at Hermes.

Another possibility is model-independent studies of neutron structure using the technique of *spectator tagging*. With a deuteron beam, one can measure protons in coincidence with the scattered electron in the process $e + D \to e' + p + X$. The spectator proton will move forward with momentum $k_p \sim k_D/2$. The neutron structure function can thus be measured on-shell, over the kinematic range $1 < Q^2 < 200$ GeV^2. With this technique, one can obtain accurate measurements of $d^p(x)/u^p(x)$ at large x. Since deuteron Fermi motion, binding and relativistic effects are all significant at large x, this quantity is surprisingly poorly known. In addition, with either polarized deuteron or ^{3}He beams, one can obtain accurate measurements of proton and neutron spin asymmetries A_1^N at large x, as measurements of this quantity appear to be crucial tests of QCD models. Spectator tagging could also be used to extend measurements of the neutron spin structure function $g_1^n(x)$ to

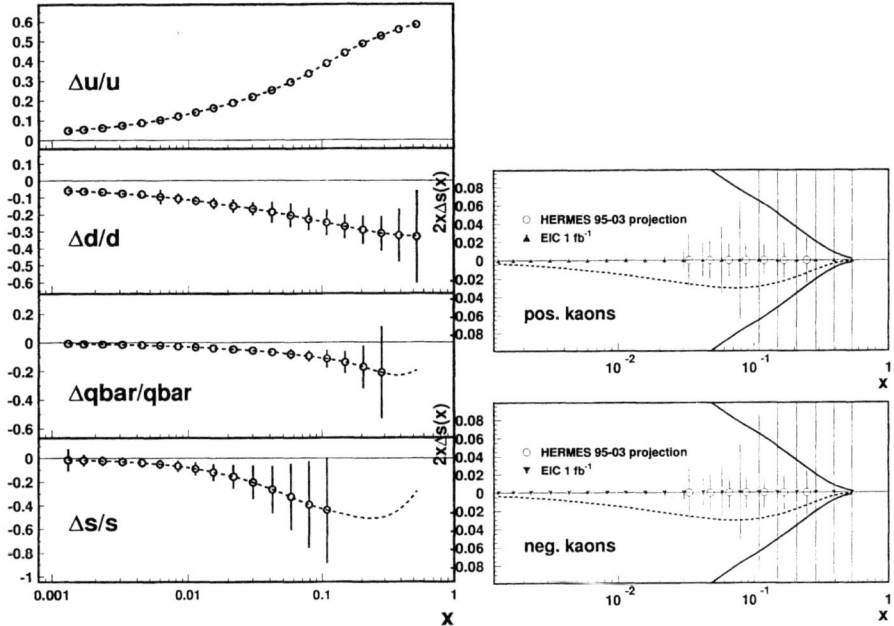

FIGURE 2. Left: statistical errors and range of x which could be covered by EIC collider for quark spin densities. Right: statistical accuracy which could be reached by EIC in one month of running at design luminosity, for strange quark spin distribution, if spin densities of non-strange quarks were accurately determined [8].

lower values of x. The additional information would be very useful in decreasing experimental errors in the determination of the Bjorken sum rule.

An electron collider would also allow precise measurements of pion and kaon structure functions. At present, pion valence quark distributions are reasonably well known in the region $x_\pi \geq 0.2$, but poorly known for smaller x. With the collider, we could obtain accurate values through the reaction $e + p \rightarrow e' + n + X$. Forward neutrons are detected in $e - p$ scattering. The photon scatters from a virtual π^+ which is very near its mass shell, and the resulting cross section is proportional to $F_\pi(x_\pi)$. Simulations by Holt and Reimer [9], in Fig. 3 show that F_π can be determined with excellent precision in the region $0.01 \leq x_\pi \leq 0.9$. At low x, one can answer the question, *"Are sea quarks and gluons as important for pions at low x as for the nucleon?"* One could then repeat these experiments with a nuclear target to determine *"How does F_π vary in the nuclear medium?"* Finally, one could obtain accurate measurements of the kaon structure functions using the reaction $e + p \rightarrow e' + \Lambda + X$. Only very sketchy information is known about the kaon structure function to date.

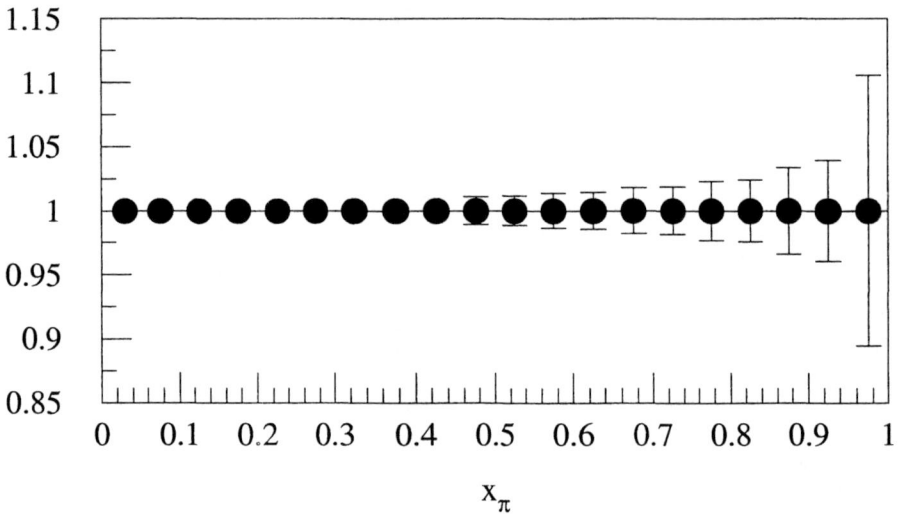

FIGURE 3. Expected precision in determination of the pion structure function F_π as a function of x_π, which could be obtained from less than one month's collider measurements of forward neutrons in $e-p$ collisions [9].

A polarized EIC facility would be able to make definitive studies of *transversity*, a third leading-twist structure function. The transversity $h_1(x)$ is proportional to $\delta q(x)$, the transverse spin difference for nucleons. One reason that transversity is interesting is that, unlike the longitudinal spin distribution g_1, gluons do not contribute to h_1. Consequently, h_1 should evolve much more like the valence quark polarization than the longitudinal spin structure function g_1, which is believed to have a very large contribution from polarized glue. The first moment of h_1 is equal to the nucleon's tensor charge, a quantity which may be calculable with lattice QCD.

As h_1 is a chiral-odd operator, it is inaccessible in inclusive DIS, but in semi-inclusive processes (e.g., $e + \vec{p} \rightarrow e' + \pi + X$) this effect could be observed by measuring the azimuthal asymmetry of leading pions. The asymmetry will be proportional to $A \sim h_1(x) H_1^\perp(z) \sin\phi$. Consequently, one measures the product of the transversity distribution times the so-called Collins fragmentation function $H_1^\perp$, another chiral-odd operator. With sufficiently precise data, one could separately determine the x-dependence of h_1 and the z-dependence of $H_1^\perp$, and hence determine h_1 to within an overall normalization. Transversity can be accessed by measuring single-spin asymmetries in semi-inclusive electroproduction on a transversely polarized target. Both HERMES and COMPASS will make exploratory

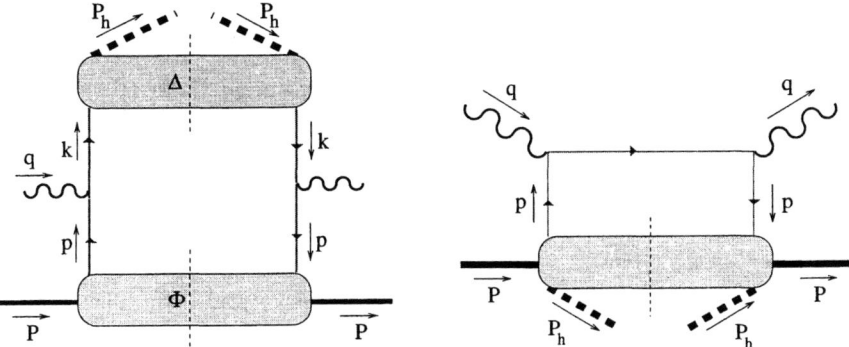

FIGURE 4. Leading contribution to fragmentation processes, in which quark q evolves to final hadron h. Left: current fragmentation; Right: target fragmentation, where the quark is struck by a virtual photon while baryon remnants fragment to final hadron.

measurements of this quantity; however the collider geometry is optimal for this type of study, since in a collider one can detect the full hadronic final state, both current and target fragmentation regimes are accessible, and one can measure complete angular distributions and hence extract moments of the parton distributions.

Parton Evolution into Hadrons: Fragmentation

An interesting partonic phenomenon which could be studied with a polarized electron-ion collider in the EIC energy regime is the *fragmentation* of quarks into hadrons. In such studies, a quark makes a transition into a final hadron which is detected. The ability of the collider to detect all final hadrons will allow detailed studies of the properties of fragmentation. The most commonly studied process is that of *current fragmentation*, shown schematically on the left in Fig. 4. In this process, the quark struck by the virtual photon decays to a final hadron h. There is also the *target fragmentation* process, where a quark is struck with a virtual photon in a lepton-induced reaction, and one observes the subsequent decay of the nucleon remnants. The kinematic situation for target fragmentation is shown schematically on the right-hand side of Fig. 4.

Target fragmentation is a largely unexplored regime of QCD. Observing such processes requires a detector capable of measuring decay fragments separated from the current jet by a large interval in rapidity. As a result, the collider geometry is probably essential for studies of the target fragmentation region. In Fig. 5 we show a plot by Mulders [10] of rapidity vs. fragmentation energy fraction z, for a $\gamma^* N$ invariant mass $W = (1-x)ys = 20$ GeV. Our experience from the EMC results suggests that both current and target fragmentation regions extend over a CM rapidity range $\Delta\eta \approx 2$, where the rapidity is defined by $\eta = 0.5\ln(P_h^-/P_h^+)$. Fig. 5 shows the z values above which current and target fragmentation could be

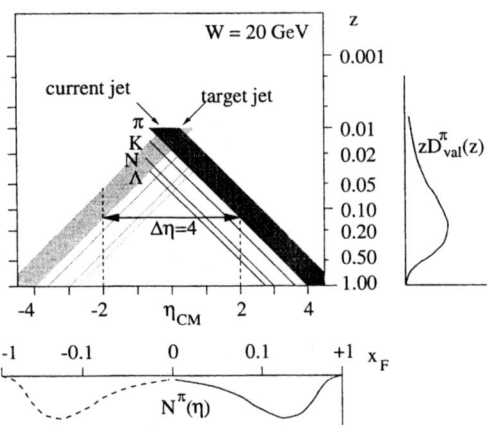

FIGURE 5. Relation between z-values in fragmentation and CM rapidity for $W = 20$ GeV. Roughly speaking, for regions separated by at least four units in CM rapidity, current and target fragmentation can be kinematically separated.

kinematically separated; for $W = 20$ GeV, these regions can be separated for all $z \geq 0.01$. Increases in W greatly lower those z-values, and demonstrates that a collider, with the properties which we have defined, has the capability of accessing and separating both the current and target fragmentation regimes.

There are two types of target fragmentation. In the first case, the momentum fraction x carried by the struck parton is small. The momentum fraction of the remnant, $1-x$, is large. In this case the subsequent decay is not correlated with the initial parton. Trentadue and Veneziano [11] described target fragmentation processes in terms of *fracture functions*. A fracture function represents the probability of finding a parton of a certain flavor i, together with a hadron h, in the target nucleon. One can subsequently determine how the fracture function evolves in Q^2, in analogous fashion to the DGLAP evolution equations for parton distributions. As the fracture functions are universal, they can be measured in other processes, for example diffraction. There exists some first experimental data for these processes from HERA. EIC could investigate this regime in detail. As this field is largely unexplored, the discovery potential is quite high. With both polarized leptons and hadrons, one could measure the extent to which the polarization of the initial state affects the quantum numbers of the final hadron h. In the current fragmentation regime, a quark which has participated in a hard scattering event evolves into a hadron. This region is "meson-rich," since the quark can become a meson by picking up a single antiquark. Naively, one expects the target fragmentation region to be relatively "baryon-rich", since after removal of a single quark the two remaining valence quarks can decay to a baryon by picking up a single quark.

There is another kinematic regime, as yet unexplored, for which an asymmetric electron-ion polarized collider at these energies would be a uniquely capable facility.

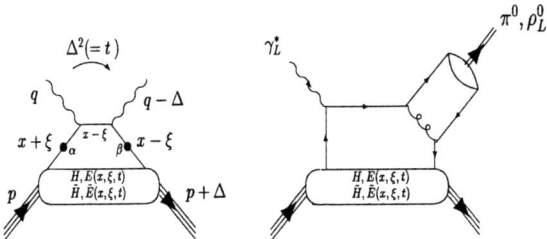

FIGURE 6. Leading order diagrams for DVCS (left) and for longitudinal electroproduction of mesons (right).

This is the regime of large x for the struck parton in target fragmentation (right side of Fig. 4). In this case the momentum fraction $1-x$ of the hadronic remnants is small. At higher-energy facilities like HERA, these slower-moving fragments cannot be analyzed as they proceed down the beam pipe. The subsequent remnant decay is correlated with the struck parton, and does not evolve with Q^2.

HARD EXCLUSIVE PROCESSES AT A COLLIDER

An asymmetric polarized beam collider provides the opportunity to study the correlations between partons in nucleons and nuclei. A particularly promising area is the study of hard exclusive processes. One of the leading terms in the amplitude for a hard exclusive process is shown schematically in Fig. 6. In this process, a virtual photon γ^* from the lepton produces either a real photon (Deeply Virtual Compton Scattering, or DVCS) or a meson (meson electroproduction). The amplitude depends on the quantity x, related to the momentum fraction carried by the parton, the "skewedness" ξ which measures the different momentum fractions carried by initial and final partons, and the four-momentum transfer $\Delta^2 = t$ at the upper vertex.

Amplitudes for this process depend upon four *generalized parton distributions* or GPD's, which are functions of the aforementioned three kinematic variables. The GPD's *interpolate* between the quantities measured in inclusive deep inelastic scattering, and the particle form factors traditionally measured in nuclear physics processes [12,13]. For example, in the forward direction, characterized by $\xi = \Delta^2 = 0$, two of the GPD's reduce to the helicity-averaged and helicity-dependent single-parton densities, the quantities measured in inclusive DIS reactions. The first moment of the GPD's with respect to x, at fixed four-momentum transfer, gives the nucleon form factors. Fig. 7 shows the kinematic range accessible for DVCS [14] by an $e-p$ collider with $\sqrt{s} = 30$ GeV, with the restriction that $|t| < 1$ GeV^2.

An asymmetric collider, which is capable of measuring all final hadrons in exclusive reactions, has the possibility of directly sampling effects of partonic correlations. One promising area would be to compare amplitudes for elastic processes,

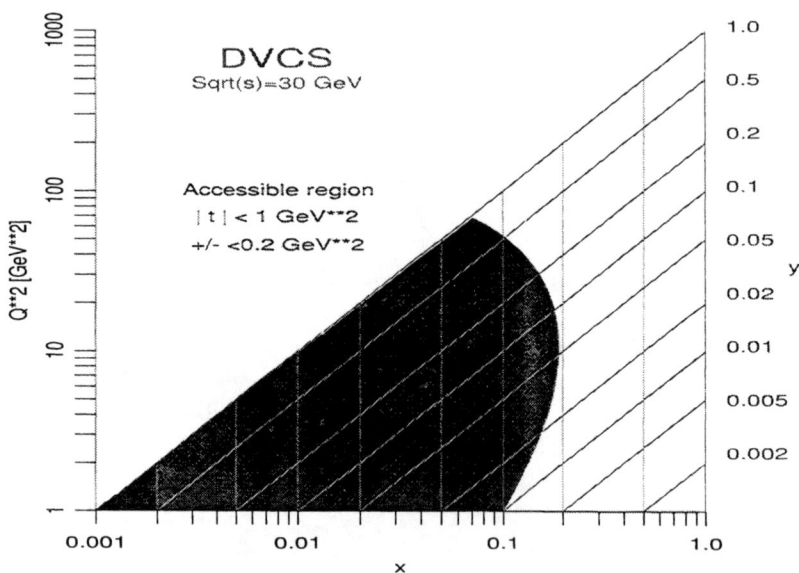

FIGURE 7. Expected kinematic range, x vs. Q^2, for DVCS processes in an $e-p$ collider with $\sqrt{s} = 30$ GeV.

where the struck nucleon remains in its ground state, with inelastic excitations resulting in a final N^*, Δ or hyperon. This would enable us to answer the question, *what is the partonic structure of light baryons?* For example, valence-quark models of baryons would suggest that diquark correlations are more important in octet than in decuplet baryons. The reasoning is that one expects a very strong attraction between quarks in the spin-isospin zero channel. In contrast, in chiral models which employ an expansion about the large-N_C limit of QCD, the octet and decuplet baryons are different rotational excitations of the same soliton, and one would not expect significant differences from diquark correlations.

The ability to control both lepton and hadron polarizations at the collider is an important element in the ability to extract information on partonic correlations. The ability to measure flavor effects in final state baryons and mesons is also important, since through the Pauli principle spin effects become correlated with flavor physics. In this regard Λ hyperon production can be extremely useful because of the self-analyzing nature of the Λ. Comparison of hyperon, octet and decuplet baryon production can answer the question, *Is SU(3) flavor symmetry valid for baryon structure?* Because u, d and s quarks are relatively light, and partly because of a paucity of experimental information on strange quark production, it is common to use SU(3)-based arguments to infer relate strange and non-strange hadronic processes. However, it is quite possible that large flavor-dependent effects may be found in parton distributions and fragmentation functions.

The area of GPD's is quite new. In 1997 Collins, Frankfurt and Strikman proved that under certain conditions factorization occurs in hard exclusive processes. Consequently such processes can be described as the product of a hard scattering term calculable from pQCD, times a soft amplitude (the GPD's) which are universal but not calculable from QCD. Both theoretical and experimental work in this area is only a few years old, but extremely rapid progress is being made in this field. Some preliminary experimental studies are being carried out at HERA and HERMES, and are proposed for Jefferson Lab at 12 GeV. In addition, theoretical studies are needed to determine which experiments most directly reveal the important physical information accessible with hard exclusive processes, e.g.: *How does the transverse momentum of partons influence the GPD amplitudes?*, and *Which observables are most sensitive to quark-quark correlations?*

HADRONIC PHYSICS WITH NUCLEI

With the addition of nuclear beams at a collider, there are a number of important questions which can be addressed:

- Measurement of the pion structure function (as described above) can also be considered for a pion in flight in a nucleus. This can answer the question, *"What role do pions play in nuclear binding?"*. Studies of nuclear effects on pion structure functions can also check measurements of nuclear anti-quark distributions, which found no enhancement over anti-quark probabilities in a nucleon [15].

- Measurement of the gluon distribution in nuclei would be relatively straightforward at EIC via charm production. One would look for the expected medium modification of the gluon distribution. Only scant data exist at present [16].

- There are indications that gluon densities, at high partonic density and low x, undergo significant change. It has been suggested that experiments in this region will reveal a *"Colored Glass Condensate"*, a collective gluonic Bose condensate analogous to effects seen in spin glasses.

- A number of important partonic phenomena can be explored with nuclei, e.g.: color transparency, color coherence, parton energy loss, hadronization in nuclei.

The physics prospects for an $e-A$ collider have been covered in detail in previous "e-RHIC" workshops; we refer the reader to the workshop proceedings for a more complete summary [6,7].

CONCLUSIONS

The scientific motivation to pursue realization of an Electron-Ion Collider is very strong. Previous experiments using hard processes indicate the essential need for a machine with a large kinematic range, high luminosity and optimal control of spin and flavor for a decisive study of hadron structure. The collider geometry offers the crucial and unique capability of complete event reconstruction in hard scattering. The collider should be available in a timely fashion to build upon the insights gained from existing programs at BNL, CERN, DESY, and SLAC. It is our strong desire that the collider concept be endorsed by the nuclear physics community in the United States and that it receive vigorous R&D funding over the next several years.

REFERENCES

1. CTEQ5 Parton Distributions, H.L. Lai *et al.*, Eur. Phys. J. C**12**, 375 (2000).
2. K. Ackerstaff *et al.*, Phys. Lett. B**464**, 123 (1999).
3. 'The Science Driving the 12 GeV Upgrade of CEBAF', JLab report, December 2000.
4. Report of the Joint DESY/GSI/NuPECC Workshop on Electron-Nucleon/Nucleus Collisions, March 3-4, 1997, Lufthansa-Zentrum, Seeheim, GSI Report 97-04.
5. *Proceedings of Workshop on Physics with a High Luminosity Polarized Electron Ion Collider*, eds. L.C. Bland, J.T. Londergan and A.P. Szczepaniak (World Scientific, Singapore, 2000).
6. *Proceedings of the 2nd eRHIC Workshop*, Yale April 6-8, 2000, (report BNL-52592, 2000).
7. *Proceedings, Nuclear Theory Summer Meeting on eRHIC*, (report BNL-52606, 2000).
8. E. Kinney and U. Stoesslein, private communication, December 2000.
9. R.J. Holt and P.E. Reimer, eprint nucl-ex/0010004.
10. P. Mulders, eprint hep-ph/0010199.
11. L. Trentadue and G. Veneziano, *Phys. Lett.* B323, 201 (1994).
12. X. Ji, *Phys. Rev. Lett.*78, 610 (1997); *Phys. Rev.* D55, 7114 (1997).
13. M. Vanderhaeghen, eprint hep-ph/0007232.
14. M. Vanderhaeghen, private communication, August 2000.
15. D.M. Alde *et al.*, Phys. Rev. Lett. **64**, 2479 (1990).
16. T. Gousset and H.J. Pirner, Phys. lett. B**375**, 349 (1996).

Learning From QCD

Frank Wilczek

Center for Theoretical Physics
Massachusetts Institute of Technology
Cambridge, MA 02139-4307

Abstract. Quantum chromodynamics, or QCD, is the modern theory of the strong interaction. It is a precisely defined theory, that cleanly embodies a few profound concepts. Its fundamental predictions have been verified experimentally with remarkable completeness and rigor. In many ways, QCD is our most perfect theory of Nature. It is, however, far from being a closed or finished story. Its implications are wide-ranging, and there are many potential applications that remain to be achieved.

INTRODUCTION

In this brief survey I'll outline in broad terms the scope and significance of QCD. I hope this provides some perspective and inspiration for the work to follow.

COMPLETING THE THEORY OF MATTER

In 1900, physics had a completely different character from what it has today. It described *how* matter will evolve, given its initial condition. Classical physics cannot explain why there are material substances with definite, reproducible properties at all, much less why there are just the particular molecules, atoms, and nuclei with the specific properties we observe in Nature. In short, classical physics cannot address questions about *what* matter is, nor *why* it is that way.

By the mid-1930s physicists had developed a very specific and fruitful model of atoms. With a few minor refinements, it is still the model we use today. According to this model, an atom consists of a very heavy, very small nucleus, containing all the positive charge of the atom and most of its mass, and a number of electrons. The electrons satisfy the rules of quantum mechanics, and the only interaction taken into account is electromagnetism. This model is adequate to provide an excellent quantitative account of the spectra of simple atoms, and a semi-quantitative account of the periodic table, chemical bonds, the behavior of metals, and much else. In a real sense it provides the foundation for, in Dirac's phrase, "all of chemistry and most of physics".

The largely mysterious nature of the nucleus was not a severe practical limitation for this work. Its size is much smaller than the size of the atom as a whole, and in calculating the influence of the nucleus on electrons, it is a good approximation to set it to zero. Its mass is so large, compared to that of the electrons, that "recoil" corrections are quite small, so the nucleus can be treated as a fixed source. In short, for most practical purposes of atomic physics, the nucleus could be treated as a black box, whose behavior was adequately described using a small number of parameters taken from experiment.

Another Layer -- and Another?

Nevertheless the fundamental problem of understanding what atomic nuclei are, and what holds them together, could hardly escape the attention of physicists. Besides their intrinsic interest, these questions were known to be tied up with the energy source of stars, the origin of chemical elements, and medical and technical applications of radioactivity. Although the early history is very confused and complicated, in hindsight it seems appropriate to mark the division between the pre-history and the history of modern thinking about the strong interaction at 1931, when James Chadwick discovered the neutron. Within a few years a working semi-phenomenological theory of nuclear structure was buiilt up. Its main idea was that nuclei are built up from protons and neutrons. These particles were supposed to obey the rules of quantum mechanics, but to be subject to a new force which, unlike atomic electromagnetism, has no macroscopic analogue. There was no deep theory of the new force, but in due course some of its properties were mapped out experimentally. It was clear that to confine protons and neutrons into a space as small as the nucleus, in opposition to electrostatic forces, the new force would have to be much more powerful than electromagnetism; on the other hand, its power evidently extended only over a short distance, becoming negligible much beyond $\sim 10^{-13}$ cm. The semi-phenomenological theory is good enough to say quite a bit about stellar burning, and to support the design of nuclear weapons and reactors -- pretty impressive!

Nevertheless from early on there were many indications that protons and neutrons are not elementary particles, but have significant internal structure and physicists still hungered for a more profound theory.

The Theory of Matter

One might have anticipated that further investigation would uncover another layer of structure beyond protons, which would have its own semi-phenomenological theory, and that there would be a continuing, incremental, step-by-step increase in our understanding of the structure of matter. Indeed, this was what seemed to be emerging from an enormous body of work in the period 1932-1970 or so.

But then developments took a dramatic and quite unexpected turn. Experiments on "ard"strong interaction processes, that is processes involving large transfers of energy and momentum, displayed an unexpected simplicity. Specifically, the simplicity emerged when one considered hard *inclusive* processes - following the total flow of energy and momentum, without distinguishing individual particles. As we shall discuss below, and in great detail in the text, it emerged that when hard strong interaction process are viewed inclusively, patterns reminiscent of elementary interactions in quantum field theory appeared. However, it was also apparent that the underlying quantum field theory, if it existed at all, would have to have some very special, unusual properties, previously unknown.

Specifically, a quantum field theory of the strong interaction would need to have the property that radiation events involving large transfers of energy and momentum are rare, whereas radiation events involving small transfers of energy and momentum are common. This is the property of *asymptotic freedom*. Translated into position space, asymptotic freedom means that the interaction turns off at short distances, or in other words that it is *antiscreened*. Screening of charge is a familiar phenomenon in electrostatics, where it is known as dielectric behavior. In quantum field theories, the vacuum acts as a medium, due to the existence of virtual particles. Screening is the usual case.

A systematic survey of quantum field theories revealed that there is a very restricted class of theories with asymptotic freedom. They all feature non-abelian gauge fields. Previous phenomenological work had hinted at a 3-valued "color" degree of freedom. By combining the general requirement of asymptotic freedom with broad phenomenological hints about quarks and color, we were led to a unique candidate quantum field theory for the strong interaction. This of course is the theory now known as quantum chromodynamics, or QCD.

QCD does not fit the pattern of incremental accommodation of new layers of structure. It is a closed, conceptual theory, that uniquely embodies a few general principles such as special relativity, locality, quantum mechanics, and local gauge symmetry. QCD cannot be significantly changed, unless one or more of these principles is compromised. Fortunately, despite many years of rigorous testing, no indication has emerged that any change is required by experiment.

We can quantify these claims, to a certain extent, as follows. QCD is not a complete theory of Nature (only, at most, of the strong interaction) and so physical events occur which are not predicted by QCD, or predicted with a different amplitude. But if QCD is correct, such deviations, if they involve interactions among quarks and gluons, must be due to non-renormalizable interactions [1] and are therefore characterized by one or more powers of a suppression factor $E\Lambda$, where E is the energy scale of the process, and Λ an energy scale where new physics (e.g., low-energy supersymmetry) might appear. The experimental data indicate $\Lambda \geq$ 1-10 Tev. By comparison the mass of the proton is 1 GeV, and the scale of nuclear binding energies is tens of MeV, so the potential errors are .1% or less.

Thus QCD is more than adequate to do the job originally intended for a theory of the strong interaction, that is, to provide a fundamental basis for completing the description of ordinary matter. QCD and QED together constitute the Theory of Matter.

EMBODYING QUANTUM FIELD THEORY

Quantum field theory was born from the marriage of two great theories of twentieth-century physics, special relativity and quantum mechanics, and has deep roots in the classical fields theories of the nineteenth century, specifically Maxwell's

---

[1] Strictly speaking, in making these comparisons we should subtract out the small, calculable effects of electroweak interactions.

electrodynamics. It has many remarkable features all its own, and brings several new basic insights into the nature of Nature.

It was not the first quantum field theory, nor is it the only successful one, but QCD occupies a very special place among quantum field theories. It displays a richness and variety of dynamical behavior far beyond that displayed by QED or electroweak theory. Since QCD is not governed by a weak coupling, its behavior cannot be treated with straightforward perturbative techniques. In QCD, intricate and profound aspects of quantum field theory are directly related to observable behavior.

The Virtues of Quantum Field Theory

Quantum field theory is the framework in which the regnant theories of the electroweak and strong interactions - which together form the Standard Model - are formulated. Quantum electrodynamics (QED), besides providing a complete foundation for atomic physics and chemistry, has supported calculations of physical quantities with unparalleled precision. The experimentally measured value of the magnetic dipole moment of the muon,

$$(g_\mu - 2)_{\text{exp}} = 233\ 184\ 600(1680) \times 10^{-11}, \qquad (1)$$

for example, should be compared with the theoretical predicition

$$(g_\mu - 2)_{\text{theor.}} = 233\ 183\ 478\ (308) \times 10^{-11}. \qquad (2)$$

In quantum chromodynamics (QCD) we cannot, for the forseeable future, aspire to to comparable accuracy. Yet QCD provides different, and at least equally impressive, evidence for the validity of the basic principles of quantum field theory.Indeed, because in QCD the interactions are stronger, QCD manifests a wider variety of phenomena characteristic of quantum field theory. These include especially running of the effective coupling with distance or energy scale and the phenomenon of confinement. QCD has supported, and rewarded with experimental confirmation, both heroic calculations of multi-loop diagrams and massive numerical simulations of (a discretized version of) the complete theory.

'Quantum field theory also provides powerful tools for condensed matter physics, especially in connection with the quantum many-body problem as it arises in the theory of metals, superconductivity, the low-temperature behavior of the quantum liquids He^3 and He^4, and the quantum Hall effect, among others. Although for reasons of space and focus I will not attempt to do justice to this aspect here, the continuing interchange of ideas between condensed matter and high energy theory, through the medium of quantum field theory, is a remarkable phenomenon in itself. A partial list of historically important examples includes global and local spontaneous symmetry breaking, the renormalization group, effective field theory, solitons, instantons, and fractional charge and statistics.

It is clear, from all these examples, that quantum field theory occupies a central position in our description of Nature. It provides both our best working description of fundamental physical laws, and a fruitful tool for investigating the behavior of complex systems. But the enumeration of examples, however triumphal, serves more to pose than to answer more basic questions: What are the essential features of quantum field theory? What does quantum field theory add to our understanding of the world, that was not already present in quantum mechanics and classical field theory separately?

The first question has no sharp answer. Theoretical physicists are very flexible in adapting their tools, and no axiomization can keep up with them. However I think it is fair to say that the characteristic, core ideas of quantum field theory are twofold. First, that the basic dynamical degrees of freedom are operator functions of space and time -- quantum fields, obeying appropriate commutation relations. Second, that the interactions of these fields are local. Thus the equations of motion and commutation relations governing the evolution of a given quantum field at a given point in space-time should depend only on the behavior of fields and their derivatives at that point. One might find it convenient to use other variables, whose equations are not local, but in the spirit of quantum field theory there must always be some underlying fundamental, local variables. These ideas, combined with postulates of symmetry (e.g., in the context of the standard model, Lorentz and gauge invariance) turn out to be amazingly powerful, as will emerge from our further discussion below.

The field concept came to dominate physics starting with the work of Faraday in the mid-nineteenth century. Its conceptual advantage over the earlier Newtonian program of physics, to formulate the fundamental laws in terms of forces among atomic particles, emerges when we take into account the circumstance, unknown to Newton (or, for that matter, Faraday) but fundamental in special relativity, that influences travel at a finite limiting speed. For then the force on a given particle at a given time cannot be deduced from the positions of other particles at that time, but must be deduced in a complicated way from their previous positions. Faraday's intuition that the fundamental laws of electromagnetism could be expressed most simply in terms of fields filling space and time was of course brilliantly vindicated by Maxwell's mathematical theory.

The concept of locality, in the crude form that one predict the behavior of nearby objects without reference to distant ones, is basic to scientific practice. Practical experimenters -- if not astrologers -- confidently expect, on the basis of much successful experience, that after reasonable (generally quite modest) precautions to isolate their experiments they will obtain reproducible results. Direct quantitative tests of locality, or rather of its close cousin causality, are afforded by dispersion relations.

The deep and ancient historic roots of the field and locality concepts provide no guarantee that these concepts remain relevant or valid when extrapolated far beyond their origins in experience, into the subatomic and quantum domain. This extrapolation must be judged by its fruits. That brings us, naturally, to our second question.

Undoubtedly the single most profound fact about Nature that quantum field theory uniquely explains is *the existence of different, yet indistinguishable, copies of elementary particles* Two electrons anywhere in the Universe, whatever their origin or history, are observed to have exactly the same properties. We understand this as a consequence of the fact that both are excitations of the same underlying ur-stuff, the electron field. The electron field is thus the primary reality. The same logic, of course, applies to photons or quarks, or even to composite objects such as atomic nuclei, atoms, or molecules. The indistinguishability of particles is so familiar, and so fundamental to all of modern physical science, that we could easily take it for granted. Yet it is by no means obvious. For example, it directly contradicts one of the pillars of Leibniz' metaphysics, his "principle of the identity of indiscernables," according to which two objects cannot differ solely in number. And Maxwell thought the similarity of different molecules so remarkable that he devoted the last part of his *Encyclopedia Brittanica* entry on Atoms - well over a thousand words - to discussing it. He concluded that "the formation of a molecule is therefore an event not belonging to that order of nature in which we live ... it must be referred to the epoch, not of the formation of the earth or the solar system ..but of the establishment of the existing order of nature ".

The existence of classes of indistinguishable particles is the necessary logical prerequisite to a second profound insight from quantum field theory: *the assignment of unique quantum statistics* to each class. Even given the existence of indistinguishability of a class of elementary particles, and complete invariance of their interactions under interchange, the general principles of quantum mechanics teaches us that solutions forming any representation of the permutation symmetry group retain that property in time, but does not constrain which representations are realized. Quantum field theory not only explains the existence of indistinguishable particles and the invariance of their interactions under interchange, but also constrains the symmetry of the solutions. For bosons only the identity representation is physical (symmetric wave functions), for fermions only the one-dimensional odd representation is physical (antisymmetric wave functions). One also has the spin-statistics theorem, according to which objects with integer spin are bosons, whereas objects with half odd integer spin are fermions. Of course, these general predictions have been verified in many experiments. The fermion character of electrons, in particular, underlies the stability of matter and the structure of the periodic table.

A third profound general insight from quantum field theory is *the existence of antiparticles*. This was first inferred by Dirac on the basis of a brilliant but obsolete interpretation of his equation for the electron field, whose elucidation was a crucial step in the formulation of quantum field theory. In quantum field theory, we re-interpret the Dirac wave function as a position (and time) dependent operator. It can be expanded in terms of the solutions of the Dirac equation, with operator coefficients. The coefficients of positive-energy solutions are operators that destroy electrons, and the coefficients of the negative-energy solutions are operators that create positrons (with positive energy). With this interpretation, an improved version of Dirac's original hole theory, with a sensible generalization to bosons and to processes where the number of electrons minus positrons changes, emerges in a straightforward way. A very general consequence of quantum field theory, valid in the presence of arbitrarily complicated interactions, is the CPT theorem. It states that the product of charge congugation, parity, and time reversal is always a symmetry of the world, although each may be - and is! - violated separately. Antiparticles are strictly defined as the CPT conjugates of their corresponding particles.

The three outstanding facts we have discussed so far: the existence of indistinguishable particles, the phenomenon of quantum statistics, and the existence of antiparticles, are all essentially consequences of *free* quantum field theory. When one incorporates interactions into quantum field theory, two additional general features of the world immediately become brightly illuminated.

The first of these is *the ubiquity of particle creation and destruction processes*. Local interactions involve products of field operators at a point. When the fields are expanded into creation and annihilation operators multiplying modes, we see that these interactions correspond to processes wherein particles can be created, annihilated, or changed into different kinds of particles. This possibility arises, of course, in the primeval quantum field theory, quantum electrodynamics, where the primary interaction arises from a product of the electron field, its Hermitean conjugate, and the photon field. Processes of radiation and absorption of photons by electrons (or positrons), as well as electron-positron pair creation, are encoded in this product. Just because the emission and absorption of light is such a common experience, and electrodynamics such a special and familiar classical field theory, this correspondence between formalism and reality did not initially make a big impression. The first conscious exploitation of the potential for quantum field theory to describe processes of transformation was Fermi's theory of beta decay. He turned the procedure around, inferring from the observed processes of particle transformation the nature of the underlying local interaction of fields. Fermi's theory involved creation and annihilation not of photons, but of atomic nuclei and electrons (as well as neutrinos) -- the ingredients of 'matter'. It began the process whereby classic atomism, involving stable individual objects, was replaced by a more sophisticated and accurate picture. In this picture it is only the fields, and not the individual objects they create and destroy, that are permanent.

The second is *the association of forces and interactions with particle exchange*. When Maxwell completed the equations of electrodynamics, he found that they supported source-free electromagnetic waves. The classical electric and magnetic fields thus took on a life of their own. Electric and magnetic forces between charged particles are explained as due to one particle acting as a source for electric and magnetic fields, which then influence others. With the correspondence of fields and particles, as it arises in quantum field theory, Maxwell's discovery corresponds to the existence of photons, and the generation of forces by intermediary fields corresponds to the exchange of virtual photons. The association of forces (or, more generally, interactions) with exchange of particles is a general feature of quantum field theory. It was used by Yukawa to infer the existence and mass of pions from the range of nuclear forces, and more recently in electroweak theory to infer the existence, mass, and properties of W and Z bosons prior to their observation, and in QCD to infer the existence and properties of gluon jets prior to their observation.

The two additional outstanding facts we just discussed: the possibility of particle creation and destruction, and the association of particles with forces, are essentially consequences of classical field theory supplemented by the connection between particles with fields we learn from free field theory. Indeed, classical waves with nonlinear interactions will change form, scatter, and radiate, and these processes exactly mirror the transformation, interaction, and creation of particles. In quantum field theory, they are properties one sees already in *tree graphs*.

Deep Structures

The foregoing major consequences of free quantum field theory, and of its formal extension to include nonlinear interactions, were all well appreciated by the late 1930s. The deeper properties of quantum field theory, which will form the subject of the remainder of this paper, arise from the need to introduce *infinitely many degrees of freedom*, and the possibility that all these degrees of freedom are excited as quantum-mechanical fluctuations. From a mathematical point of view, these deeper properties arise when we consider *loop graphs*.

From a physical point of view, the potential pitfalls associated with the existence of an infinite number of degrees of freedom showed up in connection with the problem which led to the birth of quantum theory, that is the ultraviolet catastrophe of blackbody radiation theory. Somewhat ironically, in view of later history, the crucial role of the quantum theory here was to remove the disastrous consequences of the infinite number of degrees of freedom possessed by classical electrodynamics. The classical electrodynamic field can be decomposed into independent oscillators with arbitrarily high values of the wavevector. According to the equipartition theorem of classical statistical mechanics, in thermal equilibrium at temperature T each of these oscillators should have average energy kT. Quantum mechanics alters this situation by insisting that the oscillators of frequency ω have energy quantized in units of $\hbar\omega$. Then the high-frequency modes are exponentially suppressed by the Boltzmann factor, and instead of kT receive $\dfrac{\hbar\omega e^{-\frac{\hbar\omega}{kT}}}{1-e^{-\frac{\hbar\omega}{kT}}}$.

The role of the quantum, then, is to prevent accumulation of energy in the form of very small amplitude excitations of arbitrarily high frequency modes. It is very effective in suppressing the *thermal* excitation of high-frequency modes. But while removing arbitrarily small amplitude excitations, quantum theory introduces the idea that the modes are always *intrinsically* excited to a small extent, proportional to $\hbar$. This so-called zero point motion is a consequence of the uncertainty principle. For a harmonic oscillator of frequency ω, the ground state energy is not zero, but $\frac{1}{2}\hbar\omega$. In the case of the electromagnetic field this leads, upon summing over its high-frequency modes, to a highly divergent total ground state energy. For most physical purposes the absolute normalization of energy is unimportant, and so this particular divergence does not necessarily render the theory useless.[2] It does, however, illustrate the dangerous character of the high-frequency modes, and its treatment gives a first indication of the leading theme of renormalization theory: we can only require - and

---

[2] One would think that gravity should care about the absolute normalization of energy. The zero-point energy of the electromagnetic field, in that context, generates an infinite cosmological constant. This might be cancelled by similar negative contributions from fermion fields, as occurs in supersymmetric theories, or it might indicate the need for some other profound modification of physical theory.

generally will only obtain - sensible, finite answers when we ask questions that have direct, operational physical meaning.

The existence of an infinite number of degrees of freedom was first encountered in the theory of the electromagnetic field, but it is a general phenomenon, deeply connected with the requirement of locality in the interactions of fields. For in order to construct the local field $\psi(x)$ at a space-time point x, one must take a superposition

$$\psi(x) = \int \frac{d^4k}{(2\pi)^4} e^{ikx} \tilde{\psi}(k) \qquad (3)$$

that includes field components $\tilde{\psi}(k)$ extending to arbitrarily large momenta. Moreover in a generic interaction

$$\int L = \int \psi(x)^3 = \int \frac{d^4k_1}{(2\pi)^4} \frac{d^4k_2}{(2\pi)^4} \frac{d^4k_3}{(2\pi)^4} \tilde{\psi}(k_1) \tilde{\psi}(k_2) \tilde{\psi}(k_3) (2\pi)^4 \delta^4(k_1 + k_2 + k_3)$$

(4)

we see that a low momentum mode $k_1 \approx 0$ will couple without any suppression factor to high-momentum modes k_2 and $k_3 \approx -k_2$. Local couplings are "hard", in this sense. Because locality requires the existence of infinitely many degrees of freedom at large momenta, with hard interactions, ultraviolet divergences similar to the ones cured by Planck, but driven by quantum rather than thermal fluctuations, are never far off stage. As mentioned previously, the deeper physical consequences of quantum field theory arise from this circumstance.

First of all, it is much more difficult to construct non-trivial examples of interacting relativistic quantum field theories than purely formal considerations would suggest. One finds that *the consistent quantum field theories form a quite limited class, whose extent depends sensitively on the dimension of space-time and the spins of the particles involved.* Their construction is quite delicate, requiring limiting procedures whose logical implementation leads directly to renormalization theory, the running of couplings, and asymptotic freedom.

Secondly, *even those quantum theories that can be constructed display less symmetry than their formal properties would suggest.* Violations of naive scaling relations - that is, ordinary dimensional analysis - in QCD, and of baryon number conservation in the standard electroweak model are examples of this general phenomenon. The original example, unfortunately too complicated to explain fully here [3], involved the decay process $\pi^0 \to \gamma\gamma$, for which chiral symmetry – treated classically - predicts much too small a rate. When the correction introduced by quantum field theory (the so-called `anomaly') is retained, excellent agreement with experiment results.

These deeper consequences of quantum field theory, which might superficially appear rather technical, largely dictate the structure and behavior of the Standard Model -- and, therefore, of the physical world.

---

[3] We'll do it later.

Because the strength of the interaction must be defined by a limiting procedure, its numerical value must be defined relative to a specified cutoff. This introduces an energy or distance dependence of the effective coupling. A heuristic picture of the mechanism leading to a *running coupling* is as follows. What we call empty space is filled with virtual particles. These form a cloud around any specified source, that modifies how it appears from a distance. The observed strength of a source is therefore not a single intrinsic strength, but rather an effective strength, reflecting the influence of its surrounding cloud. Depending on how far we penetrate the cloud, we see different effective strengths of the effective coupling parameter.

The deep structure of infinitely many degrees of freedom and ultraviolet divergences is reflected in the running of the coupling and the anomalous breaking of symmetry. In addition we will meet several truly non-perturbative phenomena, including confinement, spontaneous chiral symmetry breaking, instantons, and color superconductivity. Their study involves new concepts and tools.

More specifically, the apparent symmetry of the classical Lagrangian of QCD (idealized to contain two massless quarks) includes

$$SU(3)_{color} \times SU(2)_L^{flavor} \times SU(2)_{flavor}^R \times U(1)_{baryon}^{L-r} \times R(1)_{scale}^x \quad (5)$$

- color, chiral flavor, axial baryon number, and scale invariance. But in Nature what survives is merely

$$SU(2)_{flavor}^{L+R} \quad (6)$$

- isospin. The differences reflect confinement, spontaneous chiral symmetry breaking, anomalies and instantons, and asymptotic freedom. Each of these phenomena is a deep property of quantum field theory. Understanding how QCD operates in Nature involves understanding them all, not just abstractly, but in many very concrete manifestations.

EXPLOITING SIMPLICITY

There is a very powerful respect in which QCD is radically simple. At short distances, or equivalently large energy-momentum, the running coupling becomes small. This is the phenomenon of asymptotic freedom, already mentioned above.

The Jet Paradigm

Since radiations that drastically change the flow of energy-momentum are rare, an energetic quark or gluon will tend to induce, by soft radiation, a *jet* of particles moving rapidly in the direction it imprints. This phenomenon has revolutionized the design and interpretation of high-energy accelerator experiments. By doing inclusive measurements of jets, experimenters can "see" the quarks and gluons. Since these are the fundamental objects, in terms of which theoretical ideas both within and beyond the Standard Model are formulated, default expectations for jets can be reliably calculated, and consequences of theoretical conjectures can be reliably estimated.

Factorization

The ideally simple case of jet production occurs when the initial quark and antiquark are produced by electroweak currents, for example through a virtual photon in electron-positron annihilation. A much wider class of experiments involve hadrons in an intrinsic way, either because the intial state contains hadrons, or because the measurements of interest are not entirely inclusive. In many cases it is still possible to make progress, by the process of *factorization*. Roughly speaking, the idea of factorization is to combine estimates of the probability of finding a quark or gluon with given momentum inside a hadron (structure functions), or estimates of the probability of a quark or gluon to fragment into a specified hadron (fragmentation functions), with a core caculation of how the quarks and gluons themselves interact. The structure and fragmentation functions themselves are not easy to calculate, so they are taken from experiment. But once found, the same functions can be used to describe many different experiments.

The original example of factorization, which is fundamentally important and which we shall discuss in some detail, is the analysis of deep inelastic scattering. The strategy has been used successfully to estimate rates and backgrounds for dozens of very different types of experiments, including for example the W and Z boson and t quark discovery experiments.

Frontier Examples

The Higgs particle is a major target for high energy experiment. Its main coupling to ordinary matter arises through an unusual and interesting QCD mechanism, involving virtual top quarks materializing into gluons. "Gluon fusion" is the dominant production mechanism for Higgs particles at hadronic colliders, and decay into glue pairs is one of their main decay modes.

More complex, because it involves many particles with poorly characterized couplings, is the search for low-energy supersymmetry. The concepts of jets and factorization are the backbone of analyses of their production rates and decay signatures.

MANAGING COMPLEXITY

In cases where we cannot appeal to asymptotic freedom, there is no simple general algorithm to compute the consequences of QCD. There are, however, several notable ways to make progress.

Universality and Flavor Symmetries

A fundamental aspect of non-abelian gauge symmetry is that the transformation law for the gauge field fixes a value for the (classical) coupling constant. Because of this, the coupling of the gauge field to charged particles is *universal*. Whereas in an abelian gauge theory any value of the relative charge of different particles is allowed, and in a non-gauge theory, such as Yukawa theory, the couplings of mesons to particles are not constrained, in a non-abelian gauge theory the couplings must be equal, or universal. Thus quarks of different flavors - u,d,s,c,b,t - all couple to gluons in exactly the same way.

This universality has many observable, and useful, consequences. The u and d quark masses, and to a certain extent the s quark mass, are small compared to the scales associated with typical interactions in QCD. To the extent we ignore them all, we recover the famous *SU(3)* flavor symmetry of Gell-Mann. It can be used to group the hadrons in to multiplets, and to relate processes involving them. One can also reinstate the quark mass perturbatively, and obtain relations among mass splittings.

A more accurate approximation is to ignore just the u and d masses, in which case we have the *SU(2)* symmetry of isospin, which is typically accurate at the 1% level.

For c, b and t quarks the opposite approximation, treating them as infinitely massive and neglecting recoil, is more appropriate at low energy. Universality then allows one to relate the properties of one heavy quark to another; up to a simple kinematic re-scaling for the spin coupling, the basic idea is that to a gluon one heavy quark of given velocity looks just like another. This leads to a different sort of approximate flavor *SU(3)* symmetry.

CHIRAL SYMMETRY

Another fundamental aspect of gauge symmetry is that the gauge interactions preserve helicity[4]. Thus in QCD, the fundamental gauge interactions connect left-handed quarks only to left-handed quarks, and right-handed quarks only to right-handed quarks.

This is most significant for light quarks, because it leads to a new approximate symmetry, chiral symmetry. Chiral symmetry is intrinsically broken by quark mass terms, since such terms directly connect left- to right-handed particles.

If we neglect the u and d quark masses, then the *SU(2)* isospin symmetry of QCD is augmented to chiral $SU(2)_{Left} \times SU(2)_{Right}$.

A powerful phenomenology with many applications can be built up around the idea that chiral symmetry is quite accurate intrinsically, as a symmetry of the fundamental equations of QCD, but broken to a much larger extent spontaneously. That is, it is a much more accurate symmetry of the equations than of the stable solution of these equations, in the ground state of actual world. The spontaneous breaking leaves isospin intact; thus

$$SU(2)_{Left} \times SU(2)_{Right} \to SU(2)_{L=R}$$

---

[4] To be more accurate, chiral symmetry for massless fermions follows from gauge symmetry *and* the requirement of renormalizability or asymptotic freedom. An anomalous chromo-magnetic moment coupling, for example, is allowed by gauge invariance, but would violate chiral symmetry, even for massless quarks.

Whenever one has a spontaneously broken continuous symmetry, there are low-energy collective modes connected with slow space-time modulations of the direction of symmetry breaking. Indeed, in the limit of uniform rotation through all space, action of the spontaneously broken symmetry will take one from the ground state to another, physically distinct state with the same energy. Such uniform rotation is not physically realizable, and in any case would lead one to a state orthogonal to the ground state, but one can construct states of low energy that approximate the uniformly rotated ground state over a finite volume. If the gradients in space and time are small, the energy of the rotated state will be small. When one quantizes these excitations, they are interpreted as massless particles - called Nambu-Goldstone bosons, after the physicists who elucidated their existence and properties. Because the Nambu-Goldstone bosons are generated by a canonical, symmetry-based construction, many of their properties can be calculated from first principles.

In QCD, one can identify the π mesons as Nambu-Goldstone particles of spontaneously broken chiral symmetry. According to this idea, if m_u and m_d were strictly zero, the π mesons would be strictly massless. In fact, these mesons are significantly lighter than the other strongly interacting particles. Furthermore, they exhibit many other properties consistent with this interpretation. Spontaneously broken symmetry is a very powerful concept in physics generally, and it gets a workout in QCD.

Direct Calculation

Since QCD is a precisely defined theory, in the end it must be possible to compute its consequences algorithmically[5]. A very important branch of theoretical work on QCD is devoted to delivering on this promise. It has had some brilliant success, yet still poses major challenges.

The fundamental algorithm for defining QCD involves discretizing the theory, calculating vastly high-dimensional integrals, and taking a limit. In physical terms, one replaces continuous space-time by a lattice, with the quark and gluon fields restricted to taking single values on lattice points and links respectively, and then refines the lattice, keeping the points at which observations are made fixed, until the effect a limiting value is obtained. Much ingenuity and subtlety is involved in the construction. To keep the calculations manageable and accurate is important to maintain as much symmetry as possible. Quite remarkably, it is possible to maintain quite large subgroups of gauge and chiral symmetry (the latter is a quite recent discovery). The existence of the limit is by no means obvious - indeed, for a generic quantum field theory, the analogous procedure will fail. The existence of a limit is a consequence of asymptotic freedom, which also allows one to anticipate how it is approached.

For practical calculations it is necessary to use clever numerical techniques to do the integrals, typically involving some version of Monte Carlo algorithm, and to use the most powerful available computers. But the results are worth it. One highlight is a no-parameter, 10% calculation of the light hadron spectrum!

---

[5] Operationally, that is what we mean by "precisely defined".

Numerical simulation allows us to investigate situations and gather "data" on questions of great theoretical interest that are not accessible by normal experiments. We can, for example, change the number of quarks, or their masses, or the gauge group, or the temperature; or examine the typical configuration of gluon fields in detail. We'll have some fun with this.

Existing numerical techniques are totally impractical for calculating real-time behavior, scattering amplitudes, or finite density static properties. Significant progress on either of these problems would have wide ramifications well beyond QCD, since closely related problems arise in chemistry and condensed matter physics.

DESCRIBING EXTREME CONDITIONS

The behavior of matter in ultra-extreme conditions of high temperature or high density is interesting both in itself and for applications. Fortunately, these "highs" are closely related to the limits of large energy-momentum and short distance, where asymptotic freedom applies and QCD simplifies.

High Temperature

The behavior of QCD at high temperatures (and zero baryon number density) is of great intrinsic interest, since it is the answer to the child-like question, "What happens if I keep accumulating energy?" It is also of fundamental interest for cosmology, since according to Big Bang cosmology the entire Universe, in the crucial early moments of its history, was filled with matter at ultra-high temperatures (and very low baryon number density).

The fundamental result about QCD at high temperatures T is simply that many of its properties, for example the equation of state, can be calculated to a good approximation by assuming that one has an ideal gas of non-interacting quarks (and antiquarks) and gluons at temperature T.

Stated in this form, coming from the high-temperature side, the fundamental result might not seem surprising. Indeed, asymptotic freedom immediately makes it very plausible, by arguments similar to those given for the integrity of jets. On the other hand, if you considered building up the high-temperature behavior of the strong interaction from the low-temperature side, without knowing about QCD and asymptotic freedom, the emergence of any simple result would seem miraculous. For in such a phenomenological approach, as the temperature rises you must include the effects of more and more different kinds of particles (resonances), with complicated mutual interactions. Thus as late as the early 1970s there was general pessimism about the possibility of understanding conditions in early Big Bang, at temperatures above 10^{12} K, as in Weinberg's 1972 text: "However, if we look back a little further, into the first 0.001 sec. of cosmic history when the temperature was above 10^{12} K, we encounter theoretical problems of a difficulty beyond the range of modern statistical mechanics. At such temperatures, there will be present in thermal equilibrium copious numbers of strongly interacting particles - mesons, baryons, and antibaryons - with a mean interparticle distance less than a typical Compton wavelength. These particles will be in a state of continual mutual interaction, and cannot reasonably be expected to obey any simple equation of state". Nowadays, of course, we routinely contemplate much higher temperatures, and calculate the cosmological consequences of models of particle physics with confidence, based on the simplicity QCD that promises.

Existing numerical methods for calculating the consequences of QCD are well adapted to the study of finite temperature. They are consistent with the fundamental result. In addition they provide information about the onset of quark-gluon behavior. A major result from these studies is that the onset of recognizable quark-gluon behavior occurs at a surpisingly low temperature, T ≤ 200 MeV. This is surprising, because the only hadrons with masses below this scale are the pions. Thus one passes rather abruptly from a dilute, approximately ideal gas with 3 degrees of freedom (3 spin zero pions) to a different approximately ideal gas with 52 degrees of freedom (8 gluons, each with 2 helicities, plus 3 kinds of quarks u,d,s, which have 3 colors and 2 spin, and their antiquarks. $8 \times 2 + 3 \times 3 \times 2 \times 2 = 52$).

As this is written the direct experimental study of high-temperature QCD is entering a new era, with the initial operation of the relativisitic heavy ion collision facility (RHIC) at Brookhaven Lab. Very plausibly, they will produce fireballs in which temperatures T ≥ 200 MeV will be attained. As we shall discuss, there are definite quantitative and qualitative consequences of the transition to quark and gluon degrees of freedom, which ought to be observable. Indeed, there are already some indications of major changes in the nature of the highest-energy fireballs, suggestive of the quark-gluon plasma phase.

Another fascinating subject is the possibility of sharply defined thermodynamic transitions between hadronic and quark phases. An intricate web of theoretical arguments suggests that there ought to be a true critical point in the temperature-baryon number density plane. Heavy ion collisions that pass near this point ought to exhibit critical fluctuations, which may provide a tool to elucidate the QCD phase diagram experimentally.

High Density

The behavior of QCD at high baryon number density (and zero temperature) is of great intrinsic interest, since it is the answer to the child-like question, "What happens to matter if I keep squeezing?" It is also of fundamental interest for astrophysics, since this regime is plausibly realized within neutron stars, and during stellar collapse.

The fundamental result about QCD at high baryon number density is that many of its key properties, including for example the symmetry of the ground state and the energy and charge of the elementary excitations, can *not* be calculated to a good approximation by using fermi balls of noninteracting quarks. This is due to the development of a condensate of quark pairs, similar to the Cooper pairs that occur in metallic superconductors. The phenomenon of *color superconductivity* is very robust, because there is a fundamentally attractive force between quarks. It can be studied in a weak-coupling - but nonperturbative! - framework.

Color superconductivity has become an extremely active area of research over the past few years, and many surprises have emerged. Perhaps the most profound and beautiful result is the occurrence of *color-flavor locking*, a new form of symmetry breaking, in real-world (3 flavor) QCD at asymptotic densities. Color-flavor locking is a rigorous consequence of QCD. It implies confinement and chiral symmetry breaking. If the core of a neutron star is described by the color-flavor locked (CFL) phase, which seems plausible, it will be a transparent insulator that partially reflects light - like a diamond! This particular consequence of the CFL phase is unlikely to be observed any time soon, but we are working toward defining indirect signatures in observable neutron star and supernova properties.

Unfortunately, existing numerical methods for calculating the behavior of QCD very slowly at high density and low temperature. They are totally impractical, even for the biggest and best modern computers. The development of usable algorithms for this kind of problem is a most important open challenge.

SUGGESTING NEW LAWS

Since QCD is a precise, conceptually based theory, and its core has been amply verified, it is both possible and worthwhile to consider it from a logical-esthetic point of view. In other words, we can consider possibilities for perfecting the Theory, by making it depend on fewer assumptions, or increasing its beauty, independent of any additional experimental input. Such considerations have generated some remarkably concrete suggestions for extending the laws of physics.

Axions

The general framework of QCD, that is gauge symmetry and renormalizability, allows exactly one additional parameter, beyond the coupling constant (which can be traded for a mass scale) and the quark masses. This is the so-called θ parameter. The dependence of QCD on the parameter θ is unusual in several ways. This dependence is not present in the classical theory, but arises only from quantum mechanics. It breaks a continuous symmetry, for massless quark fields (axial baryon number). Physics is periodic in $\theta \to \theta + 2\pi$. The discrete space-time operations spatial inversion or parity P and time-reversal T transform $\theta \to \theta$. We shall spend quite a bit of time explaining the physics underlying these properties.

In real-world QCD the value of θ is very nearly zero. Indeed, a non-zero value of θ would imply that the strong interactions do not respect symmetry under P and T. This would induce a non-zero *electric* dipole moment of the neutron, which we will estimate. Very sensitive experiments have been performed, to search for a non-zero electric dipole moment of the neutron. They can be interpreted to yield a bound $|\theta| \leq 10^{-9}$.

For purposes of applying QCD we can simply put $\theta = 0$ and be done with it. But P and T are not exact symmetries of the world, and the question arises why QCD is so remarkably immune from being infected by their violation. When one discovers an unexpectedly simple and striking property of the world, there is a challenge to understand why it is so.

The most interesting idea for explaining why θ is small is due to Peccei and Quinn. It will be discussed in depth in the text. Roughly speaking, the idea is to promote θ from a number to a dynamical variable - a quantum field φ. Since the symmetry of QCD is enhanced at $\theta = 0$, one might expect that this is achieved at a stationary point of the energy functional for φ. If $\theta = 0$ is achieved at a minimum of the potential for φ, and φ dynamically relaxes to its minimum, then the observed smallness of θ will have been explained.

Weinberg and I pointed out that the quanta of the field φ postulated by Peccei and Quin have remarkable properties. These quanta are called *axions*. The properties of axions can be discussed in considerable detail, as a function of one unknown (but highly constrained) parameter *F*. Axions are very light, very weakly interacting particles. They are predicted to be abundantly produced in the Big Bang. They could provide the "missing mass" needed by astronomers. There are very interesting, promising, but extremely difficult experiments underway to detect axions.

Unification and Supersymmetry

The different components of the Standard Model: QCD, the *SU(3)* color gauge theory of the strong interaction, and the *SU(2)* × *U(1)* gauge theory of electroweak interactions, have a very similar mathematical structure. The core of each of these theories is interactions of vector particles with fermions, governed by the gauge principle. This structure invites us to consider the possibility that the different components of the Standard Model are different facets of a single, encompassing gauge theory.

This idea can be implemented using spontaneous symmetry breaking from a large simple group such as *SU(5)* → *SU(3)* × *SU(2)* × *U(1)* or *SO(10)* → *SU(3)* × *SU(2)* × *U(1)*. The concept of spontaneous symmetry breaking of course is borrowed from electroweak theory where *SU(2)* × *U(1)* → *U(1)*, which in turn borrowed from superconductivity where *U(1)* → Z_2. The order parameters ("Higgs fields") necessary to do this are not terribly complicated. An adjoint will do for *SU(5)*; an adjoint plus spinor **16** for *SO(10)*.

Gauge unification is extremely attractive at the level of group theory. The fermions within a family, which in *SU(3)* × *SU(2)* × U(1) fall into five distinct multiplets, are neatly unified into just two multiplets (a fundamental **5** and a two-index antisymmetric tensor **10**) in *SU(5)*, and into a single spinor **16** in *SO(10)*[6]. A major bonus is that the peculiar hypercharge assignments that are required on phenomenological grounds in the Standard Model emerge as necessary consequences of the group theory of gauge unification.

However there appears to be a fundamental barrier to consummating this unifcation. The essence of gauge symmetry is that the gauge bosons couple with universal strength. But the color gluons associated with the strong interaction *SU(3)* are observed to couple more strongly than the *W* bosons associated with *SU(2)*, which in turn couple more strongly than the photon[7].

Here the other great dynamical lesson of the Theory of Matter, that is asymptotic freedom or more generally the dependence of the strength of couplings on the distance or energy at which they are measured, comes to our rescue. Suppose that the unified gauge symmetry is spontaneous broken at some very large energy scale. At this scale, and above, the full symmetry will emerge, and the effective couplings will be equal. However by hypothesis existing experiments take place at much lower energy scales. The couplings they measure are affected by clouds of virtual particles. These dress the (universal) "bare" charges visible at short distances. Since the spectrum of virtual particles does not share exhibit the full unifed gauge symmetry, the dressing effect breaks this unifed symmetry, and the effective couplings are no longer equal.

This hypothesis predicts a non-trivial constraint among observed quantities. From two quantities - the scale of unification and the value of the coupling there - we must produce the three observed coupling constants.

The computations necessary to assess this prediction quantitively are very simple extensions of the calculation of running of the coupling in QCD. The QCD result is of course amply confirmed by experiment. Its extension to the rest of the Standard Model, and especially the extrapolation to energy scales far beyond any that have been accessed experimentally (the unification takes place at ~ 10^{16} GeV!) is very bold. Many things could upset it, such as failure of quantum field theory at extraordinary virtualities, emergence of extra dimensions, or entry into a strong-coupling regime. Probably no one would have been terribly upset if the calculation had failed. But in fact it works reasonably well. And if we include effect of the virtual particles associated with low-energy supersymmetry, then it works *extremely* well.

This result is very encouraging both for the idea of gauge unification and for the idea of low-energy supersymmetry. I think it is our best clue about the nature of physics "Beyond the Standard Model".

---

[6] The additional particle is an *SU(3)* × *SU(2)* × *U(1)* singlet. It can be interpreted as a right-handed neutrino. An attractive theory of very small neutrino masses, broadly consistent with recent observations of neutrino oscillations, can be constructed using these right-handed neutrinos

[7] To be more precise, what's most relevant is the coupling of the gauge boson coupled to hypercharge *U(1)*, which is a linear combination of the physical photon and the *Z* boson.

ADVANCING NATURAL PHILOSOPHY

Finally I'd like to mention three "applications" of QCD that are perhaps more philosophical than scientific in the usual sense. They do not make sharp falsifiable statements about natural phenomena, but rather address fundamental sources of puzzlement and wonder about Nature. You might say they go to the question "Why?", as opposed to "How?" or "What?"

Its From Bits

In his Autobiographical Notes Einstein wrote "I would like to state a theorem which at present can not based upon anything more than upon a faith in the simplicity, i.e., intelligibility, of nature: there are no *arbitrary* constants ... that is to say, nature is so constituted that it is possible logically to lay down such strongly determined laws that within these laws only rationally completely determined constants occur not constants, therefore, whose numerical value could be changed without destroying the theory"

Wheeler coined the catchy phrase "Getting Its From Bits" for the same program of producing models of Nature that are purely conceptual, with no room for introducing parameters 'from the outside' to fit observation. Within its domain, QCD achieves something remarkably close to this goal. We've emphasized that QCD contains only a few parameters; let's see how far we can be push in the direction few → none.

The coupling constant g that appears in the Lagrangian of QCD, like the corresponding constant e in QED, is a dimensionless number (in units with $\hbar = c = 1$). One might therefore be inclined to think that there are an infinite number of different versions of QCD, labeled by their differing values of g, and that it would be a fundamental problem to determine which actually held in Nature. But the real situation is more subtle, and rather different. It is simplest and most instructive to appreciate the issues by first considering a slightly idealized version of L, that contains just the two light quarks, u and d, with their masses put to zero. This is a perfectly natural thing to do, in the technical as well as the colloquial sense, since extra chiral symmetries arise in this limit. Also, as a bonus, all physical dependence on θ goes away. The resulting "QCD Lite" theory then has no mass parameters at all, and indeed no parameters at all, except for the coupling constant g, or equivalently $\alpha_s = \dfrac{g^2}{4\pi}$.

To complete the picture, we need to recognize that quantum-mechanically the value of the coupling depends on the distance, or equivalently energy, scale at which it is measured -
$\alpha_s \to \alpha_s(Q)$. This is, of course, a central feature of the theory, and one that by now has splendid experimental support. In other words, we can, within any given version of QCD Lite, measure any given *numerical* value a $= \alpha_s(Q)$, simply by choosing an appropriate Q. Thus the "different" versions of QCD Lite, with different values of the coupling parameter, in reality differ only trivially, by an overall choice of the energy scale. Of course the value of the overall energy scale makes a big difference when we come to couple QCD Lite, or of course QCD, to the rest of physics. Gravity, for example, cares very much about the absolute value of masses. But within QCD Lite

itself, if we compute mass ratios, or indeed any dimensionless quantity whatsoever, we will obtain a unique answer, independent of any choice of coupling parameter. Thus, properly understood, the value of the QCD coupling constant does not so much govern QCD itself - within its own domain, QCD is essentially unique - but rather how QCD fits in with the rest of physics. In that wider context, the actual value of the QCD coupling constant $\alpha_s(M_W)$, conventionally taken at the weak scale, is of the greatest importance. Among other things, it explains why gravity is feeble - see below.

The Origin of Mass

The fact that QCD Lite is completely free of continuously adjustable parameters makes it especially remarkable that for many purposes, notably including light hadron spectroscopy, this theory provides an excellent approximation to reality. Most of the mass of ordinary matter is concentrated in protons and neutrons. Numerical comparison of QCD Lite and full QCD shows that if we built protons and neutrons in an imaginary world with no Higgs mechanism - purely out of quarks and gluons with zero mass - their masses would not be very different from what they actually are. Their mass mostly arises from pure energy, associated with the dynamics of confinement in QCD, according to relation $m = E/c^2$. Thus it is QCD that mostly explains the origin of mass, and *not* the Higgs mechanism of electroweak theory (as is often sloppily stated).

The Feebleness of Gravity

We can attempt to extend the unification of couplings calculation further, to include gravity. Whereas the couplings of the Theory of Matter are dimensionless, and run with energy only logarithmically, due to vacuum polarization effects, the gravitational coupling has dimension $(mass)^{-2}$. Therefore it grows in importance with energy, even classically - and much faster. A simple-minded estimate, using dimensional analysis, indicates that the effective gravitational coupling becomes strong at $Q \sim 10^{18}$ GeV, the Planck mass. The other couplings unify at $Q \sim 10^{16}$ GeV, and plausibly become strong at a slightly higher energy. Thus the unification of couplings calculation, naively extended to include gravity, is not far off. This is quite a remarkable result, since the physical ingredients entering into the calculation are so disparate. The small residual discrepancy between the Theory of Matter unification scale and the Planck scale has been ascribed to the opening up of an extra spatial dimension near these scales, though of course in the present state of knowledge there are other possibilities.

A dramatic, but I think not unfair, way to state this result is that we have, within this circle of ideas, convincingly solved the central "hierarchy problem" of fundamental physics. By that I mean the question of why gravity, acting between life-size lumps of matter, is so feeble. Or, in more technical language, the probem of why the ratio of the Planck mass to the proton mass is so large. In our calculation this ratio is given as the exponential of inverses of the observed coupling constants in the Theory of Matter. No spectacularly small ("unnatural") quantities are involved. The big ratio of mass scales arises basically because the strong coupling $\alpha_s(M_U)$ at the unification scale is about 1/25, and the couplings run only logarithmically. Therefore quite a long run is required before one reaches the scale where $\alpha_s(Q)$ approaches unity, protons are assembled, and ordinary life begins.

ACKNOWLEDGMENTS

This work is supported in part by funds provided by the U.S. Department of Energy (D.O.E.) under cooperative research agreement #DF-FC02-94ER40818.

Nucleon Spin Physics: Experimental Status and Potential of HERMES, COMPASS and RHIC

Andy Miller

TRIUMF, Vancouver BC, Canada V6T 2A3

Abstract. Existing data that seem to shed the most light on the spin structure of the nucleon are summarized. An estimate is made of what important new data will emerge in the next 5 years from projects now in the construction or commissioning phase.

This is an attempt to summarize the present state of the data in the topics in nucleon spin physics that seem most important now, and to project the experimental progress that is likely to emerge in the next five years from the new experiments that are scheduled to begin running soon. These topics are: the helicity distributions of quarks and gluons, the transversity distributions of quarks, and the hard exclusive processes that promise to eventually shed light on parton orbital angular momenta. This list doesn't exhaust the field, but it can certainly exhaust my time.

THE FACILITIES

I thought that a useful first step would be to compare the experiments that will be running over the next five years, in terms of a few critical attributes. Table 1 compares the $\vec{l} + p$ experiments. One remarkable similarity among all the experiments is in the effective annual figure of merit $(P_B^2 P_T^2 f^2 I)$ for running polarized targets, after accounting for the target dilution factor f. On the other hand, we shall see that single-spin asymmetries with unpolarized targets have recently become very important. HERMES and JLAB excel in this area, and have strong potential in associated instrumentation to identify exclusive final states.

The RHIC-Spin experiments will soon be contributing strongly to spin physics at higher energy: $\sqrt{s} = 200\text{--}500\,\text{GeV}$. The polarized luminosity is expected to be $0.3\,\text{fb}^{-1}$ per year (again similar to the lepton facilities), with

Facility	beam	P_{beam}	target (dilution factor)	Luminosity/year
HERMES	27.5 GeV $\vec{e}^{\pm}$	55%	pure atomic $\vec{H}$ or $\vec{D}$(1.0)	0.16 fb^{-1} (pol)
				0.5 fb^{-1} (unpol)
COMPASS	100-200 GeV $\vec{\mu}^{\pm}$	80%	N$\vec{H}_3$ (0.18) or ^{6}Li$\vec{D}$ (0.5)	2 fb^{-1}
JLAB	6(-12?) GeV $\vec{e}^-$	90%	(un)polarized solids	"unlimited"
HERA Coll.	$\vec{e}^{\pm}$+p	55%	unpolarized protons	
	$\sqrt{s}=320$ GeV			50 pb^{-1}

TABLE 1. The $\vec{l}+p$ Experiments

colliding proton polarizations of 70%. (However, such a high value has not yet been demonstrated in a high energy ring, and it is worth noting that the figure of merit for two-spin asymmetries scales as the 4'th power in this case!) The features of the two collider detectors at RHIC are somewhat complementary. PHENIX has fine-grained EM calorimeters with very limited coverage ($|\eta| < 0.35$, $\Delta\phi = \pi$), but 2 muon arms for tagging decay of heavy flavors and W's, while STAR will eventually have an EM calorimeter with extensive coverage ($1 < \eta < 2$, $\Delta\phi = 2\pi$) for jet studies and prompt photons.

POLARIZED QUARK DISTRIBUTIONS

Most of what we know about nucleon structure has been learned from Deep-Inelastic Scattering (DIS) of high energy leptons. In DIS with polarized lepton beams and polarized nucleon targets, one probes the polarization of the quarks in a polarized nucleon by exploiting the simple fact that a quark can absorb a virtual photon only if their helicities are opposite.

A key limitation of inclusive DIS measurements is that they are sensitive only to linear combinations of flavors weighted by the *squares* of the quark charges, and hence can't distinguish quarks from anti-quarks. Thus measurements of the structure function g_1 on both proton and neutron (deuteron) targets fix only the non-singlet distribution Δq_3 (the integrand of the Bjørken Sum) among:

$$\Delta q_3(x, Q^2) = \Delta u + \Delta \bar{u} - \Delta d - \Delta \bar{d} = 6(g_1^p - g_1^n), \quad (1)$$

$$\Delta q_8(x, Q^2) = \Delta u + \Delta \bar{u} + \Delta d + \Delta \bar{d} - 2(\Delta s + \Delta \bar{s}), \quad (2)$$

$$\Delta \Sigma(x, Q^2) = \Delta u + \Delta \bar{u} + \Delta d + \Delta \bar{d} + \Delta s + \Delta \bar{s} \quad (3)$$

To the extent that $\Delta s + \Delta \bar{s}$ is negligible, $\Delta u + \Delta \bar{u}$ and $\Delta d + \Delta \bar{d}$ are also strongly constrained.

As we all know, our field took its modern shape from the combination of inclusive DIS data with baryon β-decay lifetimes, which constrain the non-singlet first moments. The admittedly questionable assumption of SU(3) flavor-symmetry leads to the famous conclusions that $\Delta\Sigma$ is small and Δs is negative — the so-called spin crisis or puzzle, depending on your taste.

As in the unpolarized case, the scaling violation of $g_1(x, Q^2)$ distinguishes between the various parton distributions — i.e. by exploiting the different Q^2-dependent behaviour of each term in:

$$g_1^{p(n)} = \frac{1}{9}\left(C_{NS} \otimes \left[\pm\frac{3}{4}\Delta q_3 + \frac{1}{4}\Delta q_8\right] + C_S \otimes \Delta\Sigma + 2N_f C_g \otimes \Delta g\right), \qquad (4)$$

one can *in principle* separate these polarized parton distribution functions. However, this has thus far yielded less of value than in the unpolarized case, for several reasons. Most importantly, the precision and range of present polarized data is insufficient. Fig. 8 compares the polarized and unpolarized world data sets. The existing NLO QCD fits to inclusive polarized DIS data strongly constrain $\Delta\Sigma$ but are inconclusive regarding either Δs or Δg. The HERMES collaboration now have in hand high-quality deuterium data that are yet to be analyzed.

A technique that can distiguish the polarizations of anti-quarks is semi-inclusive DIS, where the flavor of the struck quark is 'tagged' by a hadron h detected in coincidence with the scattered lepton. This technique has been applied by both SMC [1] and HERMES [2], with only limited success up to now mainly because of the limitations of the published semi-inclusive data on the neutron. However, this situation is also about to change radically when the existing extensive HERMES data set on the deuteron is analyzed.

What can be expected from both HERMES and COMPASS [3]? The statistical error bars will become negligible for all but the strange sea, with the precision limited by the model-dependence implicit in the fragmentation process as modelled in the Monte Carlo. The comparison of HERMES and COMPASS results at very different energies will help to define that model uncertainty.

RHIC will soon begin to provide a complementary and very elegant experimental perspective on polarized quark distributions, based on the large single-spin asymmetries that arise from the maximal parity violation in the weak processes $u\bar{d} \to W^+ \to l^+ + \nu$ and $d\bar{u} \to W^- \to l^- + \nu$. Hence only the polarization of one beam is used for each analysis (reducing the sensitivity of the figure of merit). The inferred rapidity of the W determines whether the momentum fraction x_1 from the polarized beam proton exceeds x_2 from the unpolarized proton, in which case the asymmetry is dominated by valence quarks. Otherwise, it's dominated by sea quarks. Obviously, corrections based on other types of data are needed for the contributions of the sea (valence) quarks at large (small) x. The W rapidity can be only approximately deduced from the kinematics of the detected lepton, by neglecting the P_T of the W, but this is expected to work well enough, as shown by the projections for PHENIX in Fig. 1. The STAR EM calorimeter end-caps will not only enhance the rates but also presumably extend the x-range.

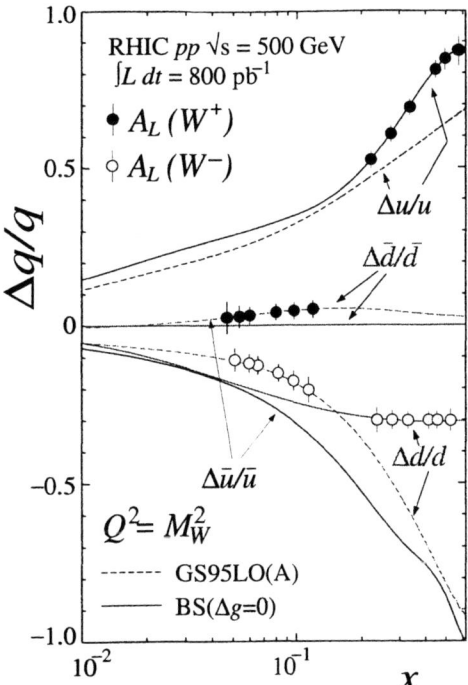

FIGURE 1. PHENIX sensitivity from $\mu^{\pm}$ decays: (8k W^+ + 8k W^-) and $e^{\pm}$ decays: (15k W^+ + 2.5k W^-). (STAR will see more electron decays: 72k W^+ + 21k W^-)

GLUON POLARIZATION

The "spin puzzle" is often expressed in terms of the latter two unknown terms in the schematic equation

$$\frac{1}{2} = \frac{1}{2}\Delta\Sigma + \Delta G + L_z \qquad (5)$$

arising from gluon helicity and the orbital angular momenta of both quarks and gluons. As we have seen, existing indirect information about ΔG is inconclusive. For a direct measurement of ΔG, one must employ processes where gluons enter in leading order. The two most important experimentally accessible processes, photon-gluon fusion (PGF) and direct photon production, are closely related, and both have large analyzing powers. Using polarized (quasi-)real incident photons at $\vec{l} + \vec{p}$ facilities, the two useful experimental signals of PGF ($\gamma + g \to q + \bar{q}$) are heavy flavor production and dual jets. At existing fixed target energies, only charm is available, and jet pairs must be replaced by pairs of hadrons with large enough P_T to provide a hard scale, subject to model-dependent interpretation.

At COMPASS and possible at HERMES, the 'golden' open charm signal is $(D^* \to \pi +) D^0 \to K^-\pi^+$, which offers the clearest interpretation. (J/ψ production is not so transparent.) However, only decays with relatively small branching fractions provide identifiable signals, so the projected statistical precision is underwhelming, even from 2.5 years running at COMPASS: $\delta(\frac{\Delta G(0.17)}{G}) \sim 0.11$. Also, x_{gluon} can't be reconstructed, only averaged, and the x coverage is limited: $0.1 < x_{gluon} < 0.3$ at 100 GeV.

Pairs of high-$p_\perp$ hadrons offer much better statistical precision and a wider x-range at HERMES and COMPASS, at the cost of larger model-dependence in the interpretation. Several competing processes can dilute the signal, although PGF has a unique negative analyzing power here. HERMES may have already seen the signal, having reported a positive extracted value for $\Delta G/G$ [5]. Additional data in hand should double their statistics. It will be possible to crudely reconstruct x_{gluon}, over the range $0.04 < x_{gluon} < 0.2$ at COMPASS (200 GeV). The projected COMPASS statistical precision is $\delta(\frac{\Delta G(4\ x-\text{bins})}{G}) \sim 0.05$ from 1 year. One possible technical complication is that the optimum COMPASS beam energy differs from that for charm production.

At RHIC with both beams polarized, the "golden signal" will be prompt photon production ($q + g \to \gamma + \text{jet}$), preferably in coincidence with the away-side jet to provide kinematic reconstruction of x_{gluon}. The quark polarization is a required input to the analysis, to be taken from the measurements discussed above. Over limited kinematic regions (small x), lepton pairs might have some advantage over real photons at a cost in statistics, as this signal is thought to be less vulnerable to competing sources of photons.

The limited calorimetric coverage of PHENIX precludes jet recognition, allowing only a smeared mean $\langle x_{gluon}\rangle$ to be calculated in the range $0.023 < \langle x_{gluon}\rangle < 0.26$, but its fine granularity helps to spatially resolve the two photons from π^0 decay, which constitutes an intense background to be identified and rejected. Only photons isolated from possible jet partners are selected, to further suppress background from π^0's, as well as fragmentation photons: e.g. $(q\bar{q})^* \to \gamma$.

The large coverage of the STAR EM calorimeter will allow recognition of the away-side jet. When the endcaps are complete, they will accept the kinematic asymmetry for large x_{quark} that provides the large $\Delta q/q$ needed to analyze the gluon helicity. A projection of the quality of kinematic reconstruction that this will provide is shown in Fig. 2. Only isolated photons were selected in the kinematic range $-1 < \eta^\gamma < 2$ and $10 < p_\perp < 20 GeV$. Initial-state k_T-smearing included in the simulation. Most reconstruction errors occur in the less interesting kinematic region $x_{quark} < x_{gluon}$, while a strong correlation allows direct reconstruction of $\Delta G(x)$ in the complementary interesting region.

The projected quality of results from the extraction of ΔG from such STAR data depends strongly on the assumed values for ΔG, as illustrated in Fig. 3.

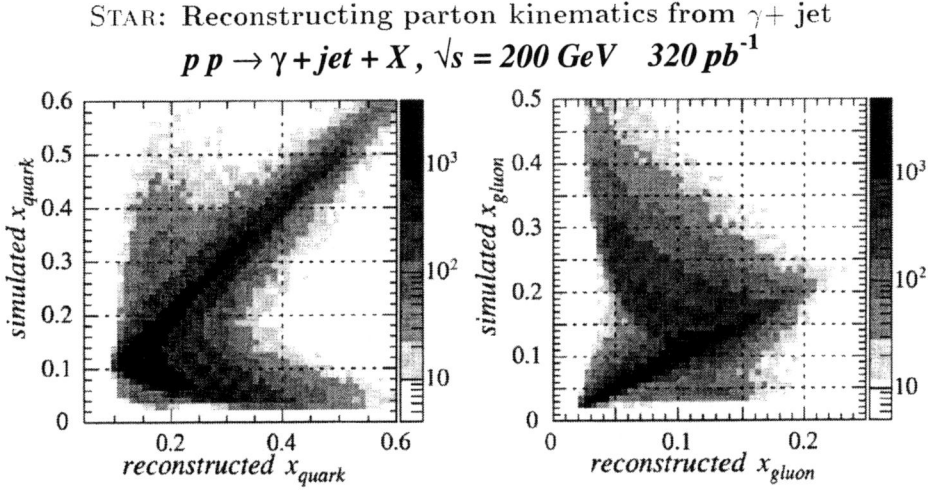

FIGURE 2. Prompt photon + jet in STAR

The reconstructed values miss the assumed input because background from $q\bar{q} \to \gamma g$, the kinematic reconstruction errors shown above, and k_T smearing are all neglected at this stage of the simulation. Data at both $\sqrt{s} = 200$ and 500 GeV are needed for full coverage in x_{gluon}. When the simulated data from the two beam energies are combined and fitted with a parametric form

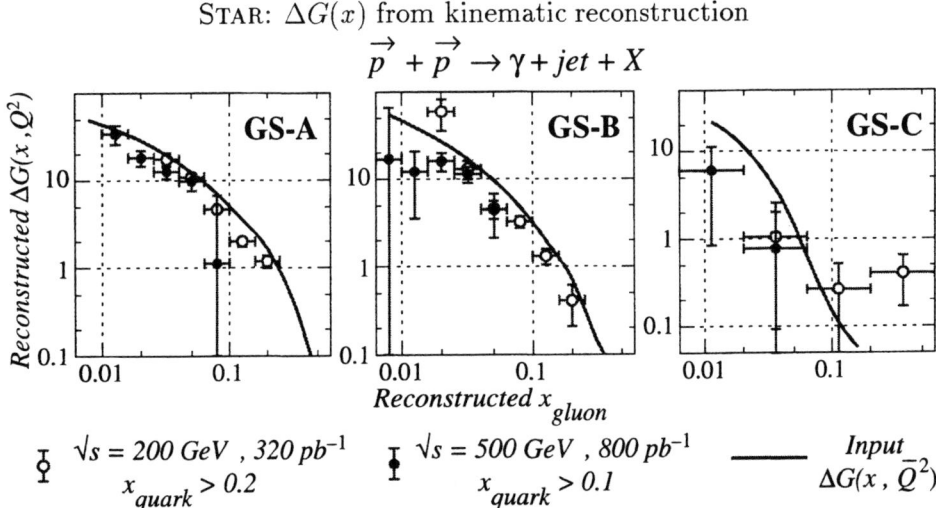

FIGURE 3. Simulated extraction of $\Delta G(x)$ from prompt photon + jet in STAR, for three different predictions for ΔG from NLO fits to existing world DIS data.

to extract the integral, the results suggest that a precision in the integral of ±0.5 including both statistical and systematic contributions could be achieved in the eventual full maturity of the program. Some theoretical progress will be required in the meantime to understand existing problems extracting $G(x)$ from unpolarized prompt-photon data.

As a final point about gluon polarization, I have attempted to compare the projected contributions of the photoproduction experiments with those of RHIC. However, this comparison is to be interpreted with caution, as the STAR error bars have been 'evolved' to $Q^2 = 2.2\,\text{GeV}^2$ by scaling with the root of the signal evolution according to Germann and Stirling. This procedure is based on the assumption that the difference between the various parameterizations, and hence the ease with which they can be distinguished, scales with Q^2 approximately as does their magnitude. Note that since it is $x\Delta G(x)$ that is plotted versus $\log(x)$, the areas under the curves represent the contributions to the intergral. (The COMPASS resolution in x_{gluon} from high P_T hadrons is such that their adjacent points are barely resolved.) One can conclude that there will be many meaningful cross-checks between the various techniques, but only the eventual STAR results can provide full x coverage.

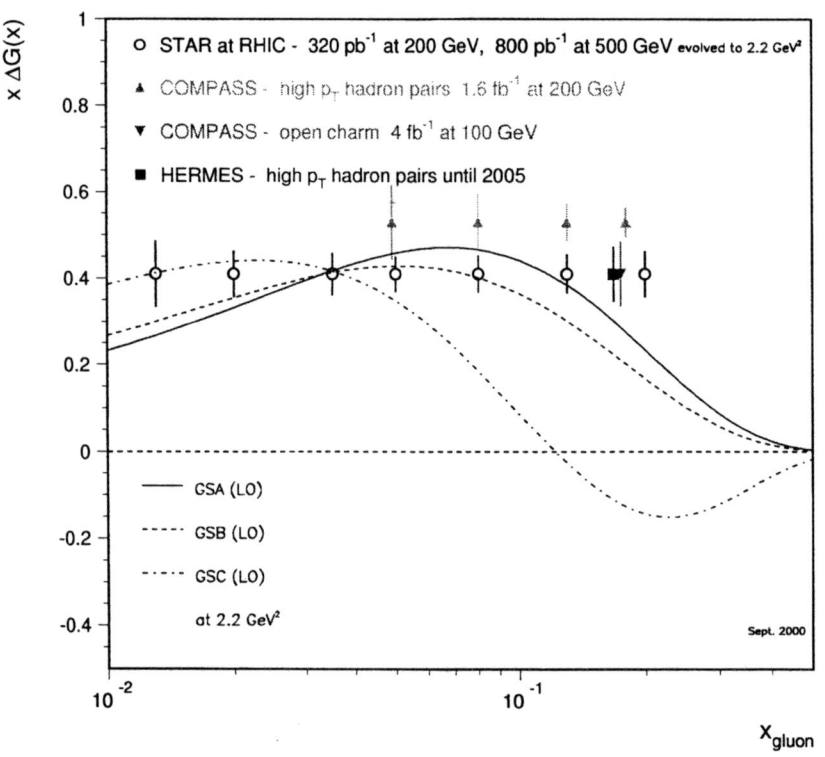

FIGURE 4. Comparing Experiments: $x\Delta G(x)$

TRANSVERSITY

A complete description in leading twist (leading order in Λ_{QCD}/Q) of the quarks in the nucleon requires three flavor-sets of distributions: the unpolarized $q(x)$ contained in $F_1(x)$, the helicity distributions $\Delta q(x)$, contained in $g_1(x)$ and the still-unmeasured set of transversity distributions $\delta q(x)$, whose structure function will here be denoted as $h_1(x)$. (Hereafter we use the singular term 'distribution' to imply that for any particular flavor.) This description can be illustrated as a spin-density matrix representation of the leading-twist quark distribution [6]:

$$\mathcal{F}(x,Q^2) = \frac{1}{2}q(x,Q^2)I^q \otimes I^N + \frac{1}{2}\Delta q(x,Q^2)\sigma_3^q \otimes \sigma_3^N$$
$$+ \frac{1}{2}\delta q(x,Q^2)(\sigma_+^q \otimes \sigma_-^N + \sigma_-^q \otimes \sigma_+^N),$$

where the Pauli matrices lie in the quark and nucleon helicity spaces as indicated. Thus in the helicity basis, δq is the helicity-flip distribution ($N^\uparrow q^\downarrow \to N^\downarrow q^\uparrow$), in contrast to the familiar $\Delta q = q^\uparrow - q^\downarrow$ as a helicity difference ($N^\uparrow q^\uparrow \to N^\uparrow q^\uparrow - N^\uparrow q^\downarrow \to N^\uparrow q^\downarrow$). In this basis, $\delta q(x)$ has no probabilistic interpretation. On the other hand, in a basis of transverse spin eigenstates, their roles are interchanged: $\delta q = q^\perp - q^\top$, This allows us to think of transversity as the distribution of quarks polarized parallel to the nucleon spin minus the distribution polarized antiparallel, in analogy to the longitudinal case of g_1 except that now the polarization of the nucleon is transverse to its (infinite) momentum, rather than parallel. In a non-relativistic quark model, this is just a simple rotation, and in fact such models predict $h_1 = g_1$. However, these models predict $g_A/g_V = 5/3$, in conflict with reality ($g_A/g_V \simeq 1.26$), reflecting the fact that in the relativistic context of current quarks, boosts and rotations do not commute. Hence differences between h_1 and g_1 can be expected from relativistic effects in the quark bound state.

What is now known as transversity was first identified in a study of the transverse asymmetry in polarized Drell-Yan reactions [7]; its general relationship to nucleon spin structure was articulated some 10 years later [8,9]. Theoretically h_1 is on an equal footing with the first two twist-2 structure functions, and is well defined as the matrix element of a bilocal quark (tensor) operator [9]. (There is no gluon transversity distribution.)

The reason that transversity is still unmeasured is that it is chiral-odd, reflecting its role as a helicity-flip distribution. (At twist-2, chirality and helicity are identical.) This chiral-odd property means that it is strongly suppressed in inclusive lepton scattering, since the hard quark couplings to the photon and gluon preserve chirality, up to quark mass effects. (In the non-perturbative context of quark distributions or fragmentation functions, chiral symmetry is spontaneously broken.) Only processes that combine two chiral-odd objects (and experiments that 'observe' two hadrons) can be usefully sensitive to h_1.

The simplest combination was mentioned above — the transversity distributions of two colliding transversely-polarized protons. The possibility of using the Drell-Yan signal at RHIC has been studied, indicating that the expected asymmetry is very small, leading to poor statistical precision [10].

The combination of a chiral-odd *fragmentation function* with the chiral-odd distribution function δq can also provide an experimental signal for transversity. Physically, we may think of such a fragmentation function as a "polarimeter" for the transverse polarization of the struck quark. Such a chiral-odd fragmentation function has been identified theoretically [11]. This 'Collins function' is often designated as $H_1^{\perp(1)}(z)$, where z represents the fraction of the struck quark's energy carried by a hadron produced by fragmentation of the struck quark. (The subscript '1' indicates leading twist, while the '$\perp$' and '(1)' superscripts indicate respectively an essential role for the intrinsic transverse momentum k_T of the quark in the hadron, and that the quantity represents the k_T^2-moment.) To use the Collins function as a transverse quark 'polarimeter', the experimenters must distinguish between hadrons produced in directions to the left and right, relative to the spin vector of the struck quark in a frame in which it is pointing 'up'. In other words, the experimental signature for transversity operating in conjunction with the Collins function is an inhomogenous distribution of the hadrons in the azimuthal angle ϕ of the hadron about the direction of the struck quarks's 3-momentum $\mathbf{q}$, relative to the plane containing both $\mathbf{q}$ and the target nucleon polarization axis. This azimuthal asymmetry is proportional to the target polarization. Such a single-spin asymmetry must involve a time-reversal odd object, which in this case is the Collins function. Its T-odd nature arises through the final-state interactions that it represents, rather than any fundamental symmetry breaking. Until recently, there was only tentative evidence that the Collins function has a significant magnitude, from both effects on jet structure due to the very small transverse quark polarization produced in Z^0 decay: $e^+ + e^- \to Z^0 \to q\bar{q} \to 2$ jets [12], and from the interpretation of large single-spin asymmetries observed in inclusive pion production from polarized p + p̄ collisions [13].

Some data exist for semi-inclusive DIS on a transversely polarized target [14], but they are limited and indecisive. Much smaller transversity-related effects based on the same Collins function are expected with a *longitudinally* polarized target. Using such a target, a single-spin azimuthal asymmetry was recently observed [15]. The sinusoidal moment of the target-related spin asymmetry corresponds to an analyzing power of $0.022 \pm 0.005(\text{stat}) \pm 0.003(\text{sys})$ for positive pions, and is consistent with zero for negative pions. It has been found that these data are consistent with models for the relevant distribution functions together with a model for the Collins fragmentation function that is also consistent with the preliminary value for this function from Z^0 decay, as well as with the large single-spin asymmtries that have been observed in inclusive pion production from p↑p or p̄↑p $\to \pi^\pm + X$ [16,17]. Hence it appears that the Collins fragmentation function is responsible for the effect and has a substantial value, thus opening the way to future measurements of transversity using transversely polarized targets. This finding has inspired firm plans to make the first measurement of at least the *shape* of the x-dependence of δu at HERMES, exploiting u-quark dominance to analyze data from only a proton target [18]. The normalization is subject to uncertainty about the magnitude of the Collins function. It can be fixed by an assumption about the equality of h_1 and g_1 at large x and relatively snall Q^2. There is also a possibility of following this with later data on the deuteron to make a flavour decomposition. However that result is expected to be severely limited by statistics, as illustrated by the simulation of the expected precisions shown in Fig. 5. COMPASS has announced no specific plans to measure transversity, but they have the potential to add the small-x region.

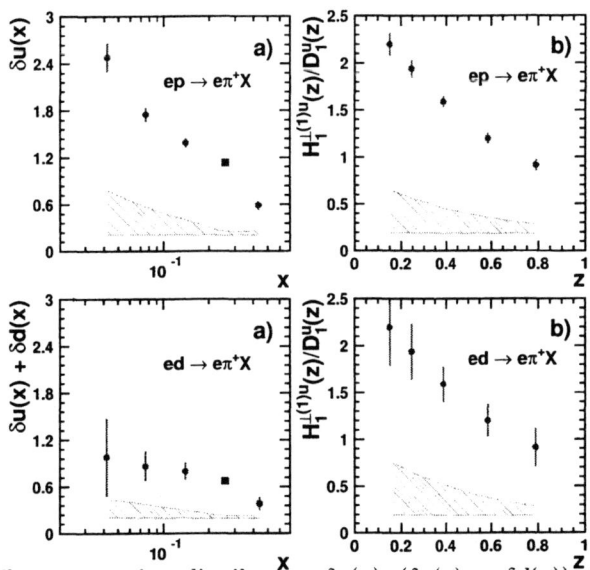

FIGURE 5. The transversity distribution $\delta u(x)$ $(\delta u(x) + \delta d(x))$ from the proton (deuteron) target, as a function of x, and the ratio of the Collins to the unpolarized fragmentation function as a function of z, as they could be measured by HERMES using 7 Million DIS events ($200\,\text{pb}^{-1}$) on *either* target. If *both* were measured, then much of the large component of the systematic uncertainty for $\delta u(x)$ arising from the assumption of u-quark dominance would presumably disappear, and the statistical uncertainties in the fragmentation functions would all be similar to those shown for the proton, with consequent improvement in the precision for $\delta u(x) + \delta d(x)$. For both targets, the d flavor of the Collins fragmentation function is neglected.

Other potential experimental signals of transversity have been identified. One is coherent production of a system of two leading pions with orbital angular momentum about the struck quark direction, in both s and p states that interfere [19]. The experimental signal is a single-spin asymmetry in the azimuthal distribution of the plane containing both pion momenta about the struck quark direction, with respect to the direction of the transverse target polarization. The asymmetry changes sign at a value of invariant mass of the two pions equal to the ρ^0 mass. The magnitude of the chiral-odd T-odd 'interference' fragmentation describing this effect is presently unknown. It has only been ascertained that its positivity limit does not rule out its usefulness, in particular in possible measurements of transversity in $\vec{p} + \vec{p}$ collisions at RHIC [10].

Several promising experimental signals of transversity are now the subjects of vigorous theoretical and experimental investigation. Evidence for the utility one of them has already appeared in experimental data, and has resulted in plans to make the first exploratory measurement in the near future. Others have also been recently identified [8,20,21]. It is likely that more signals will be found to be useful.

SKEWED PARTON DISTRIBUTIONS

As recently as 5 years ago, one aspect of the nucleon spin puzzle threatened to remain clouded in mystery. No practical experimental access to the *orbital* angular momenta of partons had been identified. However, a seminal paper [22] led to an explosion of theoretical activity, and the recent appearance of intriguing new data. This paper demonstrated that the so-called generalized or 'skewed' parton distributions (SPD's), embody information about parton orbital angular momentum, and that information about this could be extracted from exclusive processes such as Deeply Virtual Compton scattering (DVCS), illustrated in Fig. 6.

There are four SPD's, two of which appear in inclusive DIS — the unpolarized function $H(x,\xi,t)$ and the polarized one $\tilde{H}(x,\xi,t)$. In the forward limit, they reduce to familiar distributions:

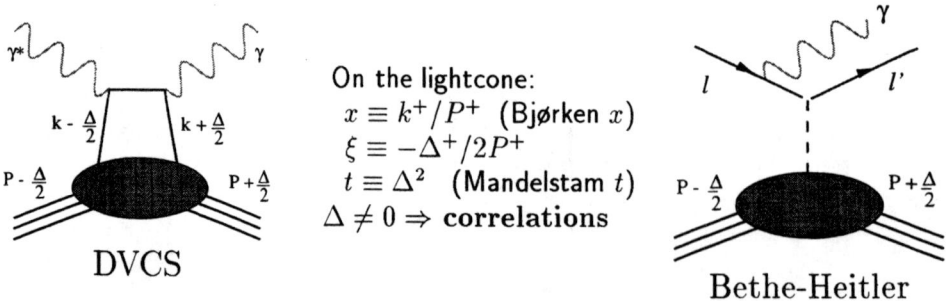

FIGURE 6. Feynman ('handbag') diagram for DVCS, and for the indistinguishable background process — radiative 'elastic' lepton scattering (Bethe-Heitler).

$$q(x) = H^q(x, \xi = 0, t = 0), \quad \text{and} \quad \Delta q(x) = \tilde{H}^q(x, \xi = 0, t = 0). \tag{6}$$

Two analogous functions are associated with target helicity flip and appear experimentally only at finite t: $E(x, \xi, t)$ and $\tilde{E}(x, \xi, t)$. Although they all are functions of three variables, their behavior is strongly constrained by symmetry principles — e.g. their first x-moments are independent of ξ and are connected with elastic form factors.

Of primary interest to us is the 'Ji Sum Rule' that relates their second moments to *total* (including *orbital*) quark angular momentum:

$$J^q = \frac{1}{2}\Delta\Sigma + L^q = \frac{1}{2}\int_{-1}^{1} x \, dx \, [H^q(x, \xi, t = 0) + E^q(x, \xi, t = 0)] \tag{7}$$

It is the second term involving E^q to which experimental access is difficult. It must be measured at finite t and extrapolated to the forward limit. Another apparent obstacle is a large background contribution from the indistinguishable Bethe-Heitler process, radiative 'elastic' lepton-proton scattering. However, this can be turned to advantage. Interference between the two processes provides direct experimental access to the DVCS amplitudes by giving rise to large asymmetries in the azimuthal distribution of the detected photons about the direction of the incident virtual photon. These asymmetries are associated with e.g. the beam lepton charge, and also in some cases with its polarization. Furthermore, each of several different moments of these azimuthal asymmetries is associated with a particular linear combination of DVCS amplitudes in the interference cross section.

HERMES kinematics are ideal for observing the DVCS/Bethe-Heitler interference. Fig. 7 shows the first measurement of lepton beam spin asymmetries in the azimuthal distribution of photons, in leptoproduction from an (unpolarized) hydrogen target. In the vicinity of $M_X = M_p$, the observed $\sin\phi$ moment of the asymmetry is large, as predicted. Several other observables can be expected to soon emerge from HERMES data, for both DVCS and exclusive production of various mesons. A recoil detector to identify truly exclusive final states is under development. The HERA collider experiments as well as Jefferson Laboratory are expected to contribute DVCS data in complementary kinematic conditions. COMPASS has initiated studies for possible future recoil detection from an unpolarized target. At this higher energy, competition from the Bethe-Heitler process is relatively small, so that the accessible observables are complementary. This field is still in its infancy, and a large program will be needed to understand and disentangle the SPD's. Transverse target polarization is expected to be particularly fruitful. Each measurement constrains only a moment of one or more functions, along a trajectory in the 3-dimensional kinematic space. Exclusive meson production by longitudinal photons can provide independent constraints. The present exploratory phase can be expected to continue for several more years.

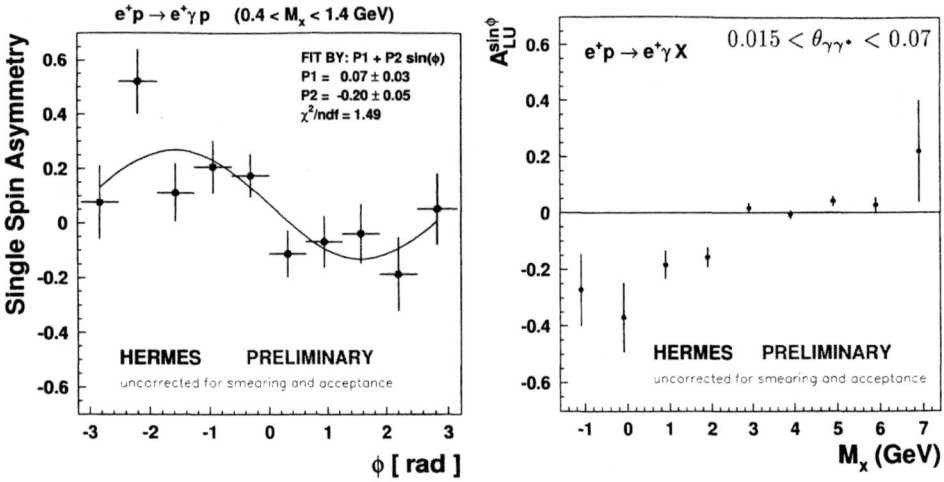

FIGURE 7. HERMES lepton beam spin asymmetries in the azimuthal distribution of photons produced from an unpolarized hydrogen target.

SUMMARY — A PERSONAL PERSPECTIVE

- Helicity distributions of quarks
  - The $\vec{l}+\vec{p}$ facilities and RHIC are complementary:
  - $\vec{l}+\vec{p}$ provides wider x-coverage and better statistics.
  - RHIC provides a definitive clean process.
  - The valence and light sea polarizations will be well defined.
  - The strange sea polarization will become as well known as the presently-published valence polarizations.

- Helicity distributions of gluons
  - HERMES and COMPASS will provide only tentative information.
  - Only if RHIC and STAR perform up to specifications will we have reliable experimental information on the integral of ΔG.
  - COMPASS and PHENIX will provide important cross-checks.

- Transversity distributions of quarks
  - HERMES should provide a respectable first measurement of at least the *shape* of δu, and perhaps δd.
  - The plans of COMPASS here have not been announced.
  - There is potential at RHIC which is under active study.

- Hard Exclusive Processes
  - This new part of the field is bursting with excitement.

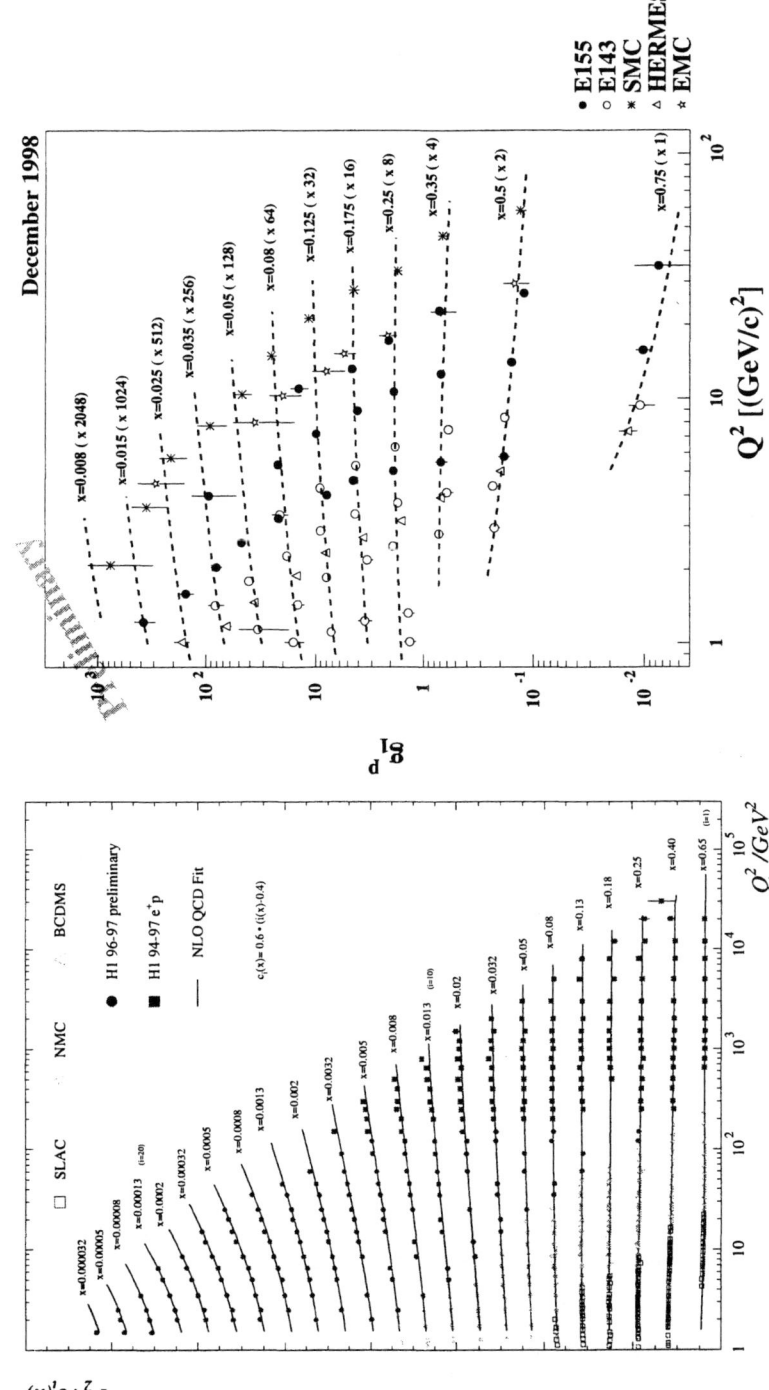

FIGURE 8. Comparison of world data for unpolarized and polarized nucleon structure functions.

- The ultimate goals are J_q, J_g.
- We're are only beginning to learn how to exploit these processes.
- HERMES, JLAB and HERA will provide lots to think about, but no final answers.
- Truly exclusive measurements will be needed.

REFERENCES

1. SMC, Adeva B. et al., *Phys. Lett.* B **420**, 180 (1998).
2. HERMES, Ackerstaff K. et al., *Phys. Lett.* B **464**, 123 (1999).
3. The COMPASS Collaboration, CERN/SPSLC 96-14 (1996).
4. Leader E., Sidorov A. and Stamenov D., *Phys. Lett.* **B488**, 283 (2000).
5. HERMES, Airapetian A. et al., *Phys. Rev. Lett.* **84**, 2584 (2000).
6. Jaffe R.L., Proc. of the 2nd Topical Workshop *Deep Inelastic Scattering off Polarized Targets: Theory Meets Experiment*, DESY 97-200, Zeuthen, 1997, p.167, ed. by Blümlein J., de Roeck A., Gehrmann T., and Nowak W.-D.; hep-ph/9710465.
7. Ralston J.P. and Soper P.E., *Nucl. Phys.* **B152**, 109 (1979).
8. Artru X., Mekhfi M., *Z. Phys.* C **45**, 669 (1990).
9. Jaffe R.L. and Ji X., *Nucl. Phys.* **B375**, 527 (1992).
10. Bunce G., Saito N., Soffer J. and Vogelsang W., *Ann. Rev. Nucl. Part. Sci.* **50**, 525 (2000).
11. Collins J., *Nucl. Phys.* **B396**, 161 (1993) and **B420**, 565 (1994).
12. DELPHI: Efremov A.V., Smirnova O.G. and Tkatchev L.G., *Nucl. Phys. (Proc. Suppl.)*, **74**, 49 (1999) and **79**, 554 (1999).
13. Boglione M. and Leader E., *Phys. Rev. D* **61**, 114001 (2000).
14. Tripet A., *Nucl. Phys.* B **(Proc. Suppl.) 79**, 529 (1999).
15. HERMES, Airapetian A. et al., *Phys. Rev. Lett.* **84**, 4047 (2000).
16. Efremov A.V. et al., *Phys. Lett.* **B478**, 94 (2000).
17. Boglione M. and Mulders P.J., *Phys. Lett.* **B478**, 114 (2000).
18. Nowak W.-D., 8th Int'l Workshop on DIS and QCD (DIS 2000), Liverpool, England, Apr, 2000, to be published in World Scientific.
19. Jaffe R.L., Jin Xuemin and Tang Jian, *Phys. Rev. Lett.*, **80**, 1166 (1998).
20. Ji X., *Phys. Rev. D* **49**, 114 (1994).
21. Bacchetta A. and Mulders P.J., *Phys. Rev. D* **62**, 114004 (2000).
22. Ji X., *Phys. Rev. Lett.* **78**, 610 (1997).

Overview of Skewed Parton Distributions

Andreas Schäfer

Institut für Theoretische Physik, Universität Regensburg, D-93040 Regensburg, Germany

Abstract: The central ideas of the intense recent discussion of Skewed Parton Distributions are reviewed.

It was always understood that a description of exclusive processes from first principles would be the next logical step for QCD, after its basic properties were established by analysing inclusive and semi-inclusive reactions and that this step would be substantially more difficult. For a few years, a new theoretical concept has attracted much attention in this context. The so called 'Skewed Parton Distributions' (SPDs) [1, 2, 3] allow for a unified QCD description of many reactions in electron-nucleon scattering and therefore make it possible to relate them in the most comprehensive way, see Fig.1 for an incomplete illustration.

While the usual parton distributions are related to hadronic forward matrix elements, SPDs are related to off forward matrix elements, see Fig. 2. The precise definition of e.g. two of the quark SPDs is:

$$\int \frac{d\lambda}{2\pi} e^{i\lambda x} \quad \langle p'|\bar{\psi}(-\lambda n/2)\gamma^\mu \psi(\lambda n/2)|p\rangle$$
$$= H_q(x,\xi,\Delta^2)\bar{U}(p')\gamma^\mu U(p)$$
$$+ E_q(x,\xi,\Delta^2)\bar{U}(p')\frac{i}{2M_N}\sigma^{\mu\nu}\Delta_\nu U(p) \quad (1)$$

where ψ, $\bar{\psi}$ are quark fields at the space-time points $\pm\lambda n/2$, n is a 4-vector defined such that $n \cdot (p+p')/2 = 1$ and $n^2 = 0$, and

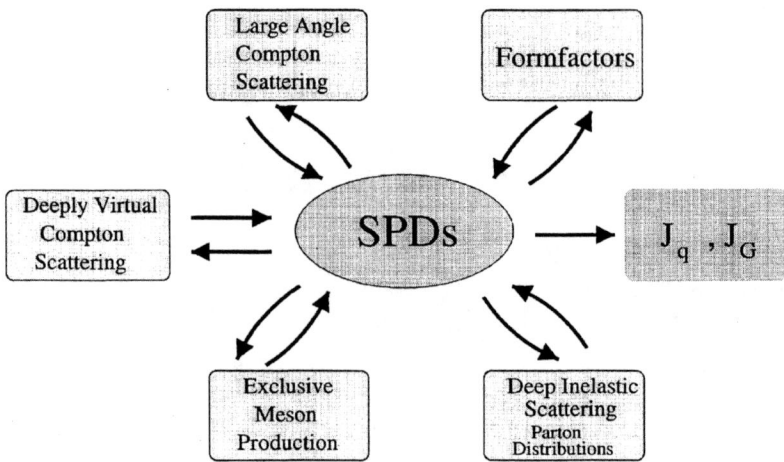

Figure 1: Skewed Parton Distributions, a unifying description of many reaction channels, determine the total angular momentum of quarks, J_q, and gluons, J_G.

$\Delta_\nu = p'_\nu - p_\nu$ is the momentum transfered to the proton. x and ξ parametrize the longitudinal momentum fraction of the quark fields, as illustrated in Fig.2 (negative values correspond to antiquarks), and $H_q(x, \xi, \Delta^2)$ and $E_q(x, \xi, \Delta^2)$ are the SPDs. Obviously $E_q(x, \xi, \Delta^2)$ can only be determined if $\Delta_\nu \neq 0$, i.e. in exclusive processes with momentum transfer to the proton.

As illustrated in Fig. 1, SPDs describe a variety of different reactions. In the forward limit, i.e. when the proton momenta p and p' become equal, one recovers the usual parton distributions, which

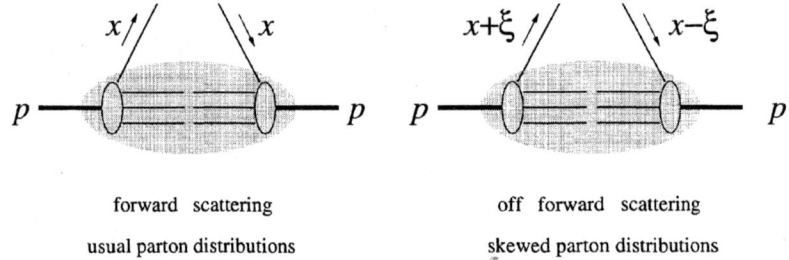

Figure 2: The relation between SPDs and usual distribution functions

thus provide boundary conditions for the SPDs. Via sum rules, i.e. equations fulfilled by integrals over the momentum fraction x, they are also connected to elastic nucleon form factors like F_1 and F_2 (Pauli and Dirac formfactor). Thus they relate the two types of quantities which so far have been the prime sources of our knowledge of hadron structure: parton distributions which tell us about the longitudinal momentum structure of a fast-moving nucleon, and form factors contain information on its transverse structure, such as its charge radius.

In addition to the situation shown in Fig.2, where a parton is extracted from the nucleon with a certain momentum fraction and returned with a different one, SPDs also have a region in x where a quark-antiquark or a gluon pair is emitted from the initial proton p, leaving it with momentum p'.

As in the case of form factors, SPDs can also describe the transition between different hadrons, allowing one to probe the overlap of their respective wave functions. This opens the way to study baryons not available as beam particles. Prominent examples of transition SPDs appear for the reactions $ep \to e\Delta^+ \gamma$ and $ep \to en\,\pi^+$.

The great advantage of SPDs is generally speaking that their large information content allows one to connect different observables and to determine quantities of physical interest which one cannot extract directly from individual observables. Let us illustrate this statement by discussing the most prominent example namely the total angular momentum of quarks, J_q:

$$\frac{1}{2}\left[\int_{-1}^{1} dx\, x\, [H_q(x,\xi,\Delta^2) + E_q(x,\xi,\Delta^2)]\right]_{\Delta^2 \to 0} = J_q \qquad (2)$$

and of gluons, J_G. Up to now only the spin fraction of the nucleon originating from quark and gluon spins could be studied. Combining such results with knowledge of J_q and J_G would obviously improve our understanding of the nucleon spin structure, although the relation of e.g. the difference $J_G - S_G$ to the expectation value of some well-defined orbital angular momentum operator is still theoretically obscure [4].

The general experimental and theoretical task to investigate SPDs is easily defined. One should measure as many reactions of the type cited in Fig. 1, determine the relevant SPDs by a combined fit to all of them and then extract J_q, J_G and any other information of interest.

In practice, however, the situation is substantially more difficult. The very fact that SPDs contain so much information on quite different processes could imply that they have complicated analytic properties. At least they could turn out to be far less smooth than the usual distribution functions implying the neccessity of a complicated many parameter fit, which in turn would require large statistics. Also, the experimental observables are not directly proportional to the SPDs but rather to convolutions of them with a hard scattering kernel. Finally the SPDs have a real and imaginary part and there are several of them, which have to be disentangled. Thus the investigation of SPDs is an experimental and theoretical challenge. The rapid progress achieved in the last years motivates, however, the general optimism encountered in this field:

Experimetally, ZEUS, H1 and HERMES have released first results on DVCS. These data are in good agreement with theory and show that the effects one is looking for are rather large. COMPASS and JLab are expected to add to these measurements soon.

Theoretically the following major steps have been achieved: Factorization proofs [3, 5] guarantee that SPDs are well-defined QCD objects. The Next-to-Leading-Order (NLO) evolution equations for SPDs have been derived [6]. Als the coeffcient functions for e.g. Deeply Virtual Compton Scattering were analysed [6]. Severe constraints have been obtained for th epossible shape of SPDs. It was shown, how different spin and angular asymmetries allow to disentangle the different SPDs [6, 7]. Specific models have been developed, which provide a good guess for how SPDs look like [8]. The coupled issues of gauge invariance and higher twist contributions starts to be understood. Finally, large parts of the extensive theoretical development just sketched has been adapted to the description of other exclusive channels, e.g diffractive meson production [9]. Along these lines most interesting relations to hadronic distribution amplitudes (often called wave-functions) wer found, etc. Thus the theoretical status looks really promising. Lattice-QCD might even provide 'measurements' of SPDs not directly accessible in an experiment.

The experimental requirements for a complete investigation of

SPDs are, however, formidable. Many different processes have to me investigated with very high luminosity, at large enough Q^2 and with high enough energy resolution to determine the final hadronic state reliably. It is highly probable that one will need different accelerators to fulfill this ambitious task. The EPIC/eRHIC collider would certainly be complementary to the fixed target experiments presently discussed, both in kinematical range and with respect to the channels easily to be studied. For a more precise statements detailed simulations would be needed.

References

[1] D. Müller, D. Robaschik, B. Geyer, F.-M. Dittes, and J. Horejsi, *Fortschr. Phys.* **42** 101 (1994)

[2] X. Ji, *J. Phys.* **G 24**, 1181(1998); *Phys. Rev. Lett.* **78**, 610 (1997).

[3] A. Radyushkin, *Phys. Lett.* **B 380**, 417 (1996).

[4] R.L. Jaffe, 'The theory of nucleon spin', hep-ph/0008038

[5] J.C. Collins, L. Frankfurt and M. Strikman, *Phys. Rev.*, **D 56**, 2982 (1997);
J.C. Collins and A. Freund, *Phys. Rev.* **D 59**, 074009 (1999).

[6] A.V. Belitsky, D. Müller, L. Niedermeier, A. Schäfer, *Phys. Lett.* **B 437**, 160 (1998); *Nucl. Phys.* **B 546**, 279 (1999); *Phys. Lett.* **B 474** 163 (2000); hep-ph/0004059

[7] M. Diehl, T. Gousset, B. Pire and J.P. Ralston, *Phys. Lett.* **B 411**, 193(1997).

[8] M. V. Polyakov and C. Weiss, *Phys. Rev.* **D 60**, 114017 (1999)
M.V. Polyakov and C. Weiss, *Phys. Rev.* **D 60** 114017 (1999);
M. Diehl, T. Feldmann, R. Jakob and P. Kroll, *Phys. Lett.* **B 460**, 204 (1999).
M. Vanderhaeghen, P. Guichon and M. Guidal, *Phys. Rev.* **D 60**, 094017 (1999).

[9] L. Mankiewicz, G. Piller and A. Radyushkin, *Eur. Phys. J.* **C 10**, 307 (1999);

The Spin Structure of the Nucleon: Theoretical Overview

Presented at the Second Workshop on Physics with a Polarized-Electron Light-Ion Collider (EPIC) September 14--16, 2000 at MIT, Cambridge, Massachusetts, USA

R.L. Jaffe

*Center for Theoretical Physics and Department of Physics
Laboratory for Nuclear Physics, Massachusetts Institute of Technology
Cambridge, Massachusetts 02139*

Abstract. I review what is known about the quark and gluon spin distributions in the nucleon. I discuss in some detail (a) the existence of sum rules for angular momentum; (b) the interpretation and possible measurement of the nucleon's transversity distributions; and (c) the uses of spin-dependent fragmentation functions.

INTRODUCTION

The modern era in QCD spin physics dates from the 1987 discovery by the European Muon Collaboration that only about 30% of the proton's spin is found on the spin of quarks [1]. Since then, particle and nuclear physicists have dreamt of facilities where QCD spin physics could be explored in detail. The recent commissioning of polarized pp physics at RHIC is the first of these to be realized [2]. Our topic - a polarized ep collider in the energy regime where perturbative QCD meets confinement - is a necessary complement to $\overline{RHIC}$ and the natural next step in unravelling the mysteries of quark confinement in QCD.

Among friends, I do not need to belabor the case for studying QCD at the boundary between the confining and perturbative domains. Two brief comments will suffice: First, quantum chromodynamics is the only nontrivial quantum field theory that we are certain describes the real world; and second, we need further experimental input to understand the highly complex QCD bound states that compose matter.

This workshop focuses on spin. While spin is an important degree of freedom, it is not the only important probe of confinement in deep inelastic processes. Flavor, twist, and quark mass dependence (through the substitution $u \to d \to s \to c \to b \to t$) yield different and complementary insights into the structure of QCD bound states. Spin, however, is today's topic.

Of course it is impossible to cover the breadth of this field in a single talk. Fortunately, others will address important subjects in detail later at this meeting. Instead, I will make a brief survey of the present situation, emphasizing our present

understanding of quark and gluon distribution functions, and then focus on three issues of current interest:
- Is there an "angular momentum sum rule" and is it experimentally testable?
- What is transversity and why is it interesting?
- Why are fragmentation functions interesting and useful in the study of spin in QCD?

THE PRESENT SITUATION

Polarization effects in QCD present a complex picture. Asymmetries need to be explained, but sometimes even if we cannot understand them, we can use them to probe other issues or isolate other important effects. Many striking asymmetries occur in the low energy or nuclear domain where we have few theoretical insights into QCD [3]. Most recent progress has occurred where the deep inelastic and soft domains overlap - the world of parton distribution and fragmentation functions. Here, spin effects help elucidate the puzzling nature of hadrons and here is where I will concentrate.

Recent Events

To set the stage for the workshop, here are lists of recent developments in experiment and theory, and a menu for expectations in the immediate future. First, experimental milestones of the past five years:
- First estimates of $\Delta g(x, Q_2)$ from evolution.
- First good look at $g_2(x, Q^2)$ from SLAC.
- First measurement of $\mu_s \equiv \langle \frac{1}{2}\vec{r} \times s^\dagger \vec{\alpha} s \rangle$ from SAMPLE.
- First fragmentation asymmetry measurements from Hermes.
- Commissioning of the polarized pp component of RHIC.

Next, theory milestones of the past five years:
- Theory of Δg measurements: Via evolution, via $\bar{c}c$ production, via $\vec{p}\|\vec{p}\| \to \gamma$ jet X.
- Development of the physics program for $\overline{RHIC}$.
- Off-forward parton distributions and their possible measurement in deeply virtual Compton scattering (DVCS).
- The theory of the nucleon's angular momentum.
- The theory of transversity and proposals to measure it.
- The classification of spin and transverse momentum effects in distribution and fragmentation processes.

And finally, prospects for the near future:
- Direct measurements of Δg.
- First measurements of transversity.
- First measurements of polarization-dependent fragmentation functions.

- Study of the inclusive/exclusive connection (i.e., higher twist), photoproduction, and DVCS at JLab.
- Flavor separation of the quark spin distributions.
- High quality measurements of $\Delta q(x,Q^2)$ at very low and very high x.

Clearly this field – the study of QCD confinement dynamics using polarized probes - requires more than a single facility. Low Q^2 and low energy are needed for DVCS and for studies of higher twist. High Q^2 is needed to study Δg via evolution. High energy is necessary to create the phase space for complex final states such as $\bar{c}c$ studies of Δg and multijet final states. Both polarized lepton beams and polarized proton beams are required. High-density polarized targets are required for high-luminosity studies of g_2 at SLAC and extraction of neutron distributions from polarized deuterium and ^{3}He scattering data.

While we enthuse about the particular subject of this workshop, we must remember that the field requires an opportunistic, even predatory mentality, ready to make use of many facilities in imaginative ways.

Bjorken's Sum Rule

Occasionally it is worth reminding ourselves what it means to "understand" something in QCD. In the absence of fundamental understanding we often invoke "effective descriptions" based on symmetries and low-energy expansions. While they can be extremely useful, we should not forget that a thorough understanding allows us to relate phenomena at very different distance scales to one another. In the case of Bjorken's sum rule, the operator product expansion, renormalization group invariance and isospin conservation combine to relate deep inelastic scattering at high Q^2 to the neutron's β-decay axial charge measured at very low energy. Even target mass and higher twist corrections are relatively well understood. The present state of the sum rule is

$$\int_0^1 dx g_1^{ep-en}(x,Q^2) = \frac{1}{6}\frac{g_A}{g_V}\left\{1 - \frac{\alpha_s(Q^2)}{\pi} - \frac{43}{12}\frac{\alpha_s^2(Q^2)}{\pi^2} - 20.215\frac{\alpha_s^3(Q^2)}{\pi^3}\right\}$$

$$+ \frac{M^2}{Q^2}\int_0^1 x^2 dx\left\{\frac{2}{9}g_1^{ep-en}(x,Q^2) + \frac{1}{6}g_2^{ep-en}(x,Q^2)\right\} \quad (1)$$

$$- \frac{1}{Q^2}\frac{4}{27}F^{u-d}(Q^2)$$

where the three lines correspond to QCD [4], target mass, and higher twist [5] corrections respectively; g_1 and g_2 are the nucleon's longitudinal and transverse spin-dependent structure functions; g_A and g_V are the neutron's β-decay axial and vector charges. F is a twist-4 operator matrix element with dimensions of [mass]^2, which measures a quark-gluon correlation within the nucleon,

$$F^u(Q^2)s^\alpha = \tfrac{1}{2}\langle PS|g\bar{u}\tilde{F}^{\alpha\lambda}\gamma u\big|_{Q^2}|PS\rangle \quad (2)$$

where g is the QCD coupling, $\tilde{F}$ is the dual gluon field strength, and $|_{Q^2}$ denotes the operator renormalization point.

The most thorough analysis of the Bj sum rule I know of is one presented by SMC in 1998 [6]. Their theoretical evaluation gives

$$\int_0^1 dx\, g_1^{ep-en}(x,Q^2)\Big|_{theory} = 0.181 \pm 0.003 \tag{3}$$

at $Q^2=5$ GeV^2. Experiment is not yet able to reach this level of accuracy. The latest data relevant to the Bj sum rule is shown in Fig. 1. The value extracted by the SMC is

$$\int_0^1 dx\, g_1^{ep-en}(x,Q^2)\Big|_{expt} = 0.174 \pm 0.05 \,{}^{+0.011}_{-0.009}\,{}^{+0.021}_{-0.006} \tag{4}$$

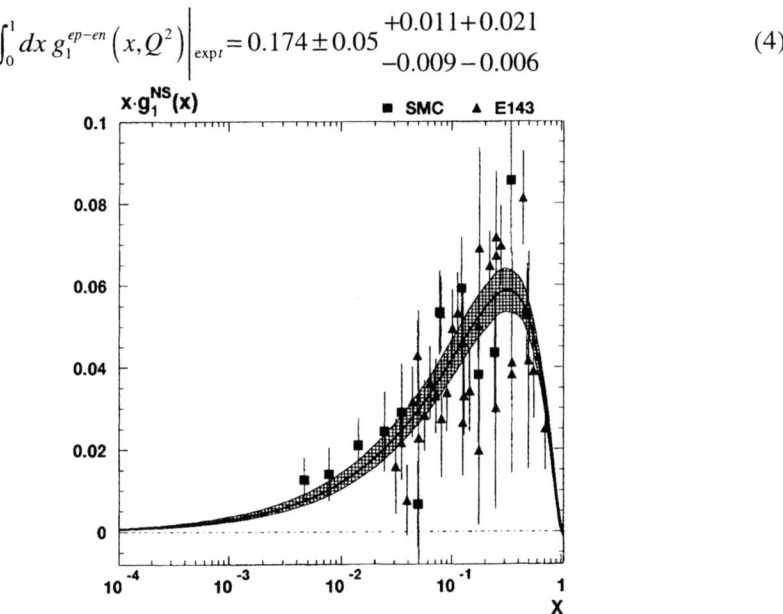

FIGURE 1. SMC analysis of data relevant to the Bjorken sum rule.

at $Q^2=5$ GeV^2, and the errors are statistical, systematic, and "theoretical" (e.g., generated by running the data to a common Q^2), respectively [6]. Further accuracy is necessary to confirm the target mass corrections and extract the twist-four contribution.

Quark and gluon distributions in the nucleon

No overview of the nucleon's spin structure is complete without a survey of the polarized quark and gluon distributions in the nucleon. These helicity-weighted momentum distributions are the most precise and interpretable information we have about the spin substructure of a hadron. The distributions are usually defined in terms of flavor-SU(3) structure,

$$\text{Singlet: } \Delta\Sigma = \Delta U + \Delta D + \Delta S$$
$$\text{Nonsinglet, isovetcor: } \Delta q_3 = \Delta U - \Delta D \quad (5)$$
$$\text{Nonsinglet, hypercharge: } \Delta q_8 = \Delta U + \Delta D - 2\Delta S$$

where $\Delta Q \equiv q^\uparrow(x,Q^2) + \bar{q}^\uparrow(x,Q^2) - q^\downarrow(x,Q^2) - \bar{q}^\downarrow(x,Q^2)$. Experimenters seem to prefer nonsinglet distributions specialized to the proton and neutron individually,

$$\text{Prooton nonsinglet: } \Delta q_{NS}(p) = \Delta U + \frac{1}{2}\Delta D - \frac{1}{2}\Delta S$$
$$\text{Neutron nonsinglet: } \Delta q_{NS}(n) = \Delta D - \frac{1}{2}\Delta U - \frac{1}{2}\Delta S \quad (6)$$

so that

$$g_1^p = \frac{2}{9}\Delta\Sigma + \frac{2}{9}\Delta q_{NS}(p)$$
$$g_1^n = \frac{2}{9}\Delta\Sigma + \frac{2}{9}\Delta q_{NS}(n) \quad (7)$$

Since the integrated quark spin accounts for only about 30% of the nucleon's spin, it is extremely interesting to know whether the integrated gluon spin in the nucleon is large. Of course the polarized gluon distribution, $\Delta g(x,Q^2)$, cannot be measured directly in deep inelastic scattering because gluons do not couple to the electromagnetic current. Instead, Δg is inferred from the QCD evolution of the quark distributions. (See Ref. (6) for details of the process and references to the original literature.) However, evolution of imprecise data only constrains a few low moments of Δg and gives only crude information on global characteristics such as the existence and number of nodes. It is clear that Δg must be measured directly elsewhere.

That said, the world's data on polarized structure functions is summarized in Figs. 2 and 3. Fig. 2 is taken from Naomi Makins's talk at DIS2000 and presents the world's data on g_1^p in the same format traditionally used for unpolarized structure function data [7]. The figure highlights the tremendous progress of the past decade as well as the need for much better data if our knowledge of polarized distributions would aspire to the same accuracy as unpolarized distributions. Note, in particular, that the entire kinematic domain over which g_1 has been measured would fit into the lower left-hand corner of the F_2 figure. Fig. 3 shows the quark and gluon distributions extracted from the world's data by SMC, together with estimates of systematic and theoretical uncertainties [6]. While the information on quark distributions is fairly precise, it is clear that we know very little about the distribution of polarized gluons in the nucleon.

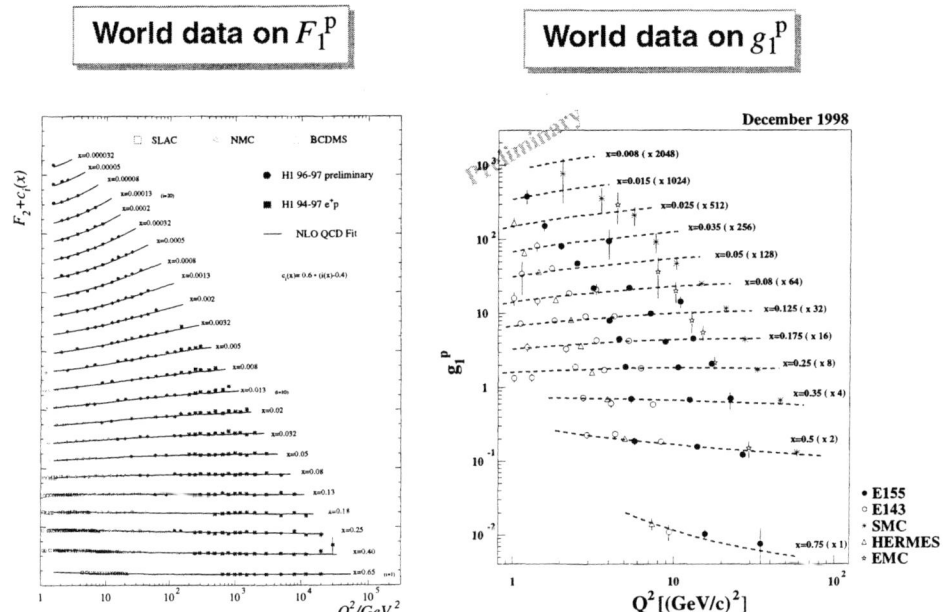

FIGURE 2. World data on spin-average and spin-dependent structure function [7].

IS THERE AN "ANGULAR MOMENTUM SUM RULE" AND IS IT EXPERIMENTALLY TESTABLE?

It has been clear for years that in some sense the nucleon's spin (projected along an axis) can be written as a sum of contributions from quark and gluon spin and orbital angular momentum [9],

$$\frac{1}{2} = \frac{1}{2}\Delta\Sigma + \Delta g + L_q + L_g \qquad (8)$$

but the interpretation and usefulness of such a relation has only recently been clarified. The principal issues are

- Are the terms separately gauge-invariant?
- Are they interaction-dependent?
- Is each separately measurable?
- Is each related to an integral over a parton x-distribution?

I believe we can now answer these questions, but the answers are not what we would like.

I want to distinguish between two different kinds of relations with the form of Eq 8. A "classic" sum rule expresses the expectation value of a local operator in a state as an *integral* (or sum) over a distribution measured in an inelastic production process involving the same state. This is the traditional definition of a "sum rule", dating back to the Thomas-Reiche-Kuhn sum rule of atomic spectroscopy. All the familiar sum rules of deep inelastic scattering - Bjorken's, Gross & Llewellyn-Smith's, etc. - are this type of relation. They are even more powerful because the distribution that is integrated has a simple, heuristic interpretation as the momentum (Bjorken-x) distribution of the observable associated with the local operator. The "spin sum rule" gives a typical example:

$$\langle P,S|\bar{q}_a\gamma^\mu\gamma_5 q_a|_{Q^2}|P,S\rangle/S^\mu \equiv \Delta q_a(Q^2)$$
$$= \int_0^1 dx\{q_{a\uparrow}(x,Q^2) + \bar{q}_{a\uparrow}(x,Q^2) - q_{a\downarrow}(x,Q^2) - \bar{q}_{a\downarrow}(x,Q^2)\} \tag{9}$$

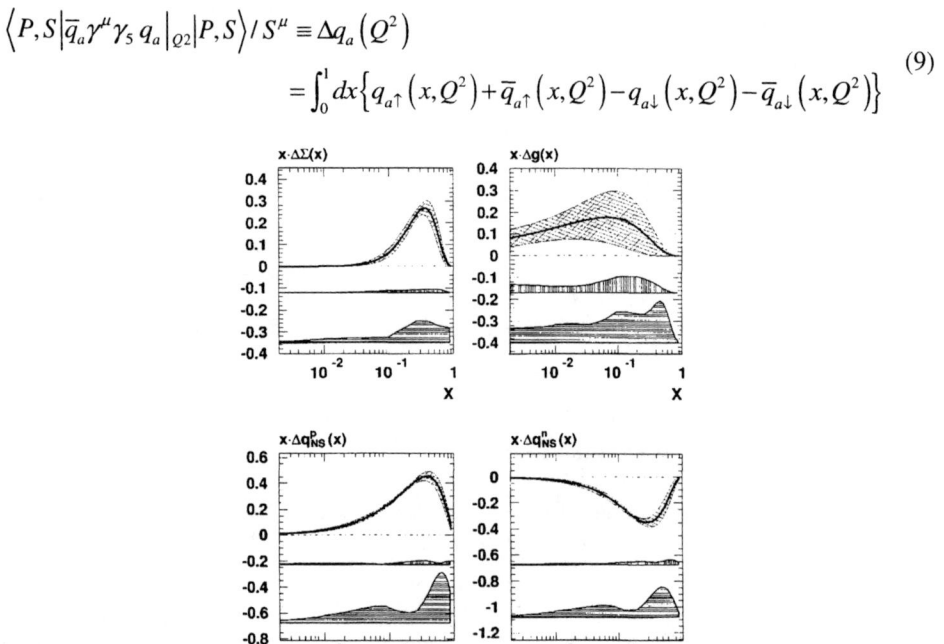

FIGURE 3. Polarized quark and gluon distribution functions. The upper figures show the distribution with a statistical error bound. The lower figures show estimates of systematic and theoretical uncertainties, respectively.

The left-hand side can be measured in β-decay or other electroweak processes. The right-hand side can be measured in deep inelastic scattering of polarized leptons from polarized targets. The meaning of the sum rule is clear because the local operator, $\bar{q}_a\gamma^\mu\gamma_5 q_a$, is the generator of the internal rotations (the "spin") of the quark field in QCD. The sum rule says that the quark's contribution to the nucleon's spin is the integral over a spin-weighted momentum distribution of the quarks.

Another, less powerful, but still interesting type of relation -sometimes called a sum rule - arises simply because an operator can be written as the sum of two (or more) other operators, $\Theta = \Theta_1 + \Theta_2$. If the expectation values of all three operators can be measured, then this relation, and the assumptions underlying it, can be tested. Such a relation exists for the contributions to the nucleon's angular momentum [9,10],

$$\frac{1}{2} = \hat{L}_q + \frac{1}{2}\Sigma + \hat{J}_g \tag{10}$$

where the three terms are *roughly* the quark orbital angular momentum, the quark spin, and the total angular momentum on the gluons. Ji has shown how, in principle, to measure the various terms in this relation [10].

A sum rule of the classic type also exists for the contributions to the nucleon's angular momentum, [13,12,14]

$$\frac{1}{2} = \int_0^1 dx \left\{ L_q(x,Q^2) + \frac{1}{2}\Delta q(x,Q^2) + \Delta g(x,Q^2) \right\} \tag{11}$$

where the four terms are *precisely* the x-distributions of the quark orbital angular momentum, quark spin, gluon orbital angular momentum, and gluon spin. However, it appears that the distributions $L_q(x,Q^2)$ and $L_g(x,Q^2)$ are not experimentally accessible. So the value of the sum rule is obscure.

Before exploring these relations for the angular momentum in more depth, let's examine the simpler and well-understood case of energy and momentum.

Sum rules for energy and momentum

One hears a lot about the "momentum sum rule" in QCD, but nothing about an "energy sum rule". The reasons are quite instructive. Energy and momentum are described by the rank-two, symmetric energy-momentum tensor, $T^{\mu\nu}$,

$$T^{\mu\nu} = \frac{i}{4}\bar{q}\left(\gamma^\mu D^\nu + \gamma^\nu D^\mu\right)q + h.c. + Tr\left(F^{\mu\alpha}F^\nu_\alpha - \frac{1}{4}g^{\mu\nu}F^2\right) \tag{12}$$

where D^μ and $F^{\mu\nu}$ are the gauge covariant derivative and gluon field strength, both matrices in the fundamental representation of SU(3). [$T^{\mu\nu}$ is ambiguous up to certain total derivatives, but these do not change the arguments presented here.]

The energy density is given by T^{00},

$$\varepsilon \equiv T^{00} = \frac{1}{2}q^\dagger\left(-i\vec{\alpha}\cdot\vec{D} + \beta m\right)q + h.c. + Tr\left(\vec{E}^2 + \vec{B}^2\right). \tag{13}$$

The expectation value of T^{00} is normalized,

$$\langle P|T^{00}|P\rangle = 2E^2 \tag{14}$$

because $|P\rangle$ is an eigenstate of the Hamiltonian, $\int d^3x T^{00}(x)|P\rangle = E|P\rangle$ This is a good start towards a sum rule. However there is no useful sum rule because there is no way to write any of the terms in Eq (13) as an integral over inelastic production data. This is not obvious, but the appearance of terms in ε that are order cubic and higher in the canonical fields is a bad sign. The first term in ε includes $\bar{q}qq$ coupling, and $\vec{E}^2 + \vec{B}^2$ involves terms cubic and quartic in the gluon vector potentials $\vec{A}$. The parton distributions of deep inelastic scattering (DIS) come from operators quadratic in the "good" light-cone components of the quark and gluon fields, q_+ and $\vec{A}_\perp$ [11].

In contrast there is a classic, deep-inelastic sum rule for P^+, where $P^+ = \frac{1}{\sqrt{2}}(P^0 + P^3)$, and the 3-direction is singled out by the gauge choice $A^+=0$. T^{++} is normalized much like T^{00},

$$\langle P|T^{++}|P\rangle = 2P^{+2}. \tag{15}$$

Unlike T^{00}, T^{++} simplifies dramatically in $A^+=0$ gauge ecause of the simplification of D^+ and $F^{+\alpha}$,

$$D^+ = \partial^+ - igA^+ \to \partial^+$$
$$F^{+\alpha} = \partial^+ A^\alpha - \partial^\alpha A^+ + g[A^+, A^\alpha] \to \partial^+ A^\alpha \tag{16}$$

As a result T^{++} is quadratic in the fundamental dynamical variables, q_+ and $\vec{A}_\perp$ and all interactions disappear,

$$T^{++} = iq_+^\dagger \partial^+ q_+ + Tr(\partial^+ \vec{A}_\perp)^2. \tag{17}$$

The two terms give the contributions of quarks and gluons respectively to the total P^+. It is straightforward to relate each to an integral over a positive definite parton "momentum" distribution,

$$iq_+^\dagger \partial^+ q \to \int dx\, xq(x)$$
$$(\partial^+ \vec{A}_\perp)^2 \to \int dx\, xg(x) \tag{18}$$

in which the parton probability density is weighted by the observable (in this case x) appropriate to the sum rule. Keeping track of renormalization scale dependence and kinematic factors of P^+, one obtains the standard "Momentum" sum rule,

$$1 = \int_0^1 dx\, x\{q(x,Q^2) + g(x,Q^2)\} \tag{19}$$

The lessons learned from this exercise generalize to the more difficult case of angular momentum:

- The time components of the tensor densities associated with space-time symmetries do not yield classic sum rules. Interactions do not drop out. They yield relations that are difficult to interpret because quark and gluon contributions do not separate. Individual terms are not related to integrals over parton distributions.
- The +-components of the same tensor densities do yield useful sum rules, which have a parton interpretation in $A^+=0$ gauge. Interactions drop out. Each term can be represented as an integral over a parton distribution weighted by the appropriate observable quantity.

Sum rules for angular momentum

The situation for angular momentum is not satisfactory. The time-component analysis yields a relation, some of whose ingredients can be measured (in principle) in deeply virtual Compton scattering. But it has no place for a separately gauge-invariant gluon spin and orbital angular momentum, no clean separation between quark and gluon contributions, and no relation to quark or gluon x distributions. The +-component analysis yields a classic sum rule with separate quark and gluon spin and orbital angular momentum contributions, each gauge invariant, each related to a parton distribution, and each free from interaction terms. Unfortunately, there does not seem to be a way to measure the terms in this otherwise perfectly satisfactory sum rule.

The tensor density associated with rotations and boosts is a three component tensor antisymmetric in the last two indices, $M^{\mu\nu\lambda}$. To extract a sum rule, we polarize the nucleon along the 3-direction in its rest frame and set $\nu=1$, $\lambda=2$ in order to select rotations about this direction. The matrix elements of M^{012} and M^{+12} are both normalized in terms of the nucleon's momentum ($P^\mu=(M,0,0,0)$) and spin ($S^\mu=(0,0,0,M)$) [9].

First consider the time component, M^{012},

$$M^{012} = \frac{i}{2} q^\dagger \left(\vec{x} \times \vec{D}\right)^3 q + \frac{1}{2} q^\dagger \sigma^3 q + 2 Tr E^j \left(\vec{x} \times \vec{D}\right)^3 A^j + Tr\left(\vec{E} \times \vec{A}\right)^3. \qquad (20)$$

The four terms look like the generators of rotations (about the 3-axis) for quark orbital, quark spin, gluon orbital, and gluon spin angular momentum respectively. Taking the matrix element in a nucleon state at rest one obtains,

$$\frac{1}{2} = \hat{L}_q + \frac{1}{2}\Sigma + \hat{L}_g + \Delta\hat{g} \qquad (21)$$

There are problems, however. There are no parton representations for $\hat{L}_g, \hat{L}_q$, or $\Delta\hat{g}$, so it is not a sum rule in the classic sense. Σ is the integral of the helicity weighted quark distribution, but $\Delta\hat{g}$ is not the integral of the helicity weighted gluon distribution. Interactions prevent a clean separation into quark and gluon contributions as they did for T^{00}. And worse still, $\hat{L}_g$ and $\Delta\hat{g}$ are not separately gauge invariant, so only the sum $\hat{J}_g = \hat{L}_g + \Delta\hat{g}$ is physically meaningful.

The most important feature of the relation, (21), is the result derived by Ji, that

$$\hat{J}_q = \hat{L}_q + \frac{1}{2}\Sigma \tag{22}$$

and $\hat{J}_g$ can, in principle, be measured in deeply virtual Compton scattering [10]. In practice, $\hat{L}_q$ may be measurable, but $\hat{J}_g$ can only be obtained by Q^2 evolution of $\hat{J}_q$, which seems beyond experimental attack for the foreseeable future. Without a handle on $\hat{J}_g$ and given the ambiguity in the definition of $\hat{L}_q$ (see below), the usefulness of Eq. [22] is unclear.

Turning to the +-component sum rule, we find a much simpler form,

$$M^{+12} = \frac{1}{2}q_+^\dagger(\vec{x}\times\vec{i}\partial)^3 q_+ + \frac{1}{2}q_+^\dagger\gamma_5 q_+ + \text{Tr}\, F^{+j}(\vec{x}\times i\vec{\partial})A^j + \text{Tr}\, \varepsilon^{+-ij}F^{+i}A^j \tag{23}$$

in $A^+=0$ gauge. [This gauge condition must be supplemented by the additional condition that the gauge fields vanish fast enough at infinity.] The four terms in M^{+12} correspond respectively to quark orbital angular momentum, quark spin, gluon orbital angular momentum, and gluon spin, all about the 3-axis. Each is separately gauge invariant[1] and involves only the "good", i.e., dynamically independent, degrees of freedom, q_+ and $\vec{A}_\perp$. Each is a generator of the appropriate symmetry transformation in light-front field theory. The resulting sum rule,

$$\frac{1}{2} = L_q + \frac{1}{2}\Sigma + L_g + \Delta g \tag{24}$$

is a classic deep inelastic sum rule. It can be written as an integral over x-distributions

$$\frac{1}{2} = \int_0^1 dx \left\{ L_q(x,Q^2) + \frac{1}{2}\delta q(x,Q^2) + L_g(x,Q^2) + \Delta g(x,Q^2) \right\} \tag{25}$$

where each term is an interaction independent, gauge invariant, integral over a partonic density associated with the appropriate symmetry generator [122-14].

Satisfying though (24) and (25) may be from a theoretical point of view. They are quite useless unless someone finds a way to measure the two new terms L_q and L_g.

---

[1] Note, however, that in any gauge other than $A^+=0$, the operators are nonlocal and appear to be interaction dependent. The same happens to the simple operators involved in the momentum sum rule, Eq. (17).

TRANSVERSITY

One of the major accomplishments of the recent renaissance in QCD spin physics has been the rediscovery and exploration of the quark *transversity distribution*. First mentioned by Ralston and Soper in 1979 in their treatment of Drell-Yan μ-pair production by transversely polarized protons [15], the transversity was not recognized as a major component in the description of the nucleon's spin until the early 1990s [16-18,11].

The transversity can be interpreted in parton language as follows: consider a nucleon moving with (infinite) momentum in the $\hat{e}_3$-direction, but polarized along one of the directions transverse to $\hat{e}_3$. $\Delta q_a(x,Q^2)$ counts the quarks of flavor a and momentum fraction x with their spin parallel the spin of a nucleon minus the number antiparallel. If quarks moved nonrelativistically in the nucleon, δq and Δq would be identical, since rotations and Euclidean boosts commute and a series of boosts and rotations can convert a longitudinally polarized nucleon into a transversely polarized nucleon at infinite momentum. So the difference between the transversity and helicity distributions reflects the relativistic character of quark motion in the nucleon. There are other important differences between transversity and helicity. For example, quark and gluon helicity distributions (Δq and Δg) mix under Q^2-evolution. There is no analog of gluon transversity in the nucleon, so δq evolves without mixing, like a nonsinglet distribution function. The lowest moment of the transversity is proportional to the nucleon matrix element of the tensor charge, $\bar{q}i\sigma^{0i}\gamma_5 q$, which couples only to valence quarks (it is C-odd). Not coupling to glue or $\bar{q}q$ pairs, the tensor charge promises to be more quark-model - like than the axial charge and should be an interesting contrast.

We now know that the transversity, $\delta q(x,Q^2)$, together with the unpolarized distribution, $q(x,Q^2)$, and the helicity distribution, $\Delta q(x,Q^2)$, are required to give a complete description of the quark spin in the nucleon at leading twist. An equation tells this story clearly:

$$A(x,Q^2) = \frac{1}{2}q(x,Q^2) I \otimes I + \frac{1}{2}\Delta q(x,Q^2)\sigma_3 \otimes \sigma_3 + \frac{1}{2}\delta q(x,Q^2)(\sigma_+ \otimes \sigma_- \otimes \sigma_+). \quad (26)$$

Here, A is the quark distribution in a nucleon as a density matrix in both the quark and nucleon helicities (hence the external product of two Pauli matrices in each term), diagrammatically equivalent to the lower part of the handbag diagram shown in Fig.4a; q governs spin average physics, Δq governs helicity dependence, and δq governs helicity flip - or transverse polarization - physics.

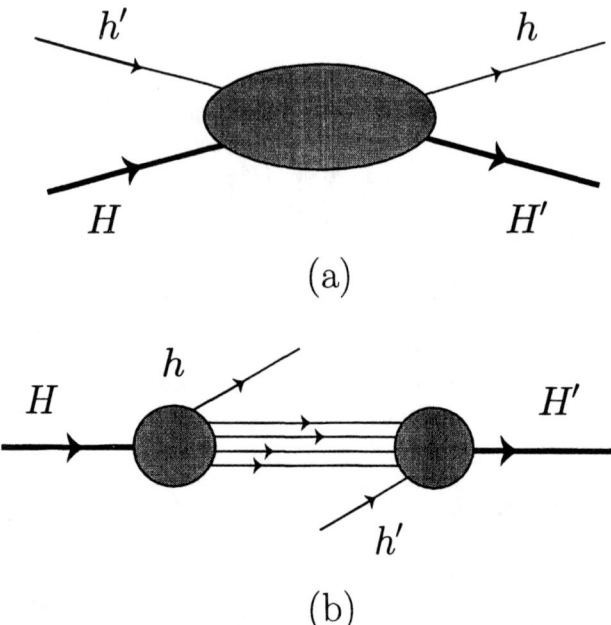

FIGURE 4. Quark hadron forward scattering. Quark helicities are labeled h and h; hadron helicities are H and H'. (a) Full scattering amplitude; (b) u-channel discontinuity, which gives the quark distribution function in DIS.

In terms of the helicity amplitude $A_{Hh,Hh'}$ in Fig. 4b, the transversity is given by $A_{++,--}$, corresponding to quark and nucleon helicity flip. The spin average (q) and helicity (Δq) distributions involve $A_{++,++}$, $A_{+-,+-}$, which preserve quark helicity. The connection between transverse spin and helicity flip is a consequence of simple quantum mechanics. The two states of transverse polarization can be written as superpositions of helicity eigenstates:; the cross section with transverse polarization has the form $d\sigma d\sigma \uparrow \alpha \langle \uparrow | ... | \uparrow \rangle$; so the difference of cross sections is proportional to helicity flip,.

At leading twist, quark helicity and chirality are identical. For this reason, the transversity distribution are called "chiral-odd", in contrast to the "chiral-even" distributions, q and Δq.

Quark chirality is conserved at all QCD and electroweak vertices; however, quark chirality can flip in distribution and fragmentation functions because they probe the soft regime where chiral symmetry is dynamically broken in QCD. This is another reason to be interested in transversity -- it probes dynamical chiral symmetry breaking, an incompletely understood aspect of QCD.

Because all hard QCD and electroweak processes preserve chirality, transversity is difficult to measure. It decouples from inclusive DIS and most other familiar deep inelastic processes. The argument is made graphically in Fig.5. In order to access transversity some second soft process must flip the quark chirality a second time. The classic example, where transversity was discovered by Ralston and Soper, is transversely polarized Drell-Yan production of muon pairs: $\vec{p}_\perp \vec{p}_\perp \to \mu^+ \mu^- X$, which is shown diagrammatically in Fig.5(b). Chirality is flipped in both soft distribution functions and the cross section is proportional to $\delta q(x_1,Q^2) \times \delta \bar{q}(x_2,Q^2)$.

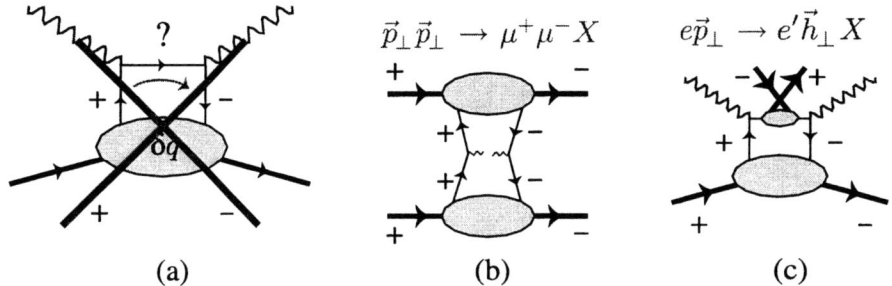

FIGURE 5. Deep inelastic processes relevant to transversity.

Transversity would not decouple from deep inelastic scattering if some electroweak vertex would flip chirality. Unfortunately (and accidentally from the point of view of QCD) all photon, $W^\pm$ and Z^0 couplings all preserve chirality. Quark-Higgs couplings violate chirality but are too weak to be of interest. Quark mass insertions flip chirality, and indeed a careful analysis reveals effects proportional to $m\delta q(x,Q^2)/\sqrt{Q^2}$ in inclusive DIS with a transversely polarized target. However the u, d, and s quarks, which are common in the nucleon, are too light to give significant sensitivity to δq.

What is needed is an insertion that flips chirality without introducing a $1/\sqrt{Q^2}$ suppression. A generic example is shown in Fig. 5(c). Much interest has been generated recently by the observation of an asymmetry at Hermes that can be interpreted as evidence for a chirality-flipping fragmentation function that couples to the nucleon's transversity. It corresponds to a particular instance of Fig. 5(c). If this effect is confirmed it suggests a bright future for transversity measurements at the next generation of polarized lepton-hadron facilities.

FRAGMENTATION AND SPIN: THE HERMES ASYMMETRY AND BEYOND

To my mind, the single most interesting development in QCD spin physics over the past two years is the azimuthal asymmetry in pion electroproduction reported by Hermes [19]. It is interesting both in itself and as an emblem of a new class of spin measurements involving spin-dependent fragmentation processes, which act as filters for exotic parton distribution functions like transversity.

Fragmentation functions as probes of unstable hadrons

Fragmentation functions allow us to access and explore the spin structure of unstable hadrons, which cannot be used as targets for deep inelastic scattering. These include the ρ, ω, and ϕ mesons, and hyperons like the Λ and Σ.

Let me give three examples:

The tensor fragmentation function of the ρ

When a quark of helicity $h = \pm\frac{1}{2}$ fragments collinearity into a ρ of helicity $H=1$, 0, or -1, there are many fragmentation functions, $F_{Hh,H'h'}$, in analogy to $A_{Hh,H'h'}$ discussed in the previous section. If we consider fragmentation of helicity eigenstates, then the fragmentation functions can be labelled by the quark helicity h and the ρ helicity H corresponding to $q_h \to \rho_H$. Parity relates three pairs, e.g., $q_{\frac{1}{2}} \to \rho_1 = q_{-\frac{1}{2}} \to \rho_{-1}$, leaving three independent combinations. These can be classified as the spin average: $q \to \rho$; the helicity difference: $(q_{\frac{1}{2}} \to \rho_1) - (q_{\frac{1}{2}} \to \rho_{-1})$; and the tensor fragmentation function, known as $\hat{b}_\rho$ in analogy to the tensor *distribution* function first analyzed in connections with the deuteron [20]: $\hat{b}_\rho = (q \to \rho_1) + (q \to \rho_{-1}) - 2(q \to \rho_0)$. The function $\hat{b}_\rho$ is independent of quark spin and has the simple physical interpretation of measuring the difference between quark fragmentation into a transverse ρ compared to a longitudinal ρ. The ππ angular distribution in ρ decay is sensitive to $\hat{b}_\rho$, so it can be measured [21]. The data are already available. The challenge to theorists is to make use of it.

The ρ double-helicity flip-fragmentation function

Consider the fragmentation of a gluon into a ρ. In addition to the fragmentation functions already discussed for quarks, a double helicity flip fragmentation function can occur. The process and the helicity labels are shown in Fig. 6. This process has a unique signature in the ππ angular distribution and no equivalent in $q \to \rho$. So it is a special probe of gluon fragmentation into the ρ.

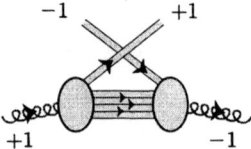

FIGURE 6. Figure showing gluon and rho helicity labels in the double helicity flip case.

Polarized quark → polarized Λ fragmentation

It should be clear that one can define longitudinal and transverse spin dependent fragmentation functions of the Λ, schematically $\vec{q}_\| \to \vec{\Lambda}_\|$ and $\vec{q}_\perp \to \vec{\Lambda}_\perp$, in direct analogy to the quark helicity and transversity distributions in a target Λ. Since the $\Lambda \to p\pi$ decay is self-analyzing, it is relatively easy to measure the spin of the Λ. By selecting Λ's produced in the current fragmentation region one can hope to isolate the fragmentation process $q \to \Lambda$. Having measured the quark spin structure of the nucleon, we can use flavor-SU(3) to estimate the way quark spins are distributed in the Λ[22]. However we do not know if this information is reflected in the fragmentation process $q \to \Lambda$. Once again the challenge is to theorists to learn how to interpret fragmentation functions in a heuristic way analogous to the quark parton model of distribution functions.

Fragmentation as a filter for novel distribution functions

Even if we do not know how to interpret fragmentation functions, we can use them as filters, to select parton distribution functions that either decouple from or are hard to extract from completely inclusive DIS. The simplest and best known example is the use of meson flavor to tag strange versus nonstrange quark distribution functions. This analysis has been developed to a high level of sophistication by the Hermes collaboration who use the felicitous term "purity" to denote the propensity for strange quarks to fragment to strange mesons and so forth [24]. They identify "favored" fragmentation processes like $u \to \pi^+$ and "disfavored" processes like $u \to \pi$pi$^-$ and set up a transfer matrix formalism to give a complete characterization of $eN \to e'$ (π, K, η) X. The interest is not principally in the various fragmentation functions, but instead to use them as filters for specific quark (and antiquark) distribution functions. Hermes and Compass hope to use these methods to extract the polarized antiquark distributions in the nucleon, $\Delta \bar{u}(x,Q^2)$, $\Delta \bar{d}(x,Q^2)$ and $\Delta \bar{s}(x,Q^2)$. Their competition in this pursuit comes from $\overline{RHIC}$, where $W^\pm$ production asymmetries can be used to trigger on specific quark flavors and extract $\Delta \bar{u}$ and $\Delta \bar{d}$.

A more complex, and potentially much more interesting example is the use of a helicity flip fragmentation function to select the quark transversity distribution. As shown in Fig. 5(c), by interposing a helicity flip fragmentation function on the struck quark line in DIS, it is possible to access the transversity. What is needed is a *twist-two*, chiral-odd fragmentation function. There are several candidates:

- $\delta \hat{q}_a(z,Q^2)$, the transverse, spin-dependent fragmentation function. This is the analog in fragmentation of transversity, and describes the fragmentation of a transversely polarized quark into a transversely polarized hadron with momentum fraction z [25,26]. To access $\delta \hat{q}$, it is necessary to measure the spin of a particle in the final state of DIS. In practice this limits the application to production of a Λ hyperon - the only particle whose spin is easy to measure through its parity violating decay.
- $\delta \hat{q}_I(z,m^2,Q^2)$, the two pion interference fragmentation function. [27-29] This describes the fragmentation of a transversely polarized quark into a pair of pions whose orbital angular momentum is correlated with the quark spin. This requires measurement of two pions in the final state. It may be quite useful, especially in polarized collider experiments [30]. I will not discuss it further here.

- $\hat{c}(z,Q^2)$, the single particle azimuthal asymmetry fragmentation function. This function, first discussed by Collins, et al. [27], describes the azimuthal distribution of pions about the axis defined by the struck quark's momentum in deep inelastic scattering.

All three of these fragmentation functions are chiral-odd and therefore produce experimental signatures sensitive to the transversity distribution in the target nucleon. Each may play an important role in future experiments aimed at probing the nucleon's transversity. Recently Hermes has announced observation of a spin asymmetry that seems to be associated with the Collins function, $c(z,Q^2)$. So although all three deserve discussion, I will spend the rest of my time on the Collins function and the Hermes asymmetry.

The Collins Fragmentation Function

The standard description of fragmentation without polarization requires a single fragmentation function usually called $D_h(z)$. It gives the probability that a quark will fragment into a hadron, h, with longitudinal momentum fraction z. [For simplicity I suppress the dependence of D on the virtuality scale, Q^2 and the quark flavor label a.] The transverse momentum of h relative to the quark is integrated out. If the transverse momentum, $\vec{p}_\perp$, is observed, then it is possible to construct distributions weighted by geometric factors. For instance,

$$c(z) \alpha \int d^2 p_\perp D_h(z, \vec{p}_\perp) \cos \chi$$

where, for comparison,

$$D(z) \alpha \int d^2 p_\perp D_h(z, \vec{p}_\perp). \qquad (27)$$

Here $D_h(z, \vec{p}_\perp)$ is the probability for the quark to fragment into hadron h with momentum fraction z and transverse momentum $\vec{p}_\perp$; χ is the angle between $\vec{p}_\perp$ and some vector, $\vec{w}$, defined by the initial state. Since we don't know the direction of the quark's momentum exactly, the transverse momentum of the hadron, $\vec{p}_\perp$, is defined relative to some large, externally determined momentum, such as the momentum of the virtual photon, $\vec{q}$, in DIS.

How can $c(z)$ figure in deep inelastic scattering? The trick is to find a vector, $\vec{w}$, relative to which χ can be defined. If the target is polarized, it is possible to define $\vec{w}$ by taking the cross product of the target spin, $\vec{s}$, with either the initial or final electron's momentum ($\vec{k}$ and $\vec{k}'$) depending on the circumstances. Generically, then, the observable associated with $c(z)$ is $\cos \chi \, \alpha \vec{k} \times \vec{s} \cdot \vec{p}$, where $\vec{p}$ is the momentum of the observed hadron in the final state. The situation is illustrated in Fig (7) from Ref. [31]. This observable is even under parity (because $\vec{s}$ is a pseudovector), but odd under time reversal. This *does not* mean that it violates time-reversal invariance. Instead it means that it will vanish unless there are final state interactions capable of generating a nontrivial phase in the DIS amplitude. This subtlety makes it hard to find a good model to estimate $c(z)$ because typical fragmentation models involve only tree graphs (if they involve quantum mechanics at all!), which are real.

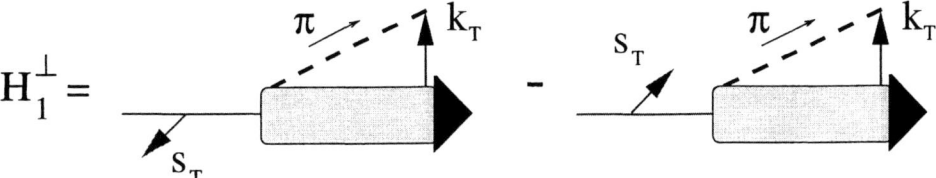

FIGURE 7. The Collins effect function $H_1^\perp$ signals different probabilities for $q(\perp/\perp) \to \pi(\vec{k}_\perp) + X$.

The Collins fragmentation function, $c(z)$, may be interesting in itself, but it is much more interesting because it is chiral-odd and combines with the transversity distribution in the initial nucleon to produce an experimentally observable asymmetry sensitive to the transversity. Two specific cases figure in recent and soon-to-be-performed experiments.

Single particle inclusive DIS with a transversely polarized target: $e\vec{p}_\perp \to e'\pi X$

If the target is transversely polarized (with respect to the initial electron momentum, $\vec{k}$), then $\vec{w} = \vec{k} \times \vec{s}$ defines a vector normal to the plane defined by the beam and the target spin. The transverse momentum of the produced hadron can be defined either with respect to the beam or the momentum transfer $\vec{q}$ - the difference in higher order in $1/Q$. $\cos\chi$ is defined by $\cos\chi = \vec{p}_\perp \cdot \vec{w}/|\vec{p}_\perp||\vec{w}|$. The kinematics are particularly simple in this case (transverse spin). Experimenters prefer to think of the effect in terms of the angle (ϕ) between two planes: Plane 1 is defined by the virtual photon and the target spin, and Plane 2 is defined by the virtual photon and the transverse momentum of the produced hadron. Then $\sin\phi = \cos\chi$ and the effect is known as a "$\sin\phi$" asymmetry. When the cross section is weighted by $\sin\phi$, the result is

$$\frac{d\Delta\sigma_\perp}{dxdydz} = \frac{2\alpha^2}{Q^2} \sum_a e_a^2 \delta_a(x) c_a(z) \qquad (28)$$

where $y = E - E'/E$ and $\Delta\sigma$, and $\Delta\sigma$ is the difference of cross sections with target spin reversed.[2] This is a leading twist effect, which scales (modulo logarithms of Q^2) in the deep inelastic limit. If $c(z)$ is not too small, it is will become the "classic" way to measure the nucleon's transversity distributions.

No experimental group has yet measured hadron production in deep inelastic scattering from a transversely polarized target, so there is no data on $\Delta\sigma_\perp$. Hermes at

---

[2] In principle, this reversal is superfluous because the $\sin\phi$ asymmetry must be odd under reversal and the rest of the cross section must be even. However, it helps experimenters to reduce systematic errors.

DESY intend to take data under these conditions in the next run. One reason for this was the observation of a $\sin\phi$ asymmetry with a *longitudinally* polarized target that Hermes announced last year [19]. It strongly suggests, but does not require, that $\Delta\sigma_\perp$ should be large.

Single particle inclusive DIS with a longitudinally polarized target: $e\vec{p}_\perp \to e'\pi X$

The possibility of a $\sin\phi$ asymmetry is more subtle in this case, and it escaped theorists' attention for a long time. The possibility of such an asymmetry was first pointed out in Ref [32]. As Q^2 and ν go to ∞, the initial and final electrons' momenta become parallel. If the target spin is parallel to $\vec{k}$, then it is impossible to construct a vector from $\vec{k}$ or $\vec{k}'$ and $\vec{s}$ in this limit. However, $\vec{k}$ and $\vec{k}'$ are not exactly parallel, so $\vec{s}$ has a small component perpendicular to the virtual photon's momentum, $\vec{q} = \vec{k} - \vec{k}'$. The vector, $\vec{w}$, can be defined as $\vec{w} = \vec{k}' = \vec{s}$, and the kinematic situation is shown in Fig. 8 from Ref. [19]. This produces an asymmetry similar to the previous case, but weighted by $|\vec{s}_\perp| \alpha\, 2Mx/Q$. Because this leading (twist two) effect is kinematically suppressed by $1/Q$, it is necessary to consider other, twist-three, effects that might be competitive. A careful analysis turns up a variety of twist-three effects, leading to a cross section of the form [31,32],

$$\frac{d\Delta\sigma_\parallel}{dxdydz} = \frac{2\alpha^2}{Q^2} \frac{2Mx}{Q} \sqrt{1-y} \sum_a e_a^2 \left\{ \delta q_a(x) c_a(z) + \frac{2-y}{1-y} h_{La}(x) c_a(z) \right\} \qquad (29)$$

where $h_L(x)$ is a longitudinal spin-dependent, twist-three distribution function analogous to g_T.

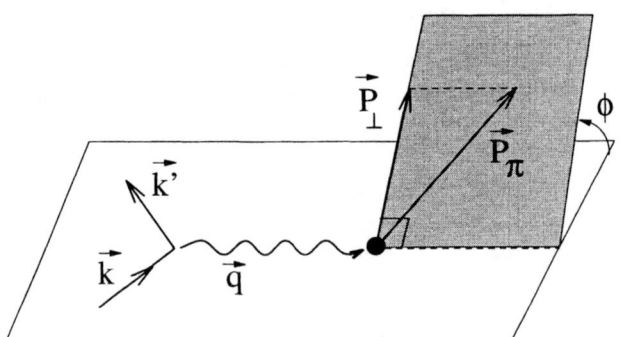

FIGURE 8. Kinematic planes for pion production in semi-inclusive deep-inelastic scattering.

By far the most interesting thing about $\Delta\sigma_\parallel$ is that Hermes has seen such an asymmetry in their π^+ data. (The Hermes data is shown in Fig. 9.) They see no effect in their π^- data. Because u quarks predominate in the nucleon, because $e_u^2 = 4e_d^2$, and because

$u \to \pi^+ \ll u \to \pi^-$, they expect no signal in π^-. They have not reported on π^0, where an asymmetry similar to π^+ would be expected.

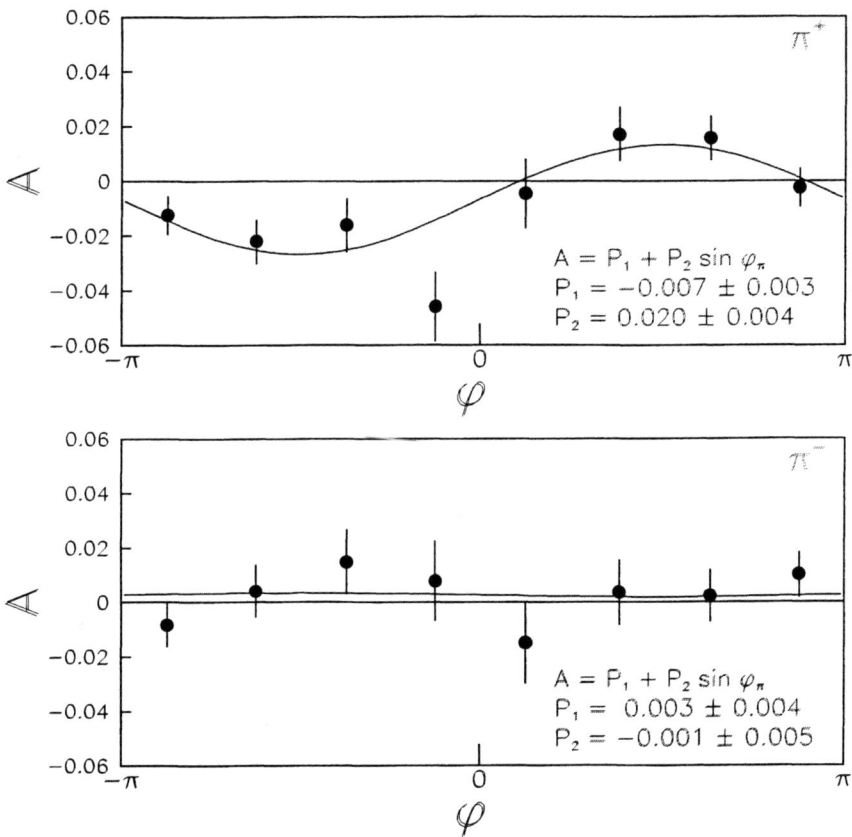

FIGURE 9. Azimuthal asymmetry $\sin\phi$ distribution for π^+ and π^- production with a longitudinally polarized target at Hermes.

If the Hermes result is confirmed, it demonstrates that the Collins fragmentation function is nonzero. Somehow the final state interactions between the observed pion and the other fragments of the nucleon suffice to generate a phase that survives the sum over the other unobserved hadrons. Whatever its origin, a nonvanishing Collins function would be a great gift to the community interested in the transverse spin structure of the nucleon. It provides an unanticipated tool for extracting the nucleon's transversity from DIS experiments. The fact that Hermes has seen a robust (2-3%) asymmetry with a longitudinally polarized target suggests that they will see a large asymmetry with a transversely polarized target (unless the effect is entirely twist three - e.g., $h_L \gg \delta q$). This in turn will lead to the first measurements of the nucleon's transversity distribution and to new insight into the relativistic spin structure of confined states of quarks and gluons.

CONCLUSIONS

A richer and more complex picture of the QCD bound states has emerged since the 1987 renaissance precipitated by the EMC observation that quarks carry only a small fraction of the nucleon spin. We know much more about the nucleon's spin than we did back then. We also know what to look for in the future: we have a clear program for future measurement and analysis of the gluon helicity distribution, δg, the quark transversity, δq_a, and the flavor decomposition of the quark spin $\Delta \bar{u}, \Delta \bar{d}$, etc.) and a host of other related subjects, which I have not had time to discuss here. This program involves several facilities and different energy regimes. The polarized *ep* collider we are considering at this workshop clearly has a central role to play.

Acknowledgments

This work is supported in part by funds provided by the U.S. Department of Energy (D.O.E.) under cooperative research agreement #DF-FC02-94ER40818.

REFERENCES

1. J.Ashman et al. [European Muon Collaboration], Phys. Lett. **B206**, 364 (1988).
2. N. Saito,"Spin Physics at RHIC", presented at the 14th International Spin Symposium, SPIN2000, Osaka, November 2000.
3. A.D. Krisch, *SPIN 98 Proceedings of the13th International Symposium on High Energy Spin Physics Protvino, Russia 8 - 12 September 1998*, N.E.Tyurin, V.L.Solovianov, S.M.Troshin, and A.G.Ufimtsev, eds. (World Scientific, Singapore, 1999)
4. S.A.Larin and J.A.Vermaseren, Phys. Lett. **B259**, 345 (1991).
5. E.V. Shuryak and A.I.Vainshtein, Nucl. Phys. **B201**, 141 (1982).
6. B.Adeva et al. [Spin Muon Collaboration], Phys. Rev. D **58**, 112002 (1998).
7. N.C.R. Makins [for the Hermes Collaboration] Talk presented at DIS2000. To be published in the proceedings.
8. G. Bunce, N. Saito, J. Soffer, and W. Vogelsang, hep-ph/0007218.
9. R.L. Jaffe and A. Manohar, Nucl. Phys. **B337**, 509 (1990).
10. X. Ji, Phys. Rev. Lett. **78**, 610 (1997) [hep-ph/9603249].
11. R.L. Jaffe, hep-ph/9602236.
12. P. Hagler and A. Schafer, Phys. Lett. **B430**, 179 (1998) [hep-ph/9802362].
13. Harindranath and R. Kundu, Phys. Rev. **D59**, 116013 (1999) [hep-ph/9802406].
14. S.V. Bashinsky and R.L. Jaffe, Nucl. Phys. **B536**, 303 (1998) [hep-ph/9804397].
15. J.P. Ralston and D.E. Soper, Nucl. Phys. **B152**, 109 (1979).
16. X. Artru and M. Mekhfi, Z. Phys. **C45**, 669 (1990).
17. R.L. Jaffe and X. Ji, Phys. Rev. Lett. **67**, 552 (1991).
18. J.L. Cortes, B. Pire, and J.P. Ralston, Z. Phys. **C55**, 409 (1992).
19. Airapetian et al. [Hermes Collaboration], Phys. Rev. Lett. **84**, 4047 (2000) [hep-ex/9910062].
20. P. Hoodbhoy, R.L. Jaffe, and A. Manohar, Nucl. Phys. **B312**, 571 (1989).
21. Schafer, L. Szymanowski, and O.V. Teryaev, Phys. Lett. **B464**, 94 (1999) [hep-ph/9906471].
22. M. Burkardt and R.L. Jaffe, Phys. Rev. Lett. **70**, 2537 (1993) [hep-ph/9302232].
23. R.L. Jaffe, Phys. Rev. D **54**, 6581 (1996) [hep-ph/9605456].
24. J.M. Niczyporuk and E.E. Bruins, Phys. Rev. **D58**, 091501 (1998) [hep-ph/9804323].
25. R.L. Jaffe and X. Ji, Phys. Rev. Lett. **71**, 2547 (1993) [hep-ph/9307329].
26. D.Boer, hep-ph/0007047.
27. J.C. Collins, S.F. Heppelmann, and G.A. Ladinsky, Nucl. Phys. **B420**, 565 (1994) [hep-ph/9305309].
28. J.C. Collins and G.A. Ladinsky, hep-ph/9411444.
29. R.L. Jaffe, X. Jin, and J. Tang, Phys. Rev. Lett. **80**, 1166 (1998) [hep-ph/9709322].
30. M. Grosse Perdekamp "Transversity Measurement at RHIC" presented at the 14th International Spin Symposium, SPIN2000, Osaka, November 2000.
31. D. Boer, RIKEN Rev. **28**, 26 (2000) [hep-ph/9912311].
32. K.A. Oganessyan, H.R. Avakian, N. Bianchi, and A.M. Kotzinian, hep-ph/9808368.

Current fragmentation in semiinclusive leptoproduction

P.J. Mulders

Division of Physics and Astronomy, Faculty of Exact Sciences
Vrije Universiteit, De Boelelaan 1081
1081 HV Amsterdam, the Netherlands

Abstract. Current fragmentation in semiinclusive deep inelastic leptoproduction offers, besides refinement of inclusive measurements such as flavor separation and access to the chiral-odd quark distribution functions $h_1^q(x) = \delta q(x)$, the possibility to investigate intrinsic transverse momentum of hadrons via azimuthal asymmetries.

LEADING QUARK DISTRIBUTION FUNCTIONS

In deep-inelastic leptoproduction (DIS), the *soft* hadron structure enters via the quark distribution functions. These distribution functions for a quark can be obtained from the lightcone[1] correlation functions [1-4].

$$\Phi_{ij}(x) = \int \frac{d\xi^-}{2\pi} \, e^{ip\cdot\xi} \, \langle P, S|\overline{\psi}_j(0)\psi_i(\xi)|P,S\rangle \bigg|_{\xi^+=\xi_T=0}, \tag{1}$$

depending on the lightcone fraction of a quark (with momentum p), $x = p^+/P^+$. In particular the At leading order, the relevant part of the correlator is $\Phi\gamma^+$

$$(\Phi\gamma^+)_{ij} = \int \frac{d\xi^-}{2\pi\sqrt{2}} \, e^{ip\cdot\xi} \, \langle P, s'|\psi_{+j}^\dagger(0)\psi_{+i}(\xi)|P,s\rangle \bigg|_{\xi^+=\xi_T=0} \tag{2}$$

where $\psi_+ \equiv P_+\psi = \frac{1}{2}\gamma^-\gamma^+\psi$ is the good component of the quark field [5].

---

[1] For inclusive leptoproduction the lightlike directions $n_\pm$ and lightcone coordinates $a^\pm = a \cdot n_\mp$ are defined through hadron momentum P and the momentum transfer q,

$$P = \frac{Q}{x_B\sqrt{2}}n_+ + \frac{x_B M^2}{Q\sqrt{2}}n_-,$$

$$q = -\frac{Q}{x_B\sqrt{2}}n_+ + \frac{Q}{\sqrt{2}}n_-.$$

Explicitly, the matrix $M = (\Phi\gamma^+)^T$ in Dirac space using a chiral representation becomes for a spin 0 target the following 4×4 matrix,

$$M_{ij} = \begin{pmatrix} f_1(x) & 0 & 0 & 0 \\ 0 & 0 & 0 & 0 \\ 0 & 0 & 0 & 0 \\ 0 & 0 & 0 & f_1(x) \end{pmatrix} \qquad (3)$$

In hard processes only two Dirac components are relevant, *one* of them righthanded and *one* lefthanded ($\psi_{R/L} = \frac{1}{2}(1 \pm \gamma_5)\psi$). Restricting ourselves to those states, the matrix for a spin 0 target becomes

$$M_{ij} = \begin{pmatrix} f_1(x) & 0 \\ 0 & f_1(x) \end{pmatrix} \begin{matrix} \text{\tiny R} \\ \text{\tiny L} \end{matrix} \qquad (4)$$

$\quad\quad\quad\quad$ R $\quad$ L

For a spin 1/2 target more quark distributions appear in the lightcone correlation function at leading order. In order to include all possible target polarizations, one can employ a spin vector[2], in which case one obtains

$$M_{ij} = \begin{pmatrix} f_1(x) + S_L\, g_1(x) & 0S_T^1 + i\,S_T^2)\, h_1(x) \\ (S_T^1 - i\,S_T^2)\, h_1(x) & f_1(x) + S_L\, g_1(x) \end{pmatrix} \begin{matrix} \text{\tiny R} \\ \text{\tiny L} \end{matrix} \qquad (5)$$

$\quad\quad\quad\quad$ R $\quad\quad\quad\quad$ L

Equivalently, and for our purposes more instructive, one can also express M as a 4×4 matrix in quark $\otimes$ nucleon spin space,

$$M^{(\text{prod})} = \begin{pmatrix} f_1 + g_1 & 0 & 0 & 2h_1 \\ 0 & f_1 - g_1 & 0 & 0 \\ 0 & 0 & f_1 - g_1 & 0 \\ 2h_1 & 0 & 0 & f_1 + g_1 \end{pmatrix} \qquad (6)$$

Note that the distribution functions exist for each quark flavor. The functions are also denoted $f_1^q(x) = q(x)$, $g_1^q(x) = \Delta q(x)$ and $h_1^q(x) = \delta q(x)$. The three functions are independent. From the fact that any forward matrix element of the above matrix represents a density, one derives positivity bounds [6],

---

[2] The spin vector is parametrized $S = S_L \frac{Q}{M\sqrt{2}} n_+ - S_L \frac{M}{Q\sqrt{2}} n_- + S_T$.

$$f_1(x) \geq 0 \tag{7}$$
$$|g_1(x)| \leq f_1(x) \tag{8}$$
$$|h_1(x)| \leq \frac{1}{2}(f_1(x) + g_1(x)) \leq f_1(x). \tag{9}$$

As can be seen $h_1(x)$ involves a matrix elements between left- and right-handed quarks, it is chirally odd [4]. This implies that it is not accessible in inclusive DIS, where the hard scattering part does not change chirality except via (irrelevant) quark mass terms.

By choosing a different basis of quark states, $\psi_{\uparrow/\downarrow} = \frac{1}{2}(1 + \gamma_5\gamma^1)\psi$ and nucleon transverse spin states (along the x-axis),

$$|N, \uparrow / \downarrow\rangle = \frac{1}{\sqrt{2}}(|N, +\rangle \pm |N, -\rangle), \tag{10}$$

one obtains the (equivalent) matrix

$$M^{(\text{prod})} = \begin{pmatrix} f_1 + h_1 & 0 & 0 & g_1 + h_1 \\ 0 & f_1 - h_1 & g_1 - h_1 & 0 \\ 0 & g_1 - h_1 & f_1 - h_1 & 0 \\ g_1 + h_1 & 0 & 0 & f_1 + h_1 \end{pmatrix} \tag{11}$$

from which one sees that $h_1(x)$ is a transverse spin density.

Leading gluon distribution functions correspond to lightcone correlators with transverse gluon fields,

$$\Gamma^{+\alpha;+\beta}(x) = \int \frac{d\xi^-}{2\pi} e^{ip\cdot\xi} \langle P, S|F^{+\alpha}(0)F^{+\beta}(\xi)|P, S\rangle\bigg|_{\xi^+=\xi_T=0}. \tag{12}$$

This can be considered as a gluon production matrix, that for a spin 1/2 hadrons is given by

$$M^{(\text{prod})} = \begin{pmatrix} G + \Delta G & 0 & 0 & 0 \\ 0 & G - \Delta G & 0 & 0 \\ 0 & 0 & G - \Delta G & 0 \\ 0 & 0 & 0 & G + \Delta G \end{pmatrix} \tag{13}$$

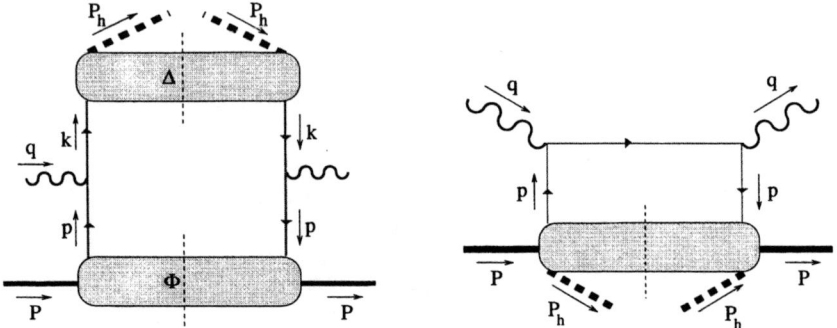

FIGURE 1. The leading contributions to current (left) and target (right) fragmentation.

Here we have used circularly polarized gluon states.

Inclusive DIS experiments have yielded a good knowledge of the unpolarized quark distributions $q(x)$ in a nucleon and via the evolution equations of $G(x)$. Polarized experiments have provided us with measurements of $\Delta q(x)$ and first indications of $\Delta G(x)$.

SEMIINCLUSIVE LEPTOPRODUCTION

Semiinclusive DIS (SIDIS), in particular one-particle inclusive DIS[3], can be and also has been used for additional flavor identification. Instead of weighing quark flavors with the quark charge squared e_q^2 one obtains a weighting with $e_q^2 \, D_1^{q \to h}(z_h)$, where $D_1^{q \to h}$ is the usual fragmentation function for a quark of flavor q into hadron h, experimentally accessible at $z_h = P \cdot P_h / P \cdot q$. possibilities to study intrinsic transverse momentum of partons, quarks and gluons, via azimuthal asymmetries and the appearance of single spin asymmetries via T-odd fragmentation functions.

Before turning to these topics, I want to address the issue of separation of current fragmentation from target fragmentation, for which the leading order description is illustrated in Fig. 1. While for current fragmentation we can use a description

---

[3] For SIDIS the lightlike directions $n_\pm$ and lightcone coordinates $a^\pm = a \cdot n_\mp$ are defined through hadron momentum P and P_h, in which case the momentum transfer q requires a transverse component

$$P = \frac{Q}{x_B \sqrt{2}} n_+ + \frac{x_B M^2}{Q \sqrt{2}} n_-,$$

$$q = -\frac{Q}{\sqrt{2}} n_+ + \frac{Q}{\sqrt{2}} n_- + q_T,$$

$$P_h = \frac{M_h^2}{Z_h Q \sqrt{2}} n_+ + \frac{z_h Q}{\sqrt{2}} n_-.$$

factorizing into distribution and fragmentation functions, target fragmentation involves a more complex soft part, namely fracture functions [7]. Here we want to mention at least one check on the precision of current fragmentation. Up to mass corrections of order M^2/Q^2 one has for current fragmentation the identities

$$x = -\frac{q^+}{P^+} \approx \frac{Q^2}{2P \cdot q} \approx -\frac{P_h \cdot q}{P_h \cdot P}, \qquad (14)$$

$$z = \frac{P_h^-}{q^-} \approx -\frac{2P_h \cdot q}{Q^2} \approx \frac{P \cdot P_h}{P \cdot q}. \qquad (15)$$

Actually incorporation of kinematical $1/Q^2$ corrections can be done by calculating the lightcone ratios (first entries in both equations) in a frame in which neither of the hadrons has a transverse momentum component.

Based on results in the EMC compilation in ref. [8] we take a rapidity interval $\Delta \eta \approx 2$ (sometimes referred to as Berger's criterium) to estimate the z-values for which one is most probably dealing with current fragmentation. For this we construct a plot using the definition of rapidity

$$\eta = \frac{1}{2} \ln \left(\frac{P_h^-}{P_h^+} \right) = \ln \left(\frac{P_h^- \sqrt{2}}{M_{h\perp}} \right) = -\ln \left(\frac{P_h^+}{M_{h\perp}} \right), \qquad (16)$$

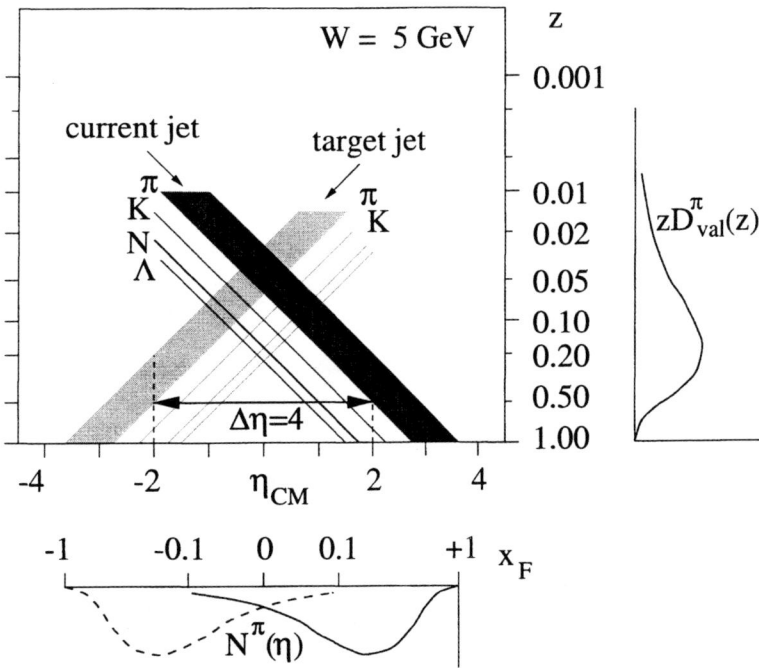

FIGURE 2. Relation between z – $values$ in fragmentation and CM rapidity for $W = 5$ GeV.

where $M_{h\perp}^2 = M_h^2 + P_{h\perp}^2$. For current fragmentation one has $z_c = P_h^-/q^-$ while for target fragmentation one is dealing with a ratio $z_t = P_h^+/(1-x)P^+$. The proportionality is all we need to deduce that for the center of mass rapidity one has

$$\eta_{cm} = \ln z_c + \ln\left(\frac{W}{M_{h\perp}}\right), \tag{17}$$

$$\eta_{cm} = \ln z_t + \ln\left(\frac{W}{M_{h\perp}}\right), \tag{18}$$

where W is the $\gamma^* N$ invariant mass, $W = (1-x)y\,s$, fixing the maximum rapidity. For two values of $W = 5$ and 20 GeV, we have indicated the relation between z and η for both current and target fragmentation for a number of hadrons in Figs 2 and 3. For light hadrons the band reflects the influence of the transverse momentum. Looking at the $\Delta\eta = 4$ difference one can estimate z-values above which current fragmentation dominates. Also indicated is how a typical (valence-like) fragmentation function produces a number density in rapidity. Clearly seen is how increased W vastly lowers the z-values where one may expect to deal with current fragments.

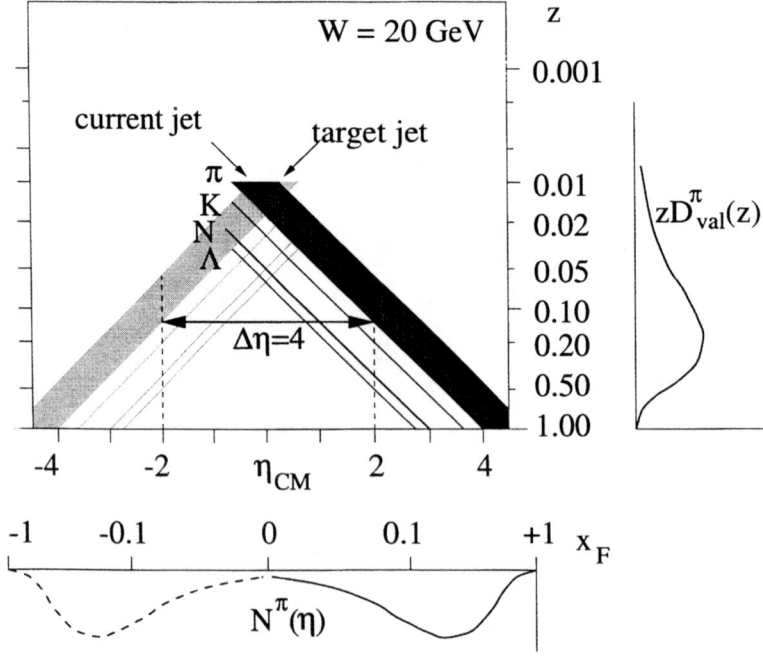

FIGURE 3. Relation between $z - values$ in fragmentation and CM rapidity for $W = 20$ GeV.

LEADING QUARK DISTRIBITION AND FRAGMENTATION FUNCTIONS IN SIDIS

While the distribution functions in DIS could be obtained from the lightcone correlation function in Eq. 1, one encounters in SIDIS two types of lightfront correlation functions, involving also transverse momenta of partons as first pointed out by Ralston and Soper [9,10] One part is relevant to treat quarks in a hadron

$$\Phi_{ij}(x, \boldsymbol{p}_T) = \int \frac{d\xi^- d^2\boldsymbol{\xi}_T}{(2\pi)^3} \, e^{ip\cdot\xi} \, \langle P, S | \overline{\psi}_j(0) \psi_i(\xi) | P, S \rangle \bigg|_{\xi^+=0}, \qquad (19)$$

depending on $x = p^+/P^+$ and the quark transverse momentum $\boldsymbol{p}_T$ in a target with $P_T = 0$. A second correlation function [11]

$$\Delta_{ij}(z, \boldsymbol{k}_T) = \sum_X \int \frac{d\xi^- d^2\boldsymbol{\xi}_T}{(2\pi)^3} \, e^{ik\cdot\xi} \langle 0 | \psi_i(\xi) | P_h, X \rangle \langle P_h, X | \overline{\psi}_j(0) | 0 \rangle \bigg|_{\xi^+=0}, \qquad (20)$$

describes fragmentation of a quark into a hadron. It depends on $z = P_h^+/k^+$ and the quark transverse momentum k_T when one produces a hadron with $P_{hT} = 0$. A simple boost shows that this is equivalent to a quark producing a hadron with transverse momentum $P_{h\perp} = -z\,k_T$ with respect to the quark.

As before we make the Dirac structure explicit and find at leading order only two relevant components, one of them righthand and one lefthanded. For fragmentation into spin 0 hadrons (e.g. pion production) this leads to the following 2×2 quark decay matrix,

$$M_{ij}^{(\text{decay})} = \begin{pmatrix} D_1(z) & i\frac{|k_T|e^{i\phi}}{M_h} H_1^\perp(z) \\ -i\frac{|k_T|e^{i\phi}}{M_h} H_1^\perp(z) & D_1(z) \end{pmatrix} \begin{matrix} \text{\textregistered} \\ \text{\textcircled{L}} \end{matrix} \qquad (21)$$

As compared to the production matrix in Eq. 4 one has the additional function $H_1^\perp(z)$, which is allowed because one cannot use time-reversal invariance to constrain the structure of Δ in Eq. 20. Such functions are referred to as T-odd. We note, furthermore, that $H_1^\perp$ is also chiral-odd, hence it will appear in a cross section in combination with a chiral-odd distribution function such as h_1. The appearance of the transverse momentum, however, has as consequence that this fragmentation function only can be measured via the dependence on the transverse momentum of the produced hadron, e.g. in azimuthal asymmetries.

For a spin 1/2 hadron one finds that the structure of Φ including transverse momentum dependence, leads to the production matrix,

$$M^{(\text{prod})} = \begin{Bmatrix} f_1 + g_1 & \frac{|p_T|}{M} e^{i\phi} g_{1T} & \frac{|p_T|}{M} e^{-i\phi} h_{1L}^\perp & 2h_1 \\ \frac{|p_T|}{M} e^{i\phi} g_{1T}^* & f_1 - g_1 & \frac{|p_T|^2}{M^2} e^{-2i\phi} h_{1T}^\perp & -\frac{|p_T|}{M} e^{-i\phi} h_{1L}^{\perp*} \\ \frac{|p_T|}{M} e^{i\phi} h_{1L}^{\perp*} & \frac{|p_T|^2}{M^2} e^{2i\phi} h_{1T}^\perp & f_1 - g_1 & -\frac{|p_T|}{M} e^{i\phi} g_{1T}^* \\ 2h_1 & -\frac{|p_T|}{M} e^{i\phi} h_{1L}^\perp & -\frac{|p_T|}{M} e^{-i\phi} g_{1T} & f_1 + g_1 \end{Bmatrix}, \quad (22)$$

to be compared with Eq. 11. Using time-reversal invariance all the distribution functions appearing in this equation are expected to be real, leaving aside mechanisms discussed in Refs [12]. For fragmentation functions, however, T-reversal cannot be used [13–15], leading to two T-odd fragmentation functions [16,17]. They are the imaginary parts of the complex off-diagonal (p_T-dependent) functions. To be precise one obtains the decay matrix with fragmentation functions after the replacements $f_1 \to D_1$, $g_1 \to G_1$, $h_1 \to H_1$, $g_{1T} \to G_{1T} + i D_{1T}^\perp$ and $h_{1L}^\perp \to H_{1L}^\perp + i H_1^\perp$.

The possibility to access the full (transverse momentum dependent) spin structure of the nucleon is in my opinion one of the most exciting possibilities offered by 1-particle inclusive leptoproduction.

Bounds

In analogy to the Soffer bound derived from the production matrix in Eq. 4 one easily derives a number of new bounds from the full matrix, such as

$$f_1(x, \boldsymbol{p}_T^2) \geq 0, \quad (23)$$
$$|g_1(x, \boldsymbol{p}_T^2)| \leq f_1(x, \boldsymbol{p}_T^2). \quad (24)$$

obtained from one-dimensional subspaces and

$$|h_1| \leq \frac{1}{2}(f_1 + g_1) \leq f_1, \quad (25)$$

$$|h_{1T}^{\perp(1)}| \leq \frac{1}{2}(f_1 - g_1) \leq f_1, \quad (26)$$

$$|g_{1T}^{(1)}|^2 \leq \frac{\boldsymbol{p}_T^2}{4M^2}(f_1 + g_1)(f_1 - g_1) \leq \frac{\boldsymbol{p}_T^2}{4M^2} f_1^2, \quad (27)$$

$$|h_{1L}^{\perp(1)}|^2 \leq \frac{\boldsymbol{p}_T^2}{4M^2}(f_1 + g_1)(f_1 - g_1) \leq \frac{\boldsymbol{p}_T^2}{4M^2} f_1^2, \quad (28)$$

obtained from two-dimensional subspaces. Here we have introduced the notation $g_{1T}^{(1)}(x, \boldsymbol{p}_T^2) \equiv (\boldsymbol{p}_T^2/4M^2) g_{1T}(x, \boldsymbol{p}_T^2)$. These bounds and their further refinements have been discussed in detail in Ref. [18]. There are straightforward extensions of transverse momentum dependent distribution and fragmentation functions for spin 1 hadrons [19] and gluons in spin 1/2 hadrons [20].

Bound on the Collins function

As an application of using the bounds, consider the Collins function $H_1^{\perp(1)}$ for which we have

$$|H_1^{\perp(1)}(z,-z\boldsymbol{k}_T)| = |\frac{\boldsymbol{k}_T^2}{2M_\pi^2} H_1^\perp(z,-z\boldsymbol{k}_T)| \leq \frac{|\boldsymbol{k}_T|}{2M_\pi} D_1(z,-z\boldsymbol{k}_T). \tag{29}$$

With the assumption

$$D_1(z,-z\boldsymbol{k}_T) = D_1(z)\frac{R_\pi^2(z)}{\pi z^2} e^{-|\boldsymbol{k}_T|^2 R_\pi^2}, \tag{30}$$

one finds for the function integrated over transverse momenta,

$$|H_1^{\perp(1)}(z)| \leq \underbrace{\frac{\sqrt{\pi}}{4M_\pi R_\pi(z)}}_{\mathcal{O}(1)} D_1(z). \tag{31}$$

Lorentz invariance relations

Since both the p_T integrated functions and the p_T dependent functions originate from (nonlocal) combinations of two quark fields, Poincaré invariance poses restrictions on the various ways we project out distribution functions. In particular we consider the inclusion of the higher-twist functions for the p_T-integrated functions, in which case the correlator $\Phi(x)$ in Eq. 1 becomes [4]

$$\Phi(x) = \frac{1}{2}\left\{f_1 \not{n}_+ + S_L\, g_1 \gamma_5 \not{n}_+ + h_1 \frac{[\not{S}_T, \not{n}_+]\gamma_5}{2}\right\}$$
$$+ \frac{M}{2P^+}\left\{e + g_T\,\gamma_5 \not{S}_T + S_L\, h_L \frac{[\not{n}_+, \not{n}_-]\gamma_5}{2}\right\}$$
$$\text{T-odd} \quad + \frac{M}{2P^+}\left\{f_T\, \epsilon_T^{\rho\sigma} S_{T\rho}\gamma_\sigma - i S_L\, e_L\, \gamma_5 + h\, \frac{i\,[\not{n}_+, \not{n}_-]}{2}\right\}. \tag{32}$$

We will compare this with the p_T-integrated result after weighing the $\Phi(x,p_T)$ with p_T, giving $\Phi_\partial^\alpha(x) \equiv \int d^2p_T\, p_T^\alpha\, \Phi(x,p_T)$, explicitly

$$\frac{1}{M}\Phi_\partial^\alpha(x) = \frac{1}{2}\left\{g_{1T}^{(1)}\, S_T^\alpha\, \gamma_5 \not{n}_+ + S_L\, h_{1L}^{\perp(1)} \frac{[\not{n}_+, \gamma^\alpha]\gamma_5}{2}\right.$$
$$\text{T-odd} \quad \left. + f_{1T}^{\perp(1)}\, \epsilon_T^{\alpha\beta} S_{T\beta} \not{n}_+ + h_1^{\perp(1)} \frac{i\,[\not{n}_+, \gamma^\alpha]}{2}\right\}. \tag{33}$$

The $\boldsymbol{p}_T^2/2M^2$ moment of the transverse momentum dependent functions turn out to be related to twist-three functions [21,22,17],

$$\underbrace{g_T - g_1}_{g_2} = \frac{d}{dx} g_{1T}^{(1)}, \tag{34}$$

$$\underbrace{h_L - h_1}_{\frac{1}{2} h_2} = -\frac{d}{dx} h_{1L}^{\perp(1)}, \tag{35}$$

$$f_T = -\frac{d}{dx} f_{1T}^{\perp(1)}, \tag{36}$$

$$h = -\frac{d}{dx} h_1^{\perp(1)}. \tag{37}$$

The above relations can for instance be used to estimate the magnitude of $g_{1T}^{(1)}$ from polarized inclusive data on g_2 [23,24].

AZIMUTHAL ASYMMETRIES

As already mentioned before, in order to experimentally investigate the full spin structure including the off-diagonal transverse momentum dependent functions (Eq. 11) one needs semiinclusive measurements. The transverse momentum dependence is probed via specific azimuthal asymmetries. We limit ourselves here to just one example, but before doing so remind the reader of the 'rules'.

- Depending on the powers t of (M/P^+) [for fragmentation functions powers of (M_h/P_h^-)] the functions show up in contributions in the cross section behaving as $(1/Q)^t$. This is sometimes referred to as a twist expansion, although it in particular for the transverse momentum dependent correlators $\Phi(x, p_T)$ and $\Delta(z, -zk_T)$ only indicates the 'lowest twist' operators that play a role, now using twist in the rigorous operator-product-expansion sense.

- Cross sections are chirally even. For instance chirally even functions like $f_{...}$ or $g_{...}$ appear together with chirally even fragmentation functions $D_{...}$ or $G_{...}$, while chirally odd functions $h_{...}$ and e appear together with chirally odd functions $H_{...}$ and E. Note that terms originating from quark mass terms multiply combinations of opposite chirality.

- The number of polarizations needed is even in the case of an even number of $T - odd$ functions combinations of distribution and fragmentation functions and it is odd in the case of an odd number of $T - odd$ functions.

The following explicit example serves to illustrate these points, namely the semi-inclusive asymmetry

$$\left\langle \frac{Q_T}{M_\pi} \sin(\phi_h^\ell + \phi_S^\ell) \right\rangle_{OTO} = \frac{2\pi\alpha^2 s}{Q^4} |S_T| 2(1-y) \sum_{a,\bar{a}} e_a^2\, x_B\, h_1^a(x_B) H_1^{\perp(1)a}(z_h), \tag{38}$$

which is the cross section weighted with the magnitude $Q_T = |P_{h\perp}|/z_h$ and involving the angles of the transverse momentum of the produced hadron, ϕ_h^ℓ (with repect to lepton scattering plane) and the transverse spin of the target, ϕ_S^ℓ. Since the Collins functions $H_1^\perp$ is T-odd and chirally odd, it can appear together with the chirally odd distribution function h_1, but since the latter is T-even, the combination appears in a single spin asymmetry: unpolarized lepton, transversely polarized target, production of a spinless particle. Many other examples have been discussed in the literature [25,17,22,26], some of them will be discussed at this meeting [27]. Also recent experimental indications of nonvanishing azimuthal asymmetries exist [28–30]

QCD DYNAMICS

The study of distribution and fragmentation functions is interesting since it identifies well-defined quantities that can be extracted from experiment by using high energy (expansion in powers of $1/Q$ with calculable $\ln Q^2$ perturbative corrections) and identified as specific matrix elements of quark and gluon fields. We will illustrate below how the QCD dynamics enters here. In Eq. 32 the quark-quark correlation function was expanded including (higher-twist) terms proportional to (M/P^+). These terms are the leading terms in a correlator involving $\overline{\psi}(0)\, D^\alpha \psi(\xi)$ where D^α is the covariant derivative. For transverse indices one can use the QCD equations of motion, $(i\slashed{D} - m)\psi = 0$ to show that

$$\frac{1}{M}\Phi_D^\alpha(x) = \frac{1}{2}\left\{\left(xg_T - \frac{m}{M}h_1\right)S_T^\alpha\,\gamma_5\slashed{n}_+ + S_L\left(xh_L - \frac{m}{M}g_1\right)\frac{[\slashed{n}_+,\gamma^\alpha]\gamma_5}{2}\right.$$
$$\left. - xf_T\,\epsilon_T^{\alpha\beta}S_{T\beta}\slashed{n}_+ - xh\,\frac{i[\slashed{n}_+,\gamma^\alpha]}{4}\right\} \qquad (39)$$

One can identify the socalled *interaction dependent* pieces via $\Phi_A \equiv \Phi_D - \Phi_\partial$

$$\frac{1}{M}\Phi_A^\alpha(x) = \frac{1}{2}\left\{\underbrace{\left(xg_T - g_{1T}^{(1)} - \frac{m}{M}h_1\right)}_{x\tilde{g}_T}S_T^\alpha\,\gamma_5\slashed{n}_+\right.$$
$$+ S_L\underbrace{\left(xh_L - h_{1L}^{\perp(1)} - \frac{m}{M}g_1\right)}_{x\tilde{h}_L}\frac{[\slashed{n}_+,\gamma^\alpha]\gamma_5}{2}$$
$$\left. - \underbrace{\left(xf_T + f_{1T}^{\perp(1)}\right)}_{x\tilde{f}_T}\epsilon_T^{\alpha\beta}S_{T\beta}\slashed{n}_+ - \underbrace{\left(xh + 2h_1^{\perp(1)}\right)}_{x\tilde{h}}\frac{i[\slashed{n}_+,\gamma^\alpha]}{4}\right\} \qquad (40)$$

Relations

The relations following from Lorentz invariance and the equations of motion can be combined to relate the functions discussed above. In particular consider the 'leading' functions g_1 and $g_{1T}^{(1)}$ appearing in the matrix in Eq. 11 and the 'subleading' functions g_T and $\tilde{g}_T$ discussed in the previous section. From the equations of motion and Lorentz invariance, respectively, we get (omitting quark mass terms),

$$g_T = \frac{g_{1T}^{(1)}}{x} + \tilde{g}_T$$
$$= g_1 + \frac{d}{dx} g_{1T}^{(1)} \qquad (41)$$

from which it is straightforward to derive the Wandzura-Wilczek relation [31,17],

$$g_T = \int_x^1 dy \, \frac{g_1(y)}{y} + \underbrace{\left(\tilde{g}_T - \int_x^1 dy \, \frac{\tilde{g}_T(y)}{y} \right)}_{\bar{g}_T}. \qquad (42)$$

Using this also $g_{1T}^{(1)}$ can be expressed in g_1 and $\tilde{g}_T$. Often this relation is used to make further assumptions, e.g. the assumption that the interaction-dependent part $\tilde{g}_T \approx 0$ (and hence also $\bar{g}_T \approx 0$) or the assumption that the p_T-weighted function $g_{1T}^{(1)} \approx 0$. Although such assumptions are at the present time still fairly *ad hoc*, they allow us to obtain order of magnitude estimates of the functions from just the leading twist function g_1.

The equivalent relations for the h-functions are

$$h_L = -2 \frac{h_{1L}^{\perp(1)}}{x} + \tilde{h}_L$$
$$= h_1 - \frac{d}{dx} h_{1L}^{\perp(1)}, \qquad (43)$$

from which one obtains [4,17]

$$h_L = 2x \int_x^1 dy \, \frac{h_1(y)}{y^2} + \underbrace{\left(\tilde{h}_L - 2x \int_x^1 dy \, \frac{\tilde{h}_L(y)}{y^2} \right)}_{\bar{h}_L}. \qquad (44)$$

For the T-odd functions, we will present the equivalent relations for the Collins fragmentation functions,

$$H(z) = -2z \, H_1^{\perp(1)}(z) + \tilde{H}(z)$$
$$= z^3 \frac{d}{dz} \left(\frac{H_1^{\perp(1)}}{z} \right), \qquad (45)$$

from which one obtains [17]

$$H(z) = \tilde{H}(z) + 2 \int_z^1 dz' \, \frac{\tilde{H}(z')}{z'}, \qquad (46)$$

i.e. this T-odd function is purely interaction-dependent as one might have expected for such functions.

CONCLUDING REMARKS

In this talk I have tried to indicate new opportunities in semiinclusive leptoproduction. In a collider with sufficient energy one can reliably study current fragmentation. This allows first of all a better flavor separation of the 'ordinary' unpolarized and polarized distribution functions $f_1^q(x)$ and $g_1^q(x)$. In principle, it also allows access to the chiral-odd distribution function $h_1^q(x)$, but the measurements require a chiral-odd fragmentation function, which for the case that one integrates over all transverse momenta, requires polarimetry in the final state. Measurement of transverse momenta of the produced hadron opens a rich new field, e.g. the existence of chiral-odd fragmentation function $H_1^\perp$ for spin 0 particles (Collins function). Since this function is also T-odd, it enables access to h_1^q via a single spin asymmetry. Last but not least one must realize that the transverse momentum dependent functions carry the information on the nonperturbative structure of the nucleon, often in a way complimentary to higher-twist functions.

REFERENCES

1. D.E. Soper, Phys. Rev. D 15 (1977) 1141; Phys. Rev. Lett. 43 (1979) 1847.
2. R.L. Jaffe, Nucl. Phys. B 229 (1983) 205.
3. A.V. Manohar, Phys. Rev. Lett. 65 (1990) 2511.
4. R.L. Jaffe and X. Ji, Nucl. Phys. B 375 (1992) 527.
5. J.B. Kogut and D.E. Soper, Phys. Rev. D 1 (1970) 2901.
6. J. Soffer, Phys. Rev. Lett. 74 (1995) 1292.
7. L. Trentadue and G. Veneziano, Phys. Lett. B 323 (1994) 201; D. de Florian and R. Sassot, Phys. Rev. D 56 (1997) 426; M. Grazzini, G.M. Shore and B.E. White, Nucl. Phys. B 555 (1999) 259.
8. T. Sloan, G. Smadja and R. Voss, Phys. Rep. 162 (1988) 45.
9. J.P. Ralston and D.E. Soper, Nucl. Phys. B 152 (1979) 109.
10. R. D. Tangerman and P.J. Mulders, Phys. Rev. D 51 (1995) 3357
11. J.C. Collins and D.E. Soper, Nucl. Phys. B 194 (1982) 445.
12. Possible T-odd effects could arise from soft initial state interactions as outlined in D. Sivers, Phys. Rev. D 41 (1990) 83 and Phys. Rev. D 43 (1991) 261 and M. Anselmino, M. Boglione and F. Murgia, Phys. Lett. B 362 (1995) 164. Also gluonic poles might lead to presence of T-odd functions, see N. Hammon, O. Teryaev and A. Schäfer, Phys. Lett. B 390 (1997) 409 and D. Boer, P.J. Mulders and O.V. Teryaev, Phys. Rev. D 57 (1998) 3057.
13. A. De Rújula, J.M. Kaplan and E. de Rafael, Nucl. Phys. B 35 (1971) 365.
14. K. Hagiwara, K. Hikasa and N. Kai, Phys. Rev. D 27 (1983) 84.
15. R.L. Jaffe and X. Ji, Phys. Rev. Lett. 71 (1993) 2547.
16. J. Collins, Nucl. Phys. B 396 (1993) 161.
17. P.J. Mulders and R.D. Tangerman, Nucl. Phys. B 461 (1996) 197; Nucl. Phys. B 484 (1997) 538 (E).

18. A. Bacchetta, M. Boglione, A. Henneman and P.J. Mulders, Phys. Rev. Lett. 58 (2000) 712
19. A. Bacchetta and P.J. Mulders, hep-ph/0007120.
20. P.J. Mulders and J. Rodrigues, hep-ph/0009343.
21. A.P. Bukhvostov, E.A. Kuraev and L.N. Lipatov, Sov. Phys. JETP 60 (1984) 22.
22. D. Boer and P.J. Mulders, Phys. Rev. D 57 (1998) 5780.
23. A.M. Kotzinian and P.J. Mulders, Phys. Rev. D 54 (1996) 1229; A.M. Kotzinian and P.J. Mulders, Phys. Lett. B 406 (1997) 373.
24. P.J. Mulders and M. Boglione, Nucl. Phys. A 666&667 (2000) 257c.
25. A.M. Kotzian, Nucl. Phys. B 441 (1995) 234; R.D. Tangerman and P.J. Mulders, Phys. Lett. B 352 (1995) 129.
26. D. Boer, R. Jakob and P.J. Mulders, Nucl. Phys. B 564 (2000) 471.
27. M. Boglione, contribution to this meeting, hep-ph/0010166; M. Boglione and P.J. Mulders, Phys. Rev. D 60 (1999) 054007; M. Boglione and P.J. Mulders, Phys. Lett. B 478 (2000) 114.
28. A. Bravar, Nucl. Phys. Proc. Suppl. B 79 (1999) 521.
29. H. Avakian, Nucl. Phys. Proc. Suppl. B 79 (1999) 523.
30. E. Efremov, O.G. Smirnova and L.G. Tkatchev, Nucl. Phys. Proc. Suppl. B 79 (1999) 554.
31. S. Wandzura and F. Wilczek, Phys. Rev. D 16 (1977) 707.

Probing "Generalized Parton Distributions" with JLab at 12 GeV

S. Stepanyan
Thomas Jefferson National Accelerator Facility, Newport News, VA 23606, USA
Christopher Newport University, Newport News, VA 23606, USA

Abstract

We discuss plans for Jefferson Lab to explore GPDs in hard exclusive reactions. A broad experimental program is proposed with up to 11 GeV polarized electron beams. Such an upgraded CEBAF, with high precision, and high performance detectors, will be a suitable place to study deep virtual production of mesons and photons in the range of $Q^2 < 8$ GeV/c^2. These studies are proposed as a key program for the Hall B CLAS detector.

1 Introduction

One of the fundamental topics of modern high energy physics is the understanding of nucleon structure. Studies with leptonic beams in the deep inelastic scattering region (DIS) led to discovery of the quark-gluon structure of the nucleon. It was found, for example, that quarks carry about half of the nucleon momentum, and about 25% of the spin of the nucleon.

To have a more complete picture of nucleon structure, new information is needed, particularly information on quark-quark and quark-gluon interactions in the nucleon. Recent developments in the theory showed that such information can be obtained in hard exclusive leptoproduction experiments. Formalism, developed by Ji [1], Radyushkin [2] and others, for the QCD description of hard exclusive reactions introduces new Generalized Parton Distributions, (or Skewed Parton Distributions, or Off-Forward Parton Distributions). GPDs contain information on quark-quark correlations, on transverse and angular momentum distributions, and provide a unified description of a wide range of inclusive and exclusive reactions.

A broad program for studying GPDs with longitudinally polarized electron beams is proposed for JLab with the upgraded 12 GeV machine [3]. First explorations of the most promising channels will begin even at lower beam energies.

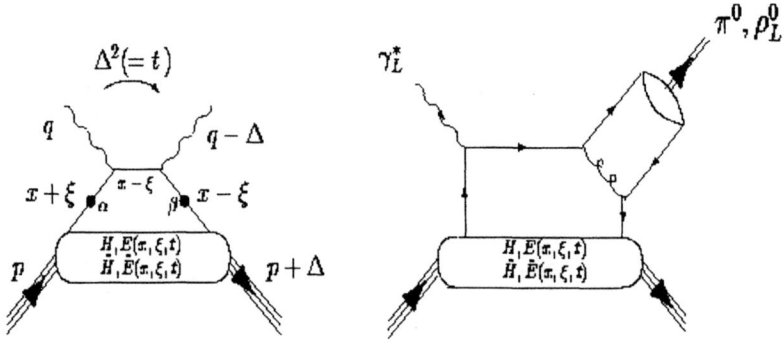

Figure 1: Handbag diagrams corresponding to Deep Virtual Compton Scattering (a) and Deep Virtual Meson Production (b).

In this report an overview of the experimental program proposed for CEBAF 12 GeV upgrade and some details of key reactions are presented.

2 Physics Motivation

It has been shown that in leading-order pQCD the amplitude of Deep Virtual Meson Production with longitudinally polarized photons (generally called Deep Exclusive Scattering, DES), and the amplitude of Deep Virtual Compton Scattering (DVCS) in the forward direction can be factorized into a hard-scattering part (exactly calculable in pQCD) and a non-perturbative nucleon structure part as illustrated in Figure 1. In these so-called "handbag" diagrams, the lower blob represents the structure of the nucleon and can be parametrized in terms of four structure functions, known as the GPDs.

The GPDs are defined as $H, \tilde{H}, E,$ and $\tilde{E}$, and depend upon three kinematic variables: $x, \xi,$ and t. x is the momentum fraction of the struck quark in the quark loop and, as such, is not directly accessible experimentally. ξ is the longitudinal-momentum fraction of the transfer Δ with $\xi = x_B/(2 - x_B)$. $t = \Delta^2$ is the standard momentum transfer between the final-state meson and the virtual photon. H and E are spin-independent and $\tilde{H}$ and $\tilde{E}$ are spin-dependent functions. The GPDs are also quark flavor dependent.

The GPDs H and $\tilde{H}$ are generalizations of the parton distributions measured in deep inelastic scattering. In the forward direction (defined by $q = q'$), H reduces to the quark distribution $q(x)$ and $\tilde{H}$ to the quark-helicity distribution

$\Delta q(x)$ measured in deep inelastic scattering. Furthermore, at finite momentum transfer, there are model-independent sum rules that relate the first moments of these GPDs to the standard elastic form factors. Also Ji [1] has shown that the second moment of these GPDs gives access to the sum of the quark spin and the quark orbital angular momentum to the nucleon spin.

3 Experimental Program

The proposed experimental program includes production of mesons in Deep Exclusive Scattering (DES) and the Deep Virtual Compton Scattering (DVCS). Systematic studies are feasible in a wide range of kinematics, as shown in Figure 2. The interests are X_B, Q^2 and t dependence of exclusive reactions like:

$$\vec{e}p \to ep\gamma$$

$$ep \to ep(\rho^{+,o,-})$$

$$ep \to ep(\pi^{+,o,-})$$

$$ep \to e\Delta(\pi^{+,o,-})$$

$$ep \to e(\Lambda\Sigma)K$$

In the case of DES the GPD formalism is valid for longitudinally polarized photons and therefore a L/T separation is essential. In the scaling regime cross sections for DES should scale as $\sigma_L \sim 1/Q^6$, for DVCS $\sigma \sim 1/Q^4$. It is important to measure cross sections and the spin observables. These will give access to the real and the imaginary parts of GPD amplitudes.

The spin dependent ($\tilde{H}$ and $\tilde{E}$) and the spin independent (H and E) GPDs can be separated by measuring pseudo-scalar and vector meson productions, respectively. Comparison of production of different final mesonic states ($\rho^{+,o,-}$, ω, $\pi^{+,o,-}$, η, etc) will allow quark flavor decomposition of GPDs.

The CLAS detector in Hall B will allow simultaneous measurements of many multiparticle final states. Currently CLAS has been operated at luminosity $L = 10^{34}$ cm^{-2} sec^{-1} and has wide angular coverage of final state particles (see Figure 3.a). The upgraded CLAS will have increased acceptance and improved resolution in the forward scattering region, and will have the capability to run at luminosity $L = 10^{35}$ cm^{-2} sec^{-1}, Figure 3.b [4]. High luminosity is very important for high Q^2 exclusive measurements.

In the following chapter the experimental expectations of some of the key channels are presented.

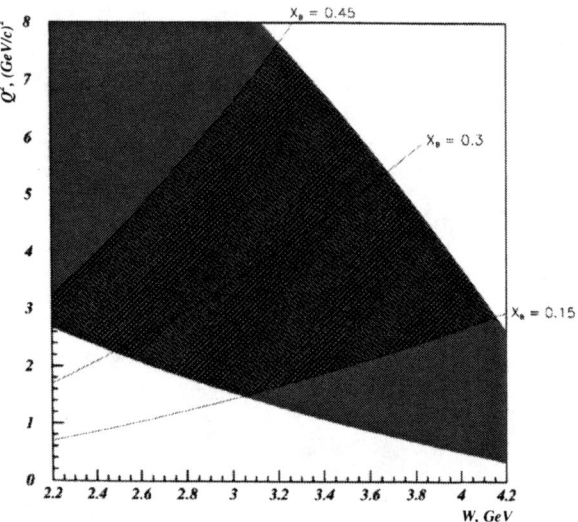

Figure 2: The accessible range of Q^2 and W with 11 GeV beam energy. Lines represent fixed values of X_B.

3.1 Deep exclusive ρ^o production

The cross section of ρ_L^o production is sensitive to the H and E GPDs. In the past, the total cross section of longitudinally polarized ρ^os has been measured in a number of experiments [6], [7], [8]. The data suggest that at high energies and Q^2 above a few GeV/c^2 the perturbative two gluon exchange mechanism (PTGEM) is dominant. However in the low energy region ($W \leq 10$ GeV) this mechanism significantly underestimates the data (see Figure 4). It is evident that at $W \sim 3$ GeV to 4 GeV, the quark exchange mechanism (QEM) becomes the dominant contribution. This region will be accessible with the upgrade of CEBAF accelerator.

The analysis of angular distributions of charged poins, after ρ^o decay, will allow reliable L/T separation assuming S-channel helicity conservation.

To estimate the expected sensitivity at 11 GeV, pseudo-data were generated according to the L/T ratio shown in Figure 5. It was assumed that $\sigma_L \sim 1/Q^6$ and $\sigma_T \sim 1/Q^8$. The detector response was simulated according to the design characteristics. In Figure 6 the reconstructed longitudinal, transverse and the total cross sections for exclusive ρ^o production are shown. Errors correspond to 500 hours of CLAS running at a luminosity $L = 10^{35}$ cm^{-2} sec^{-1}. The obtained slopes of the Q^2 dependence reproduce the generated spectra (lines on the plot) well.

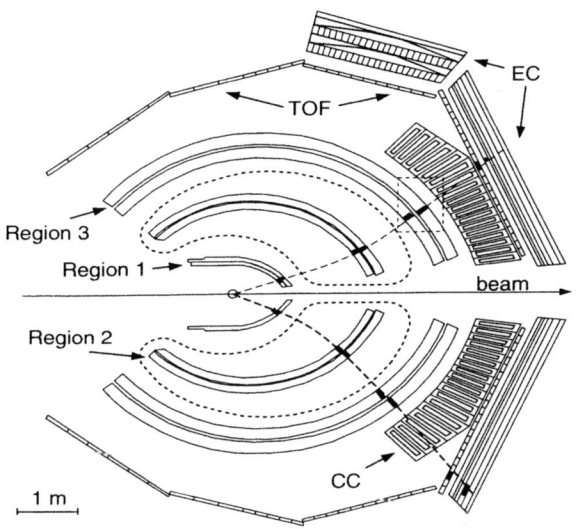

a) Current CLAS detector. Operating luminosity $L = 10^{34}$ cm^{-2} sec^{-1}, detection angular ranges: electrons $\theta = 10^o - 50^o$, charged hadrons $\theta = 8^o - 135^o$, neutrals $\theta = 8^o - 45^o$.

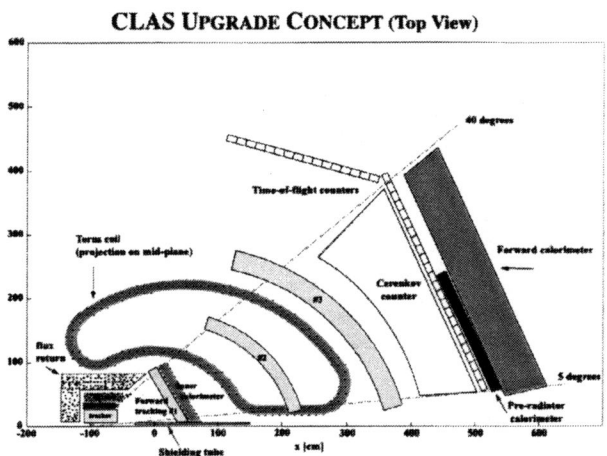

b) CLAS upgrade concept. Design luminosity $L = 10^{35}$ cm^{-2} sec^{-1}, detection angular ranges: electrons $\theta = 5^o - 40^o$, charged hadrons $\theta = 5^o - 110^o$, neutrals $\theta = 5^o - 110^o$.

Figure 3: Top view of current and upgraded CLAS

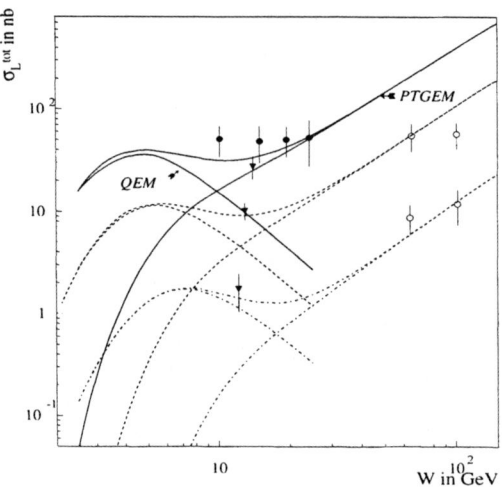

Figure 4: Total longitudinal cross section for ρ_L^o electroproduction as calculated in [5]. Data from NMC [6] (triangles) at $Q^2 = 5.5$ GeV/c^2 (highest point), 8.8 GeV/c^2 and 16.9 GeV/c^2 (lowest point), E665 [7] (black circles) at $Q^2 = 5.6$ GeV/c^2 and ZEUS [8] (open circles) at $Q^2 = 8.8$ GeV/c^2 (upper points) and 16.9 GeV/c^2 (lower points). Calculations are shown at $Q^2 = 6$ GeV/c^2 (full lines), $Q^2 = 9$ GeV/c^2 (dashed lines) and $Q^2 = 17$ GeV/c^2 (dashed-dotted lines). The curves which grow at high W correspond to gluon exchange whereas the curves which are peaked below W $\approx$ 10 GeV correspond to quark exchange. The incoherent sum of both mechanisms is also shown.

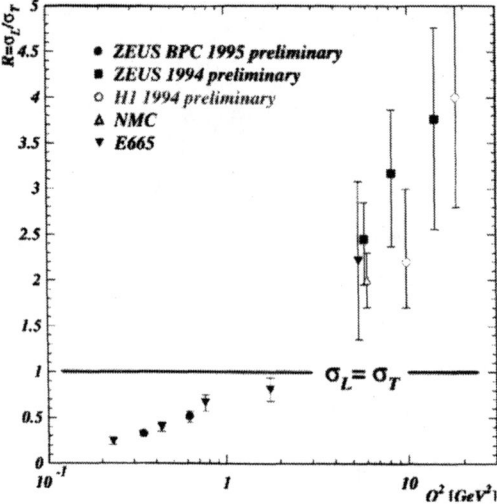

Figure 5: L/T ratio for exclusive ρ^o electroproduction as a function of Q^2.

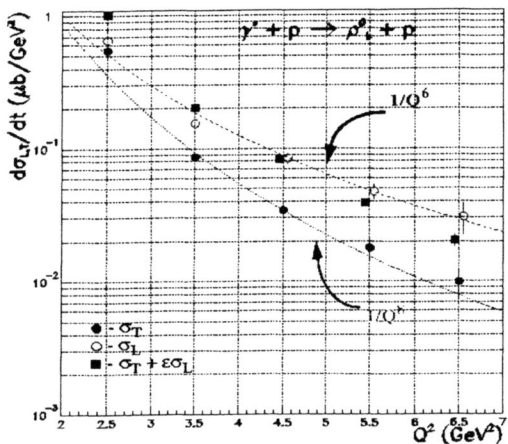

Figure 6: Reconstructed from pseudo-data σ_L, σ_T and total cross section of ρ^o electroproduction. Lines are simulated Q^2 dependences for σ_L and σ_T. Error bars correspond to 500 hours of CLAS running at luminosity $L = 10^{35}$ cm^{-2} sec^{-1}.

3.2 Deep exclusive π^+ production

The cross section of π^+ production is sensitive to the spin-dependent GPDs, $\tilde{E}$ and $\tilde{H}$. There are two diagrams contributing to π^+ production. First is the pseudo-vector (PV) contribution, when a virtual photon interacts with a quark in a nucleon, which after gluon exchange, combines with a second quark to form the pion (see Figure 1.b). Second is the pseudoscalar (PS) contribution (pion pole), when a virtual photon knocks out a pion from the meson cloud (Figure 7). The PV mechanism is sensitive to $\tilde{H}$ and PS is sensitive to $\tilde{E}$.

Relative contributions of these two diagrams depend on t and X_B. At small t and large X_B the PS contribution is large, while at large t and small X_B the PV component is dominant (see Figure 8 [9]). (Also the π^+/π^o comparison will contribute to separation of $\tilde{H}$ and $\tilde{E}$, since in the case of π^o production there is no pion pole effect.)

For identification of the longitudinal part of the π^+ cross section Rosenbluth separation will be done using measurements at several beam energies. In Figure 9 reconstruction of cross sections from simulated pseudo-data are presented. In order to perform Rosenbluth separation pseudo-data at beam energies 6 GeV and 8 GeV are simulated. Only statistical errors are shown, corresponding to 1000 hours of running at $L = 10^{35}$ cm^{-2} sec^{-1}.

Reasonable accuracy of separation is achievable at Q^2 up to 5.5 GeV/c^2. One should point out that if measurements show dominance of the longitudinal part

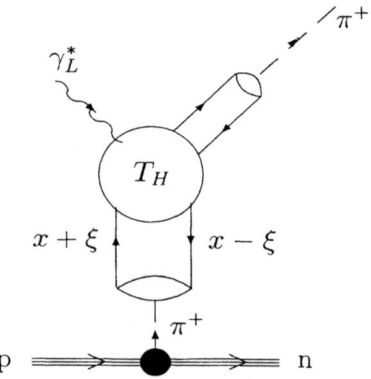

Figure 7: Pseudoscalar contribution to "hard" π^+ electroproduction. The virtual photon knocks out a pion from the meson cloud of the nucleon.

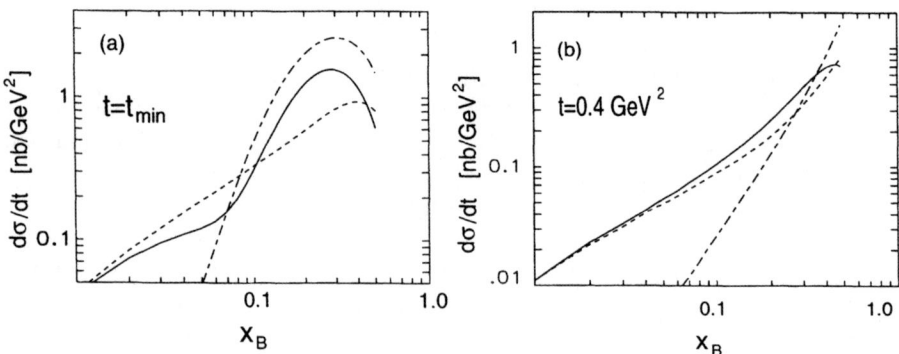

Figure 8: Cross section for exclusive π^+ production through the scattering of longitudinally polarized photons off the proton at $Q^2 = 10$ GeV/c^2 for (a) $t = t_{min}$ and (b) $t = -0.4$ GeV/c^2. The dashed and dot-dashed curves show the pseudovector and pseudoscalar contributions, respectively.

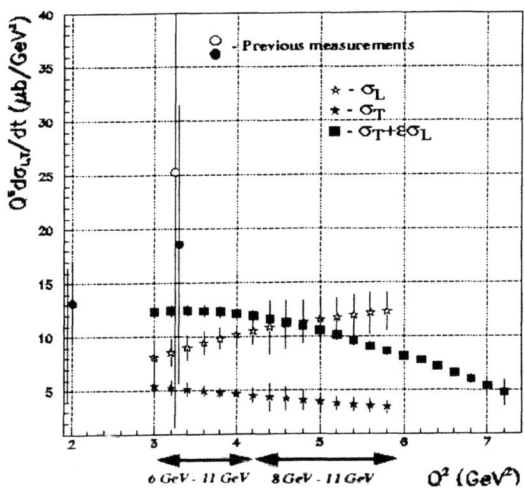

Figure 9: Separated L, T and unseparated differential cross section for π^+ production.

of the cross section at high Q^2, then for studying GPDs the total cross section, measured with higher accuracy and higher values of Q^2, can be used.

Using π^+ production it is possible also to measure transverse target spin asymmetry in the reaction $e\vec{p} \to en\pi^+$. Although CLAS will have a longitudinally polarized target, there is a transverse component to the direction of the virtual photon. Measured asymmetry will be sensitive to the product of the imaginary parts of $\tilde{E}$ and $\tilde{H}$.

In Figure 10 the expected results (data point with error bars) are plotted with model calculations. Errors correspond to a 1000 hours of CLAS running at luminosity $L = 10^{35}$ cm^{-2} sec^{-1}. There is reasonable accuracy and sensitivity for Q^2 up to ~ 4 GeV/c^2.

3.3 DVCS

DVCS is the most promising channel for studying GPDs. The validity of the handbag diagram is expected to be at lower Q^2 than in the case of DES. This is supported by measurements of the $\gamma^*\gamma\pi^o$ form factor on e^-e^+ colliders. In leading order pQCD, the DVCS process and the production of π^o by two photons, where one of the photons is highly virtual, have the same handbag diagram. In Figure 11 the recent measurements of $F_{\gamma^*\gamma\pi^o}$ from CLEO [11] are shown. The curves correspond to the leading order and the next to leading order calculations. As can be seen from the figure, $F_{\gamma^*\gamma\pi^o}$ starts to scale as $1/Q^2$ already at $Q^2 \sim 2$ GeV/c^2 - 3 GeV/c^2. When accounting for the internal quark motion the data are

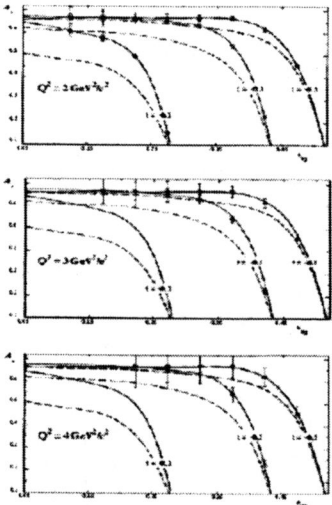

Figure 10: Transverse target asymmetry in the reaction $e\vec{p} \to en\pi^+$ as a function of X_B at different t and pseudo-data are generated assuming 1000 hours of CLAS running at luminosity $L = 10^{35}$ cm^{-2} sec^{-1}.

well described for $Q^2 > 1$ GeV/c^2

One complication with experimentally studying DVCS is the interference with the Bethe-Heitler (BH) process (see Figure 12).

$$\frac{d\sigma}{dQ^2 dX_B dtt\phi} = |T^{VCS} + T^{BH}|^2 \qquad (1)$$

DVCS contributions can be extracted from measurements in various ways:

- Direct measurements of the absolute DVCS amplitude in the region where the BH contribution is small and can be neglected

- Extracting the imaginary part of the DVCS amplitude by measuring Single Spin Asymmetry (SSA) with longitudinally polarized beam

- Extracting the real part of the amplitude by measuring the beam charge asymmetry (e$^-$ and e$^+$ beams)

For a complete understanding of GPDs, studies of all three processes are desirable in a wide range of kinematics.

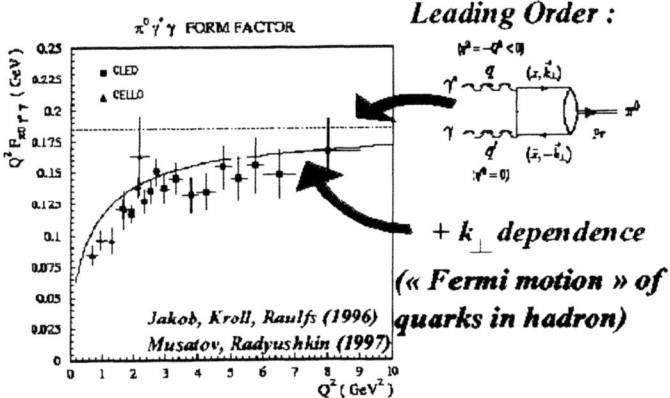

Figure 11: Experimental data on $F_{\gamma^*\gamma\pi^0}$ from [11] with predictions in the leading order and the next to leading order pQCD

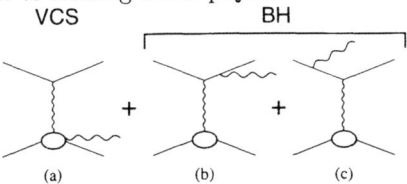

Figure 12: Feynman diagrams for VCS and Bethe-Heitler processes contributing in the amplitude of $ep \to ep\gamma$ scattering.

The first two measurements can be achieved with electron beams as will be available after the CEBAF 12 GeV upgrade, while the third one requires use of positron beams.

In Figure 13 the cross sections are shown for the DVCS and BH processes as a function of $\theta_{\gamma^*\gamma}$, the polar angle between virtual and real photons, for beam energies of 6 GeV, 12 GeV, 27 GeV and 200 GeV. Calculations are done at $Q^2 = 2.5$ GeV/c^2 and $X_B = 0.3$ when real photons are produced in the lepton scattering plane [12]. At 12 GeV beam energy in the range of $\theta_{\gamma^*\gamma} > -2.5^o$ the BH contribution is large, while at $\theta_{\gamma^*\gamma} < -5^o$ ($-t > 0.4$ GeV/c^2) the DVCS contribution dominates in the cross section.

The measurements of $ep \to ep\gamma$ in the range $\theta_{\gamma^*\gamma} < -5^o$ ($-t > 0.4$ GeV/c^2) will be sensitive to the square of the DVCS amplitude. In Figure 14 the t-dependence of the differential cross section calculated by [12] is presented. The solid and dotted curves are calculations with different model assumptions. The points are pseudo-data generated assuming 500 hours of CLAS running at luminosity of $L = 10^{35}$ cm^{-2} sec^{-1}. One can see reasonable sensitivity to the model assumptions at this kinematics.

The availability of the highly polarized electron beam will allow also to mea-

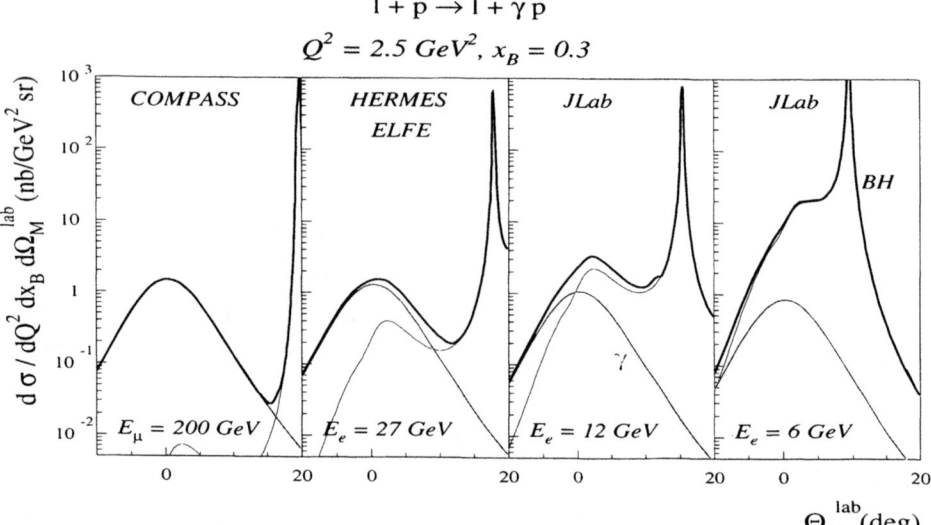

Figure 13: Cross section of $ep \to ep\gamma$ as a function of the angle between virtual and real photons at beam energies 6 GeV, 12 GeV, 27 GeV and 200 GeV. Kinematics of the reaction is fixed at $Q^2 = 2.5$ GeV/c^2 and $X_B = 0.3$. The green curve is the contribution of DVCS, the red curve is Bethe-Heitler and the blue curve is the total. Cross sections are calculated according to [12].

sure the Single Spin Asymmetry (SSA) in DVCS. The asymmetry arises as a result of an interference between real and imaginary parts of the longitudinal and transverse amplitudes.

$$\frac{d^5\sigma^+}{dQ^2 dX dt d\phi} - \frac{d^5\sigma^-}{dQ^2 dX dt d\phi} \sim Im(T^{VCS}) \times Re(T^{BH}) \qquad (2)$$

where "+" and "-" denote positive and negative helicities of the beam.

Similar estimations of the expected statistical errors for measurements of beam spin asymmetries with longitudinally polarized beam have been done as well. In Figure 15 the asymmetries are shown with projected (statistical) errors for 500 hours of running CLAS in the same conditions as above. The curves correspond to the different model approaches and show sensitivity to the model parameters.

4 Summary

A broad experimental program is proposed for JLab with CEBAF 12 GeV.

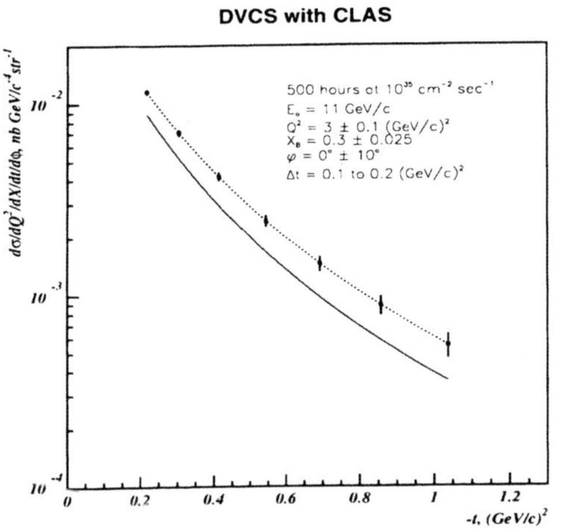

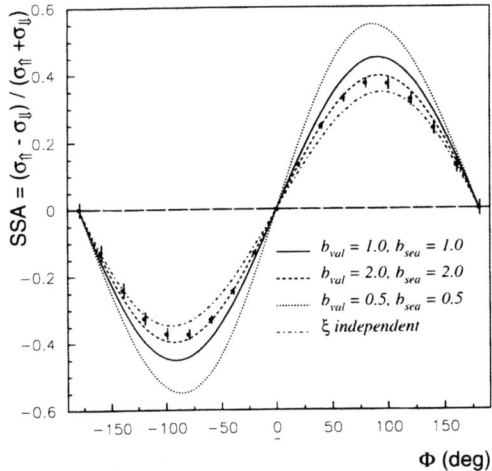

Figure 14: t-dependence of the cross section of the $ep \rightarrow ep\gamma$ process at beam energy 11 GeV. Errors are evaluated according to 500 hours of running CLAS at luminosity 10^{35} cm^{-2} sec^{-1}. Pseudo-data were integrated in the bins of $Q^2 = 3 \pm 0.1$ GeV/c^2 and $X_B = 0.3 \pm 0.025$ and $\phi_{\gamma^*\gamma} = 0 \pm 10^o$. The solid line is the calculation without ξ dependence. The dotted line is the calculation with ξ dependence.

Figure 15: The Single Spin Asymmetry in the $ep \rightarrow ep\gamma$ reaction measured with longitudinally polarized 11 GeV electron beam. Points are pseudo-data generated for 500 hours of running on CLAS at luminosity 10^{35} cm^{-2} sec^{-1}. Pseudo-data were integrated in the bins of $Q^2 = 3$ GeV/$c^2 \pm 0.1$ GeV/c^2 and $X_B = 0.2 \pm 0.05$ and $-t = 0.2 \pm 0.1$ GeV/c^2.

Data on deep virtual meson and photon production will be collected in the range of Q^2 up to 8 GeV/c^2 and X_B from 0.1 to 0.5.

- L/T separation in the production of mesons is feasible up to $Q^2 \sim 6$ GeV/c^2
- Unseparated cross sections can be measured up to $Q^2 \sim 8$ GeV/c^2
- Beam asymmetry in DVCS can be measured up to $Q^2 \sim 7$ GeV/c^2
- Target asymmetry for π^+ production is feasible at $Q^2 \sim 5$ GeV/c^2

A large amount of data will be collected in essentially an unexplored domain. If the scaling regime is reached then measurements will have high sensitivity to GPD modes. Otherwise data will be very valuable for understanding pre-asymptotic effects.

The author thanks V. Burkert, L. Elouadrhiri and M. Guidal for help in preparing the material and for very helpful discussions. Also thanks to all authors of the proposal "Deep Virtual Electroproduction of Mesons and Photon with JLab at 12 GeV" for 12 GeV upgrade for valuable input. Many thanks to organizers of the "Second Workshop on Physics with EPIC" for an invitation to the workshop.

References

[1] X. Ji, Phys. Rev. Lett. **78**, 610 (1997); Phys. Rev. D **55**, 7114 (1997).

[2] A.V. Radyushkin, Phys. Lett. B **380**, 417 (1996); Phys. Rev. D **56**, 5524 (1997).

[3] Deep Virtual Electroproduction of Mesons and Photons with JLab at 12 GeV. http://www.jlab.org/div_dept/physics_division/skewed/

[4] Hall B plans for energy upgrade. http://www.jlab.org/div_dept/physics_division/skewed/

[5] M. Vanderhaeghen, P.A.M. Guichon, and M. Guidal, Phys. Rev. Lett. **80**, 5064 (1998).

[6] M. Arneodo et al., Nucl.Phys. **B429** (1994) 503.

[7] M.R. Adams et al., Z.Phys.C **74** (1997) 237.

[8] M. Derrick et al., Phys.Lett.B **356** (1995) 601.

[9] L. Mankiewicz, G. Piller, and A. Radyushkin, Eur. Phys. J. C **10**, 307, (1999).

[10] A. Radyushkin "Measuring Skewed Parton Distributions", JLab-THY-00-DRAFT.

[11] J. Gronberg et al. (CLEO Collaboration), Phys.Rev. **D 57**, 33 (1998)

[12] M. Guidal "Computer Code for DVCS and BH Calculations", Private communications.

e-A Physics at a Collider

G. T. Garvey

Los Alamos National Laboratory, P-25, MS H846, Los Alamos, NM 87545

INTRODUCTION

An electron-nucleus (e-A) collider with center-of-mass energy in excess of 50 GeV per electron-nucleon collision will allow the physics community to obtain unprecedented new knowledge of the partonic structure of nuclei. If reliable information is to be extracted on these partonic densities, it is essential to realize that with our current level of understanding of QCD, momentum transfers to the struck partons greater than 1 GeV/c are necessary. This requirement puts a priority on high center-of-mass energy if partonic densities are to be measured over a wide range.

Comparing the partonic structure of the free nucleon to that of bound nucleons and measuring the systematic changes in that structure as a function of nucleon number (A) will provide deeper insight into the origins and dynamics of nuclear binding. In addition, e-A collisions will allow the exploration of partonic densities appreciably higher than is accessible in e-p collisions. An e-A collider will allow one to measure the gluonic structure functions of nuclei down to $x \sim 10^{-3}$, information valuable in its own right and essential to a quantitative understanding of highly relativistic A-A collisions.

The time-space evolution of partons can only be investigated by studying the modifications of hard collisions that take place when nuclear targets are employed. In a hard collision the partonic fragments interact, hadronize, and reinteract on their way to the distant detectors without revealing their evolution into the hadrons finally detected. Nuclear targets of differing A place varying amounts of nuclear matter in proximity to the hard collision producing unique information about the quantum fluctuations of incident projectile prior to the collision and on the early evolution of the produced partons.

Using charged leptons (e,μ) to investigate this physics has been the richest source of information to date and extending the reach of these investigations by the constructing an e-A collider is the best opportunity within reach of the US nuclear science community. This work will potentially affect all areas of strong interaction research. A host of additional issues of considerable interest to nuclear scientists can be investigated; some of which are discussed below.

DIS KINEMATICS AND THE ADVANTAGES OF A COLLIDER

Research with e-A collider has two major advantages over fixed target studies. First, a collider delivers more energy to the center of mass, providing greater range in x and Q^2 in the primary collisions, and second, final states of the target may be observed that are typically inaccessible to fixed target experiments.

The energy in the center of mass squared (s) of a two-body collision is a relativistic invariant given by

$$s = (p_L + p_N)^2, \qquad (1)$$

which for an energetic lepton (p_L) incident on a nucleon (p_N) at rest is

$$s_F^{1/2} = (2 M_N p_L)^{1/2}. \qquad (2)$$

Thus, for a 25 GeV lepton, $s_F^{1/2}$ = 6.8 GeV. In the collider mode $s^{1/2}$ is given by,

$$s_C^{1/2} \approx (4 p_L p_N)^{1/2}. \qquad (3)$$

Thus, a 10 GeV electron colliding with a 30 GeV proton produces $s_C^{1/2}$ = 34.6 GeV. The increase in energy by employing a collider is evident. It is impractical, to say the least, to provide a 600 GeV electron beam to achieve the equivalent center-of-mass energy on a fixed target.

Figure 1 depicts the kinematics of a deep inelastic scattering (DIS) between a nucleon and a charged lepton. E and E' are the COM energies of the scattered lepton. The fraction of the nucleon's momentum, carried by the struck parton is x. The fraction of lepton energy given up in the collision is y. Requiring that $Q^2 > 1$ GeV^2, the lowest value useful value of x that can be accessed is $x_{min} \approx 1/0.9s$. At e-RHIC with 10 GeV electrons and nuclear beams of 100 GeV/nucleon, $x_{min} \approx 3 \times 10^{-4}$.

A collider also allows much more effective measurement of the final states of the target than is the case for fixed target experiments. Fixed target exclusive and semi-inclusive measurements are forced to employ "thin" targets (less than an interaction length) if nuclear fragments are to escape the target. The use of requisitely thin targets reduces the luminosity, making the acquisition of sufficient statistics a serious problem. Further, the boost acquired by target fragments in the collider mode makes them readily detectable when they separate from the beam.

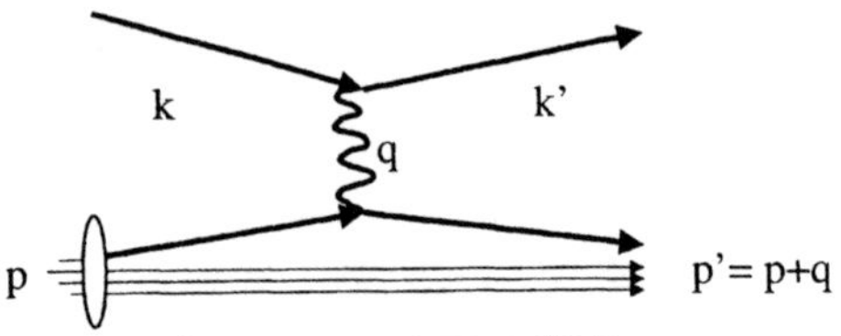

$Q^2 = -q \cdot q = 4EE' \sin^2\theta/2$

$x = Q^2/2p \cdot q \quad y \equiv p \cdot q/p \cdot k$

$x \approx Q^2/yS$

Useful range:

$x_{min} = 1/0.9S = 3 \times 10^{-4}$ (e-A)

$= 10^{-4}$ (e-p)

Figure 1. Collider DIS kinematics.

PREVIOUS FIXED TARGET EXPERIMENTS

To better understand how an e-A collider can advance our knowledge of the partonic structure of nuclei, it is useful to briefly examine the e-A and μ-A DIS experiments that have been carried out in the fixed target mode.

The highest energy DIS experiments on nuclei used tertiary beams of muons on fixed nuclear targets. The NMC [1,2] series of experiments at CERN were carried out at $s^{1/2} = 23.7$ GeV, and those of E-665/FNAL [3,4] at $s^{1/2} = 30.6$ GeV. The muon intensities were typically 2×10^6 μ/s and employing thick nuclear targets of 600 g/cm^2 targets achieved luminosities of 6×10^{32} cm^{-2}s^{-1}nucleon^{-1}. These thick targets are only practical for inclusive experiments. In order to carry out exclusive experiments, the target thickness must be reduced by approximately an order of magnitude, making the acquisition of adequate statistics a serious problem. The fixed target e-A experiments have much greater incident beam intensity, however the beam energy has been limited to 30 GeV or less. The resulting COM energy is therefore below 7.5 GeV, greatly limiting the range of x and Q^2 that can be investigated. An appropriate e-A collider overcomes all of these difficulties. Figure 2 shows the range of x and Q^2 that have been covered by previous fixed target experiments. The vast range covered at HERA is only for *e-p* collisions, while the figure on the right shows the range that could be covered by an e-A collider with $s^{1/2} = 63$ GeV ($E_e = 10$ GeV, $E_A = 100$ GeV/A).

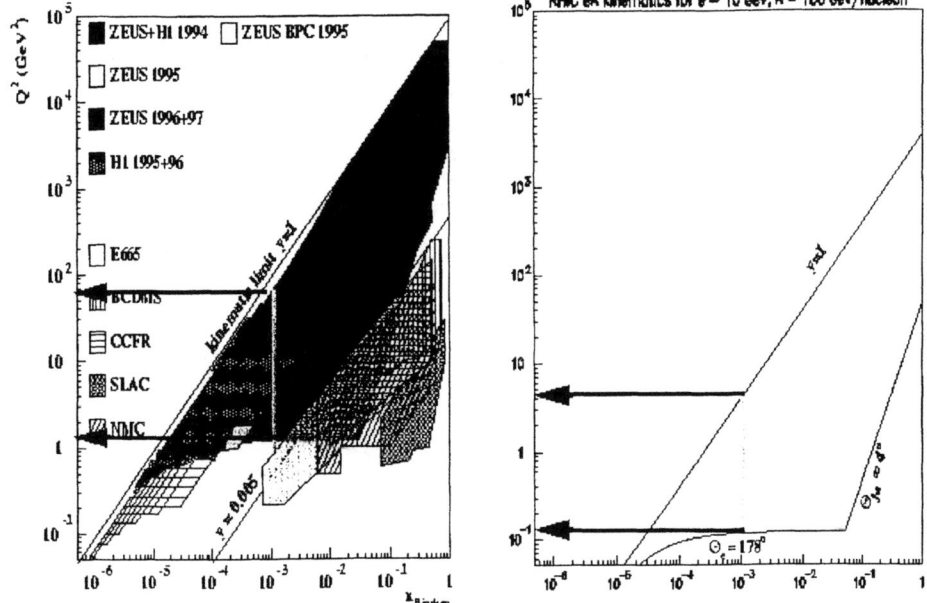

Fig. 2. On the left is shown the x and Q^2 coverage of previous experiments. The ZEUS and HERA measurements were only e-p. The rest are fixed-target experiments. The range of e-RHIC is shown on the right.

PREVIOUS STUDIES AND LUMINOSITY

Table I lists some of the previous studies of potential e-A colliders. A serious obstacle encountered in the previous studies has been the projected limited luminosity. The luminosities that had been projected for energetic e-A colliders were in the range of $l = 6 \times 10^{31}$ cm^{-2}s^{-1} per electron-nucleon. This luminosity was evaluated as interesting and useful but not compelling. The recent plans for both versions (EPIC and e-RHIC) of an electron-ion collider (EIC) propose to achieve $l = 10^{33}$ cm^{-2}s^{-1} per electron-nucleon [5]. This corresponds to 86 inverse pb per day and as previous studies established that significant physics results are obtained at 200 (pb)$^{-1}$ this is easily in reach with the proposed colliders. Achieving these luminosities requires electron cooling of the ion beam and an intense electron beam; the latter requirement is achieved already at the present day B-factories.

Table I. Previous attention to e-X colliders.

- HERA 30 GeV e on 800 GeV p
- Future Physics at HERA (1995-96) (e-A, e-p)
- GSI ENC (1997) (Seeheim Proceedings)
- Indiana University (1999) (EPIC 99)
- BNL e-RHIC 3 Workshops (December 1999, April 2000, June 2000)
- MIT-Indiana University, EPIC (September 14–16, 2000)

PARTONIC DISTRIBUTIONS IN NUCLEI

Table II lists some of the research areas that can be pursued at an e-A collider of sufficient COM energy. Due to the limited space available only a few these topics can be even briefly discussed. Recall that the conventional description of a nucleus is that of a collection of nucleons, weakly bound in a potential created by their mutual interaction. The underlying interaction is believed to arise from the exchange of virtual mesons between the nucleons. Thus it was a surprise when the NMC [6] found a systematic nuclear dependence to the nucleon's $F_2(x,Q^2)$ structure function. Their measurement of the structure function per nucleon as measured in Fe and ^{2}H showed definite nuclear dependence, only some of which is understood as of today. A host of dedicated fixed target experiments followed [7], confirming the existence of nuclear dependence but with some significant modifications of the original EMC results. This body of research has produced relative F_2 structure functions over a broad range of A, x, and Q^2. As the measured structure functions show regular and smooth dependence, only a few precise measurements of relative structure functions are required to characterize the nuclear behavior over the interval 2< A < 200. Figure 3(a) is a conceptual presentation of the current experimental knowledge of the $F_2^A(x)$ structure functions relative to the $F_2^d(x)$ structure function of deuterium. The rise at the largest values of x is generally ascribed to the nucleon's increased Fermi momentum. The enhancement and EMC regions presumably arise from nuclear binding effects but a quantitative explanation has yet to be developed [7]. The decrease observed for $x <$ 0.05 is termed shadowing and is the object of considerable attention as it has a strong quantitative impact on the outcome of relativistic heavy ion collisions [8,9].

Table II. Some research areas with an e-A collider.

- Extending x and Q^2 of nuclear parton distributions, $g^A(x)$ in nuclei
- Investigation of the saturation of parton density ?? (e-A, $x < 10^{-3}$)
- Diffractive processes, rapidity gaps, "Pomeron" (e-A, e-p)
- Time-space evolution of partonic processes (e-A)
- DIS semi-exclusive, exclusive underlying meson-baryon (MB) structure of nucleon, nuclear modifications of MB structure, skewed parton distributions
- Spin structure of the nucleon, extend $g_1(x,Q^2)$, $\Delta g(x)$ via 2-jet production

The γ–p cross section for photon energies in excess of 2 GeV is only ~ 0.1 mb [10], which corresponds to a mean-free path in nuclear matter greatly in excess of 100 fm. Thus, one expected to observe the high-energy γ-A cross sections that were directly proportional to A. However, such was not the case as the cross section increases appreciably less steeply than A times the γ-p cross section. This is because the photon sometimes fluctuates into a $q\bar{q}$ pair which has a cross section typical of strong interactions (20 mb) and as such is often adsorbed and thus "shadowed" from the rest of the nucleus (even high energy neutrino-nucleus interactions evidence shadowing!). In QCD, shadowing in DIS is described in terms of the interaction of the $q\bar{q}$ color singlet fluctuations of the photon with target gluons. Most of the shadowing arises from pairs with large relative separation as they have the largest cross sections because their color is not well screened. These cross sections cannot be reliably calculated in perturbative QCD and hence some degree of phenomenology must be introduced to characterize the interaction. Shadowing effects are both significant and universal. They are best studied with DIS as the light-cone wave function of the photon's $q\bar{q}$ pair can be characterized with high confidence. The coherence length of the $q\bar{q}$ fluctuation is $l_c \approx (2m_n x)^{-1}$. As the effects of shadowing set in as soon as the coherence length exceeds the internucleon separation distance in a nucleus of ~ 2 Fm, its onset is expected at $x \cong 0.05$. Based on such a picture, shadowing should reach a limiting value when the coherence length becomes long compared to the nuclear diameter ($l_c > 2.4 A^{1/3}$). The observation of this limit in heavy nuclei requires $x \leq 10^{-3}$, which in turn requires $s^{1/2}$ of at least 60 GeV. Figure 3(b) shows the present data and makes it clear that better and more extensive data are required to investigate this region. The lowest x data shown in Fig. 3(b) comes from E665 and have Q^2 well below 1 GeV^2, and hence must be regarded with suspicion with regard to their interpretation.

Compared to the present knowledge of the u, d, and s quark distributions, the nuclear gluon structure functions ($g_A(x,Q^2)$) are effectively unmeasured. This is because leptonic probes do not directly couple to glue. The gluon distribution must be inferred from the change of $F_2(x,Q^2)$ with $\ln Q^2$ using the DGLAP evolution equations. This requires measurement of the relative $F_2^A(x,Q^2)$ of some characteristic nuclei to the better than 1% over a fair range in Q^2. In addition to providing direct information on shadowing, the relative yields are used to extract the ratio

$$R_{A,d}(x) = \frac{d\left[\frac{F_2^A(x,Q^2)}{F_2^d(x,Q^2)}\right]}{d\ln Q^2}, \qquad (4)$$

which yields the relative gluon distributions. Such measurements are very difficult and have only been carried out in a single instance by the NMC [11]. Figure 4 shows the result of the measurement and the ratio of the gluon distributions extracted by Pirner and Gousset [12]. Figure 5 shows the projected statistical accuracy [13] that can be achieved in a 5-day measurement of relative gluon distributions between C and

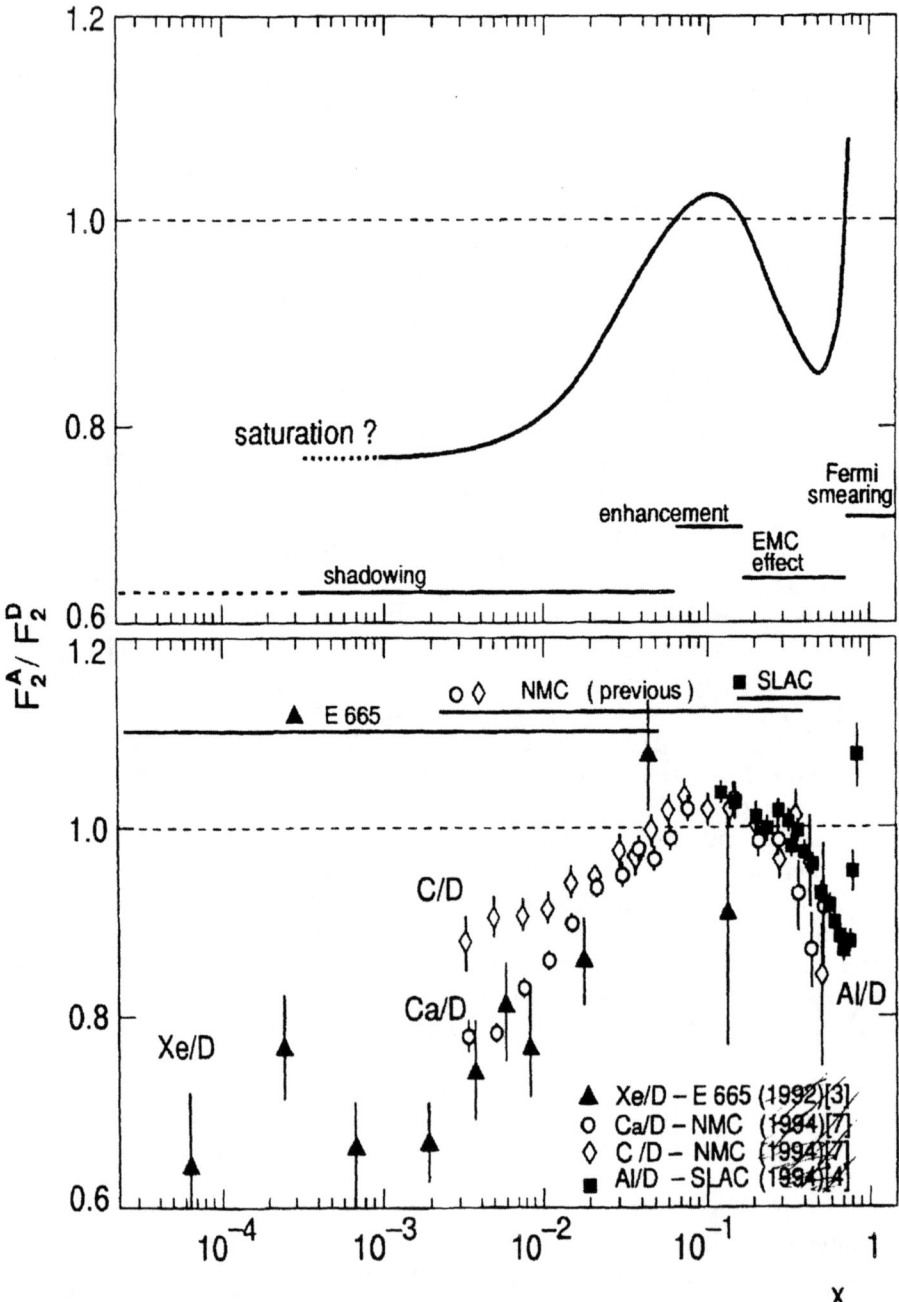

Fig. 3. The upper figure shows an idealized version of nuclear effects on the $F_2(x)$. The lower shows actual data for four cases.

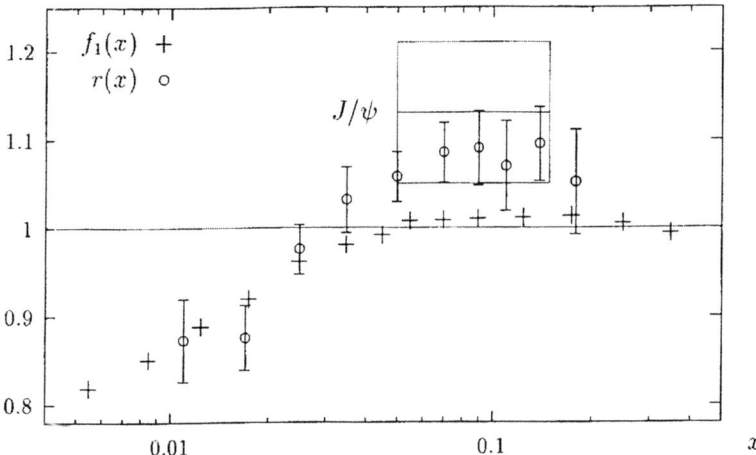

Fig. 4. The ratio $r(x) = G^{Sn}(x)/G^{C}(x)$ of tin to carbon gluon density as a function of x together with the ratio of structure functions $f_1(x) = F_2^{Sn}(x)/F_2^{C}(x)$. The box represents the extraction of r from J/ψ electroproduction data.

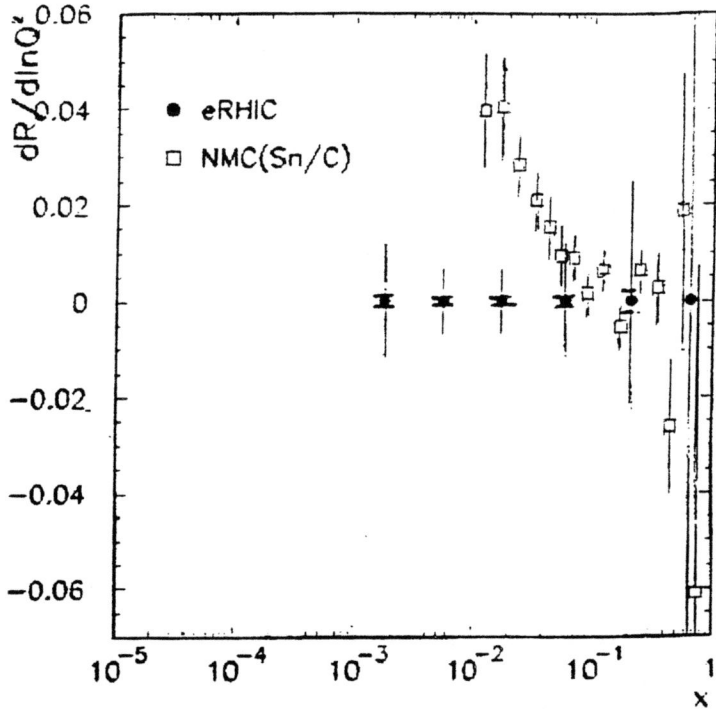

Fig. 5. Shows the statistical improvement that can be achieved at e-RHIC, with a 5-day run on $d(Sn/C)/d\ln^2 Q$.

111

Sn. The issue in measuring this ratio to the required accuracy is not in obtaining adequate statistics, which is readily achievable with the projected luminosity, but rather reducing the systematic error below 1%. This challenge is common to many of the experiments envisioned for the EIC.

It is difficult to overemphasize the importance of shadowing on all high-energy processes occurring in nuclei. Shadowing plays a role in determining the final-state particle multiplicities and energy densities achieved in relativistic heavy ion collisions. Shadowing produces opacity effects that can mask color transparency in vector meson production, and the inability to properly characterize shadowing presently limits the extraction of partonic energy loss in the nuclear medium.

There is the possibility of reaching a new state of partonic matter – a saturated parton density – using an e-A collider. As shown in Fig. 6, studies at HERA [14] have shown that the gluon density grows rapidly at small x and high Q^2. This growth leads naturally to the question as to whether partonic densities eventually saturate. A great deal of theoretical work has been done on this subject, but there is no convincing evidence that such saturation has been observed in any experiment. At the saturation value the partonic matter would become black to strongly interacting projectiles. The DGLAP equations would become nonlinear due to parton-parton recombination. It is also clear that this saturated partonic matter is universal in the sense that it is independent of the system that spawns it. Figure 7 illustrates where in x and Q^2 such condensed partonic matter might exist. In the regime depicted in Fig. 7 the strong coupling constant α_s may be small enough that perturbative procedures may be employed. Mclerran and Venogopolan [15] have investigated a classical gluon field that models the saturated state. They find it has many of the properties of a glass, so they term the state a "colored glass condensate". e-A collisions play an important role in the search for and study of such a state by exploiting the fact that the longitudinal gluon density in a heavy nucleus is enhanced relative to that in a single nucleon. In the nuclear rest frame, an incident relativistic electron sees the nucleus contracted to a longitudinal diameter, $D_A = \gamma^{-1} 2.4 A^{1/3}$. The valance quarks in the nucleons are contracted by roughly the same amount, as they carry equivalent momenta. However the low x partons in the nucleons carry much smaller momenta and hence are contracted to a much smaller degree. In fact, the gluons from different nucleons overlap in the longitudinal direction and present a gluon density that exceeds that of a single nucleon. For a nucleus with A nucleons the average enhancement factor is conservatively, $<R_A>/2 = 3/8 A^{1/3}$, which for A = 200 is an enhancement of 2.2. Using the HERA data shown in Fig. 8, one sees that for $Q^2 = 5$ GeV^2 that $xg(x)$ increases by a factor of 2 between $x = 10^{-2}$ and 10^{-4}. Thus, increasing the gluon density by a factor of 2.2 is equivalent to the gluon density that would be found in the proton at an x that is $6 \cdot 10^{-3}$ times smaller. Thus, an e-Pb collision at $x = 10^{-3}$ at e-RHIC sees a gluon density that occurs at $x = 6 \cdot 10^{-6}$ in an e-p collision. Such a collision would require achieving $s = 2 \times 10^5$ GeV. Thus it is clear that an e-A collider with beams of nuclei with large A offer a considerable enhancement in investigating the interesting and important physics that can occur in a high parton density regime.

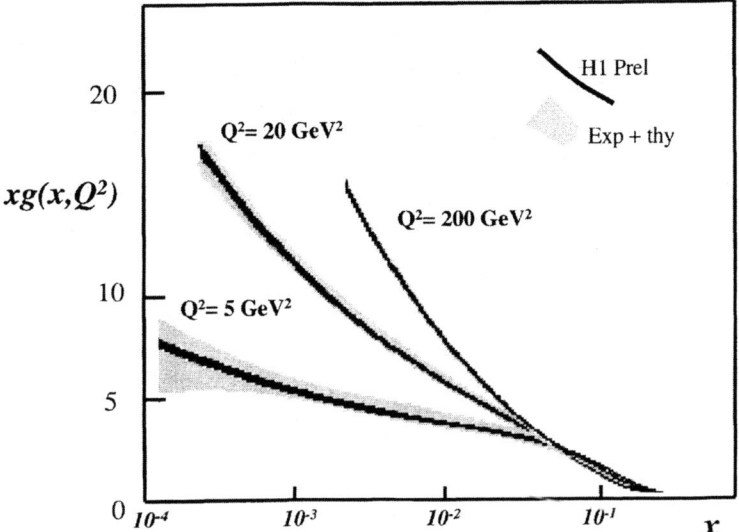

Fig. 6. $xg(x,Q^2)$ in the proton as a function of x for selected Q^2.

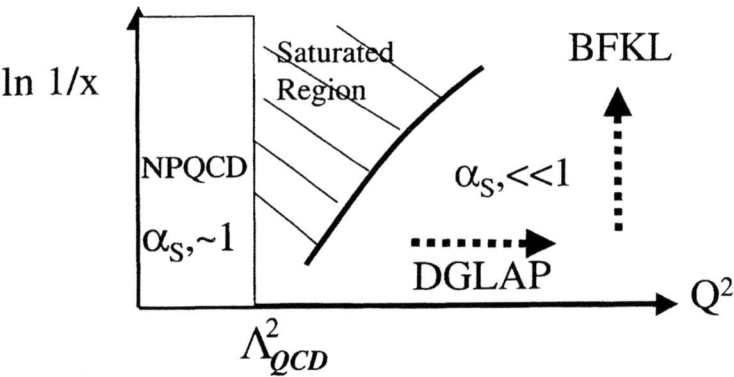

Fig. 7. Shows the region in x and Q^2 where a gluon condensate might occur.

DIFFRACTION

Diffraction is an important, many-faceted, and somewhat controversial subject. However, it is readily observable experimentally and represents a relatively large fraction (~10%) of the total cross section. There are several important issues to be investigated in the diffractive sector of e-A collisions. The structure of the Pomeron in nuclear matter can be studied, a posited intimate relationship between diffraction and shadowing [16,8] can be investigated, the nuclear gluon distribution can be determined via measurement of exclusive vector meson production, and various modes used to establish color transparency. The signatures for the identification of diffractive processes discussed below illustrate why a collider is usually essential to study many of its aspects.

Diffraction generally refers to hadron-hadron scattering where one of the hadrons more or less retains its identity. For example, a reaction such as p + p, leading to final states such as $x + p$, $x + n$, $x + \Delta$, $x + \Lambda$, etc., are regarded as diffractive when the leading baryon retains a large fraction (>95%) of the momentum of one of the colliding protons. The fact that the adsorbing hadron effectively retains its momentum leads to a coherence requirement that limits the invariant mass (M) of the diffracted state. For a particle of mass m and momentum p to diffract from a stationary target into a state of mass M, the minimum squared momentum transfer is

$$t_{min} = \left(\frac{M^2 - m^2}{2p}\right)^2 . \tag{5}$$

Thus, to retain coherence over a target of radius R, wave mechanics requires that

$$M^2 - m^2 \leq \frac{2p}{R} . \tag{6}$$

For targets with $R = 1$ fm this requirement produces pronounced forward peaking (e^{-bt}) with $b = 8$ GeV^{-2}.

As a result of this peaking, the diffracted particle is well separated in rapidity from the target. Typically, an examination of the final state of the target is usually necessary to clearly identify diffractive processes. The exception to this observation is diffractive vector meson production. In this case the virtual photon either fluctuates into a vector meson which is diffracted off the target or into a $q\bar{q}$ pair that scatters onto mass shell by virtue of two gluon exchange with the target. One distinguishing feature of diffractive e-N scattering is the large rapidity gap between the target and any associated fragments. This is not expected in conventional deep inelastic scattering; in QCD the exchanged objects are always colored. Such an exchange leads to fragmentation making large rapidity gaps most unlikely.

Figures 8(a) and (b) show the signatures of diffractive processes observed in e-p scattering at HERA [14]. The signatures are found in the fraction of the beam momentum carried by the leading baryon. The diffractive events are well separated, carrying more than 97% of the beam momentum. Figure 8(b) shows that the extent of the rapidity gap associated with non-diffractive DIS falls rapidly below $\eta_{max} = 2$, so events with η_{max} less than 2 may safely be taken to be diffractive.

Diffractive processes are typically described by the exchange of a quantity with vacuum quantum numbers, often referred to as a Pomeron. The fact that the Pomeron is colorless (color singlet) accounts for the observed rapidity gap. If a colored object (gluon, quark) were exchanged the rapidity gap would filled by string fragmentation. As the Pomeron is not part of perturbative QCD, its structure must be determined by

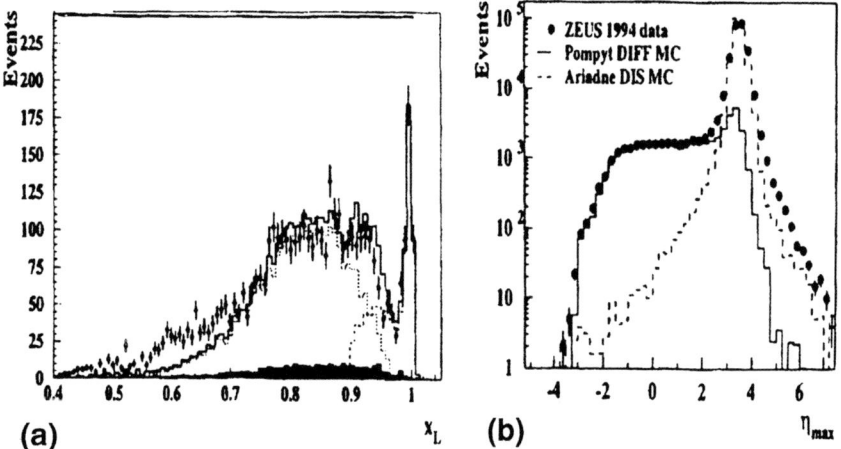

Fig. 8. (a) Shows the fraction of beam momentum carried by final-state protons. The diffractive events are at $x_L \leq 0.97$. (b) Shows the distribution of the maximum rapidity gap in DIS e-p events at ZEUS. For $\eta_{max} < 2$, the diffractive events dominate.

experiment. The Pomeron's Regge trajectory is expected to also contain the elusive glueball. Figure 9 depicts a diffractive e-p DIS where the proton does not change its state, the dotted double line represents the Pomeron. In the total set of DIS events there are many that also exhibit the properties of diffractive scattering. The set of parton densities associated with these diffractive events are labeled as

$$f_{j/B}^D\left(\beta, Q^2, x_{IP}, t\right) , \qquad (7)$$

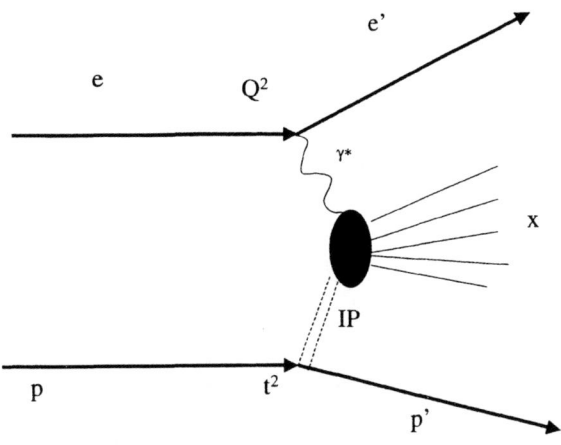

Fig. 9. Diagram of DIS off a Pomeron.

115

where $\beta = x/x_{IP}$. The outgoing baryon carries a momentum fraction $1-x_{IP}$ and squared momentum transfer t. These events are used to measure the partonic content of the Pomeron. At HERA it is found [14] that the Pomeron is ~ 90% glue at $Q^2 = 4.5$ GeV^2, which decreases to ~ 80% at $Q^2 = 75$ GeV^2. It will be fascinating to see if there is any modification of this dominantly gluonic structure in the nuclear medium.

The principal interaction of the virtual photon's $q\bar{q}$ color singlet pair in the nuclear medium is with the nuclear gluons. The cross section for this interaction in the target rest frame is [17],

$$\sigma^{inel}_{q\bar{q},N}(E_{inc}) = \frac{\pi^2}{3}b^2\alpha_s(Q^2)xg_N(x,Q^2), \qquad (8)$$

where $x = Q^2/2mE_{inc}$ and $b \sim 3/Q$. This cross section grows rapidly with E_{inc} because of the rapid increase in the gluon density at small x, and increasing Q^2 as shown in Fig. 6.

In the special case where the final state is exclusive vector meson production, the photon's $q\bar{q}$ pair must couple to the target via the exchange of <u>two gluons</u> (see Fig. 10) and is thus sensitive to the square of the gluon density. Thus these experiments should be pursued because of their particular sensitivity to the gluon density. The different vector mesons (ρ, ϕ, ψ) provide a variety of final states that can be used in such studies. Of course, when nuclear targets are employed, final-state interactions must be considered. These final-state interactions are interesting in and of themselves as they involve issues such as color transparency. This is a rich area for study that extends all the way from diffractive di-jet production to exclusive vector meson production. Again the role of a high-energy e-A <u>collider</u> is critical to making progress.

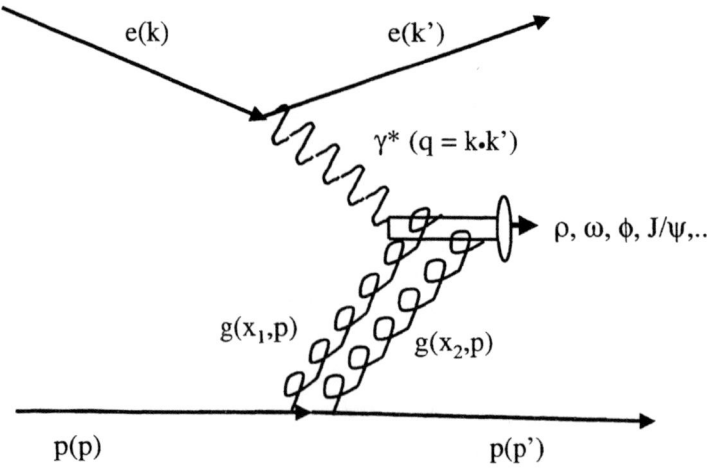

Fig. 10. Exclusive vector meson production in a QCD description.

TIME-SPACE EVOLUTION

As mentioned earlier, nuclear targets offer the only possibility of investigating the time-space evolution of partonic processes. This note has been full of examples of such evolution. Two further examples will illustrate the importance of the partonic interactions in the medium and the time space evolution of partonic state. The first example is the very large difference observed in the A dependence of acquired $< P_T^2 >$ of the final state di-muons from the Drell-Yan process as compared to those coming from J/Ψ and Y production and decay [18]. Figure 11 shows this difference. The existence of a difference is not surprising, as only the incident quark undergoes strong interactions in the Drell-Yan process, while the vector mesons formed for the most part by gluon fusion suffer strong interaction both on the incident gluon and on the resulting $c\bar{c}$ pair. However, the fact that the difference is as large as a factor of 5 is difficult to account for. It is also noteworthy that the J/Ψ and Y show the same effects, because the Y is appreciably smaller than the J/Ψ and hence should experience weaker interactions in the medium. As a further example of time space evolution, Fig. 12 shows the A^α dependence of the J/Ψ and Ψ' cross sections observed in p $\rightarrow$ A reactions, as a function of x_F [19]. At small or negative x_F the velocity of the $c\bar{c}$ system is small enough that the J/Ψ and Ψ' have the time to form within the nuclear medium. The radius of the Ψ' is twice that of the J/Ψ so is more readily adsorbed as reflected in the Fig. 12. At larger x_F the $c\bar{c}$ emerges from the nucleus before either state has formed so they display similar nuclear dependence. Thus one can see that formation and coherence times are critical to understanding observed reaction yields. The study of such time space evolution is greatly enhanced in e-A collisions where the initial state can be characterized in its color content and transverse size much more effectively than in a hadronic collision. Low luminosity and low center-of-mass energy have greatly hampered the leptonic production and subsequent study of the evolution of $c\bar{c}$ pairs in nuclei. In addition to its intrinsic interest, the value it would add to the understanding the observed attenuation of $c\bar{c}$ states in p-A collisions and its use as a signature of QG plasma formation cannot be over emphasized at this time. Such studies become straightforward at e-RHIC and would greatly extend capability of investigating the time space evolution of many $q\bar{q}$ systems.

Limited luminosity and center of mass energy has not allowed a convincing demonstration of color transparency [4]. The present evidence is shown in Fig. 13. The notion of color transparency is fundamental to the concept of Bjorken scaling so it is important to demonstrate its existence and to quantify it. The increased luminosity and center of mass energy of e-RHIC will permit an investigation of a far greater range of Q^2, can use other vector mesons in addition to the rho ($\phi, J/\psi$), and can provide far greater statistics than the earlier studies.

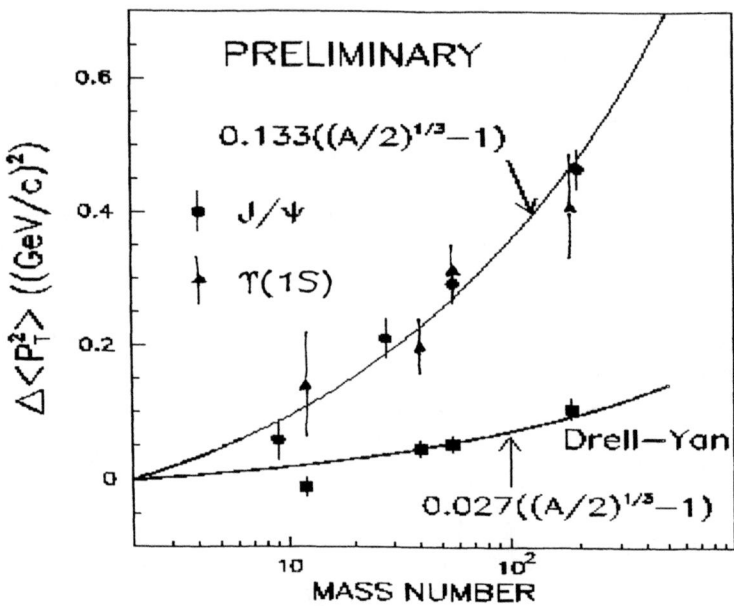

Fig. 11. The A dependence of the observed $\langle p_T^2 \rangle$ of the final dilepton pairs in the Drell-Yan process relative to those from vector mesons.

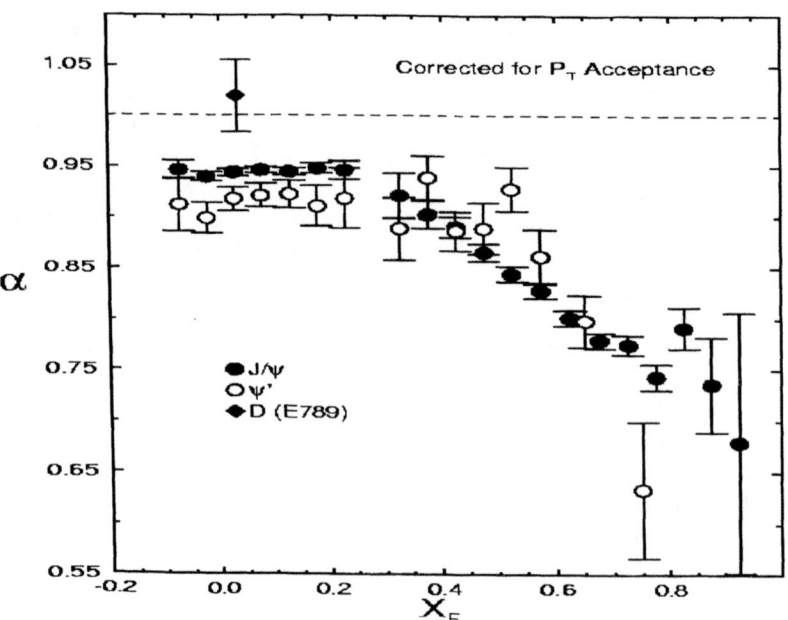

Fig. 12. Shows the A^α dependence of the J/Ψ and Ψ' yield as a function of $x_F \equiv x_1 - x_2$.

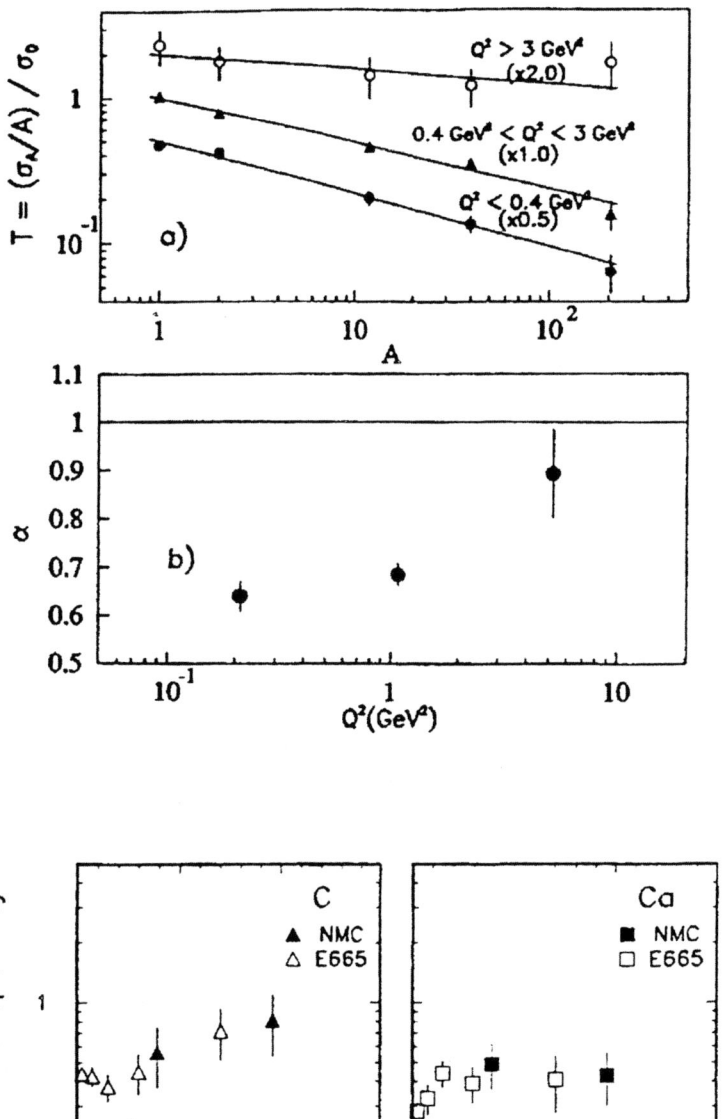

Fig. 13. The upper two diagrams show the evidence for color transparency from E665. The lower two figures compare results from E665 and NMC.

CONCLUSION

The above amply demonstrates the important, indeed critical role that a e-A collider can play in the investigation of the effects of the nuclear medium on the partonic structure of the nuclear constituents and on the time-space evolution of partonic processes. It should be obvious that such a collider is the most fundamental tool available for the investigation of this physics. Research at this facility should be full of surprises but results amenable to quantitative interpretation and incorporation into the rest of high-energy nuclear physics.

REFERENCES

1. Ameodo, M. et al., NMC Collaboration, *Nucl. Phys.* **B441 3**, 12 (1995).
2. Ameodo, M. et al., NMC Collaboration, *Nucl. Phys.* **B487 3** (1997).
3. Adams, M.R. et al., E665 Collaboration *Z. Phys. C* **67**, 403 (1995).
4. Adams, M.R. et al., E665 Collaboration, *Phys. Rev. Lett.* **74**, 1525 (1995).
5. Ben-Zvi, Ilan, 2nd e-RHIC Workshop, Yale University, April 6–8, 2000; Schwant, Peter, Workshop on Physics with EPIC, MIT, September 14–16, 2000.
6. Aubert, J.J. et al., EMC Collaboration *Phys. Lett.* **B123**, 275 (1983).
7. Geesaman, D.F., K. Saito, and A.W. Thomas, *Ann. Rev. Nucl. Part. Sci.* **45**, 337 (1995).
8. Frankfurt, L., and M. Strikman, *Eur. Phys. J. A* **5**, 293 (1999).
9. Ayala, A.L., M.B. Gay Ducati, and E.M. Levin, *Nucl. Phys.* **B493**, 305 (1997).
10. Review of Particle Physics, *Eur. Phys. J. C* **15**, 1 (2000).
11. Ameodo, M et al., NMC Collaboration, *Nucl. Phys.* **B481**, 23 (1996).
12. Glousset, T. and H.J. Pirner, *Phys. Lett.* **B375**, 349 (1996).
13. Sloan, T., private communication, reported at the 1st eRHIC Workshop, BNL, December 3–4. 1999.
14. Abramowicz, H. and A.C. Caldwell, *Rev. Mod. Phys.* **71**, 1276 (1999).
15. Iancu E., A. Leonidov, and L. McLerran, hep-ph/0011241.
16. Gribov, V.N., *Sov. J. Nucl. Phys.* **9**, 369 (1969), *Sov. Phys. JETP* **29**, 483 (1969).
17. Frankfurt, L and M. Strikman, hep-ph/9907221.
18. Leitch, M.L. et al., E866 Collaboration, APS Long Beach Meeting May 2, 2000.
19. Leitch, M.L. et al., E866 Collaboration, *Phys. Rev. Lett.* **84**, 3256 (2000).

Deeply inelastic scattering off nuclei at RHIC

Raju Venugopalan

Physics Department and RIKEN-BNL Research Center, BNL, Upton, NY 11973, USA

Abstract. We discuss the physics case for an electron--nucleus collider at RHIC.

INTRODUCTION

A high energy electron--nucleus collider, with a center of mass energy $\sqrt{s} = 60-100$ GeV, presents a remarkable opportunity to explore fundamental and universal aspects of QCD. The nucleus, at these energies, acts as an amplifier of the novel physics of high parton densities—aspects of the theory that would otherwise only be explored in an electron—proton collider with energies at least an order of magnitude greater than that of HERA. An electon--nucleus collider will also make the study of QCD in a nuclear environment, to an extent far beyond that achieved previously, a quantitative science. In particular, it will help complement, clarify, and reinforce physics learnt at high energy nucleus--nucleus and proton—nucleus collisions at RHIC and LHC over the next decade. For both of these reasons, an eA collider facility represents an important future direction in high energy nuclear physics.

We will summarize here the physics arguments that support both the key points above. We will also briefly discuss experimental observables in deeply inelastic scattering (DIS) and signatures of novel physics. Accelerator and detector issues have been discussed elsewhere. Details on these, on the physics issues, and references to an extensive literature can be found in proceedings [1],[2] of two of the three eRHIC workshops that were held in the last year [1] and in the earlier proceedings of eA HERA workshops [3].

The physics arguments can be separated according to the kinematic regions of interest [2]. Very roughly, these are
- the small x_{Bj} region $x_{Bj} < 1/(2m_N R_A) \approx 0.01$ for a large nucleus), where the virtual photon interacts coherently with partons in a nucleus over a region exceeding its longitudinal extent $2R_A$.

---

[1] More information on eRHIC and on previous eA studies for HERA can also be found at the website: http://quark.phy.bnl.gov/raju/eRHIC.html
[2] m_N below is the nucleon mass.

- intermediate x_{Bj} region ($1/(2m_N R_A) < x_{Bj} < 1/(2m_N R_N) \approx 0.1$ for a large nucleus), where the virtual photon interacts coherently over longitudinal distances larger than the longitudinal size of the nucleon $2 R_N$, but smaller than the longitudinal size of the nucleus $2 R_A$.
- the large x_{Bj} region ($x_{Bj} > 1/(2m_N R_N) \approx 0.1$ for a large nucleus) where the virtual photon/W or Z boson is localized within a longitudinal distance smaller than the nucleon size.

In this talk, we will cover only the physics of the small x_{Bj} region. Due to space limitations, we will not cover the interesting physics at intermediate x_{Bj} that can be studied with an eA collider. This covers the region in x_{Bj} from where inter-nucleon forces become important to coherent effects involving several nucleons. A nice discussion of these issues (in the context of the HERA eA collider proposal) can be found in Ref. [3]. We will not discuss the physics of the large x_{Bj} region either - this topic has been covered by other participants at this meeting [4].

eA Physics At Small x_{Bj}: $x_{Bj} < 1/(2 m_N R_A)$

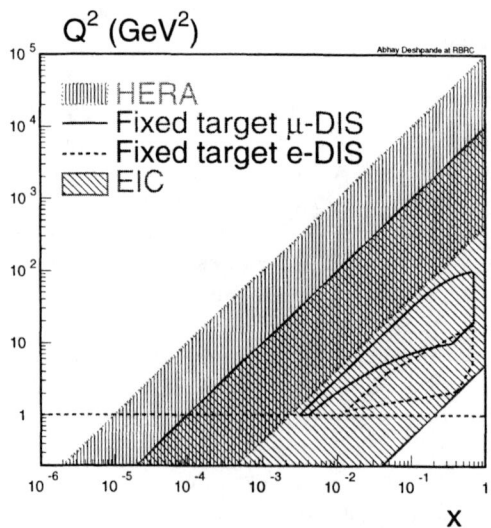

FIGURE 1. The x-Q^2 range of the electron ion collider (EIC) compared to that of the HERA ep collider and fixed target experiments. The EIC's reach would encompass the fixed target regime as well as part of the HERA regime.

This regime of small x_{Bj}'s ($x_{Bj} \leq 0.01$) is easily accessed by an electron-heavy ion collider in the energy range $\sqrt{s} \approx 60\text{-}100$ GeV. Fig 1 is a plot of the x-Q^2 plane delineating the range mapped for $\sqrt{s} = 63$ GeV (10 GeV electrons on 100 GeV heavy ions at RHIC). What is novel about these energies is that for the first time one can study the physics of $x_{Bj} \ll 0.01$ in a nucleus for $Q^2 \gg \Lambda^2_{QCD}$, where $\Lambda_{QCD} \sim 200$ MeV. Previous (fixed target) experiments such as NMC and E665 and current ones such as HERMES and COMPASS could only access small x_{Bj} at small Q^2's.

Some questions that may come immediately to mind are:
i) why is it important to simultaneously have large Q^2 at small x_{Bj}?
ii) Hasn't HERA explored this x_{Bj} - Q^2 range already? What then can one learn by studying the same regime with an eA collider?
iii) In the nuclear context, havn't the fixed target experiments at CERN, DESY and Fermilab, studied the small x_{Bj} regime?
iv) Does the collider environment have a compelling advantage in the study of small x_{Bj} physics?

In the following, we will address in detail the physics issues that underly these queries. The pithy answer to all of these however is that *an eA collider in the desired energy range may probe a hitherto inaccessible regime of QCD, where the properties of strongly interacting matter are radically different from those studied previously.* Understanding the properties of QCD in this regime may provide us the answer to fundamental questions about the strong interactions that remain unanswered. A brief list of these open questions is: a) what is the nature of multi—particle production? b) how do cross--sections behave at high energies? Are the bulk features of the cross-section computable in QCD? c) Are the properties of hadrons universal at very high energies? d) what is the nature of confinement-in particular, as probed in striking phenomena such as hard diffraction? and e) what are the initial conditions for heavy ion collisions, and how do they affect the formation of a quark gluon plasma?

We will also emphasize that mapping the relevant x-Q^2 regime with the proposed collider, at the high luminosities considered, will provide measurements of several physical quantities, with a much higher degree of precision, and of course in a wider kinematic range. Aside from their intrinsic interest, these quantities will be extremely important for the physics goals of other current and future experiments.

Why is it important to simultaneously have large Q^2 and small x_{Bj}?

In deeply inelastic scattering (DIS), one has the exact kinematic relation

$$y\, x_{Bj} = Q^2/s.$$

All of these variables are invariants - they are frame independent. The invariants x_{Bj} and Q^2 are of course well known - they are simply related, respectively, to the fraction of the momentum of a hadron or nucleus carried by a parton, and to the momentum transfer squared from the electron to the hadron. The invariant y, in the rest frame of the target, is the ratio of the energy transferred to the hadron to the energy of the electron. It has the kinematic range $0 \leq y \leq 1$. For the purposes of this discussion, we will assume that $y \sim 1$, or $x_{Bj} \sim Q^2/s$ [3].

The physics of small x is the physics of high energies [4] The total cross--section in strong interaction physics can be parametrized by the power law behavior $\sigma(s) \sim s^\varepsilon$,

---

[3] How large a value of y can be obtained without being swamped by uncertainties in the radiative corrections is a very important technical issue we will not address here. It has been addressed previously in proceedings of the eA at HERA workshops. These can be accessed on the World Wide Web at the URL: http://www.desy.de/~heraws96/proceedings/
[4] For a review of recent theoretical developments, see Ref.[5]..

where $\varepsilon \sim 0.08$. Thus the cross-section grows with decreasing x and, at high energies, is dominated by small x. This behaviour is explained in Regge phenomenology via Pomeron exchange - the t-channel exchange of an object with vacuum quantum numbers. Though the Pomeron hypothesis has had some striking success [7] in explaining high energy data, and has been the paradigm for understanding non--perturbative multi--particle production, it is not clear that it can be interpreted as an actual particle and understood as arising from the fundamental theory. A popular construction, first postulated by Francis Low and Shmuel Nussinov [6] is that the Pomeron is two gluon exchange with vacuum quantum numbers in the t channel. However, since total cross-sections at lower energies than available currently were dominated by very soft transverse momenta, it proved very hard to come up with a robust QCD based theory of the Pomeron (or more generally, that of the behavior of the bulk of the cross-section at high energies), that would also have predictive power.

The situation has changed with the advent of colliders at very high energies. With the Hadron Electron Ring Accelerator (HERA) at DESY, where 27.5 GeV electrons collide with 920 GeV protons, corresponding to a center of mass energy $\sqrt{s} \sim 300$ GeV, one can have $Q^2 = 1\text{-}10$ GeV^2 for $x_{Bj} \sim 10^{-4}$. In prior experiments, at these low $x_{Bj} \sim 10^{-4}$, only $Q^2 \ll \Lambda^2_{QCD}$ could be accessed. Thus even though one was probing the small x_{Bj} regime of the Pomeron, the coupling constant was too large to make predictions and therefore test/extract information about the theory in this regime. Since QCD is enormously complex, the lack of a small parameter in this regime was problematic. It hobbled progress in small x_{Bj} physics even though a wealth of tantalizing small x_{Bj} data [3] exists at small Q^2. A large number of models were constructed to understand the data, but their connection to the fundamental theory is still tenuous.

At HERA, the coupling $\alpha_S(Q^2) \ll 1$ in a significant portion of the small x_{Bj} regime of interest. Since the coupling is weak, computations can be made in pQCD and tested against the data. It lead, for instance, to the resurrection of the idea of the perturbative Pomeron developed by Lipatov and colleagues in the late 1970's-now known by the acronym BFKL Pomeron [9]. In QCD, a Pomeron can be constructed from the exchange of gluon ladders--the so--called hard Pomeron. The leading order BFKL result predicts rising cross--sections that rise more rapidly than the HERA data support. The next to leading order correction is very large and negative, thereby causing great confusion (and interest) in the QCD community [10][11]. One possibility is a more subtle resummation of next-to-leading order small x effects in the BFKL framework [12]; another is to formulate the problem of QCD at small x_{Bj} in the language of high parton densities - thereby performing a different sort of resummation [13],[14],[15]. This topic will be addressed in the following sub-section.

The ability to probe large Q^2's at small x_{Bj} has thus given us a handle on understanding, in a quantitative way, the hitherto inaccessible small x_{Bj} regime of QCD. This represents tremendous progress since it is this regime of the theory that controls the bulk of high energy cross--sections. Without understanding this regime, one cannot claim reliably that one completely understands the theory.

In what follows, we will discuss what we have learnt from the HERA experiments in the small x_{Bj} and large Q^2 regime, and how these experiments point to novel physics that may be fully explored with an eA collider.

From HERA Towards A New Regime of High Parton Densities

The wide kinematic range in x and Q^2 of the HERA collider can be seen in Fig. 1. One of the striking results from HERA is that the gluon distribution, extracted from scaling violations of F_2, grows rapidly at small x and high $Q^2 \gg \Lambda^2_{QCD}$. This is shown in Fig. 2. This tells us that at high energies the proton is not a simple object with three valence quarks and a few gluons that bind together the quarks. At a fixed external scale $Q^2 = 20$ GeV^2, one finds 25-30 gluons, per unit rapidity, at $x_{Bj} = 10^{-4}$ in the proton. The proton is therefore very rapidly growing dense as the resolution scale in x is shifted to smaller x's.

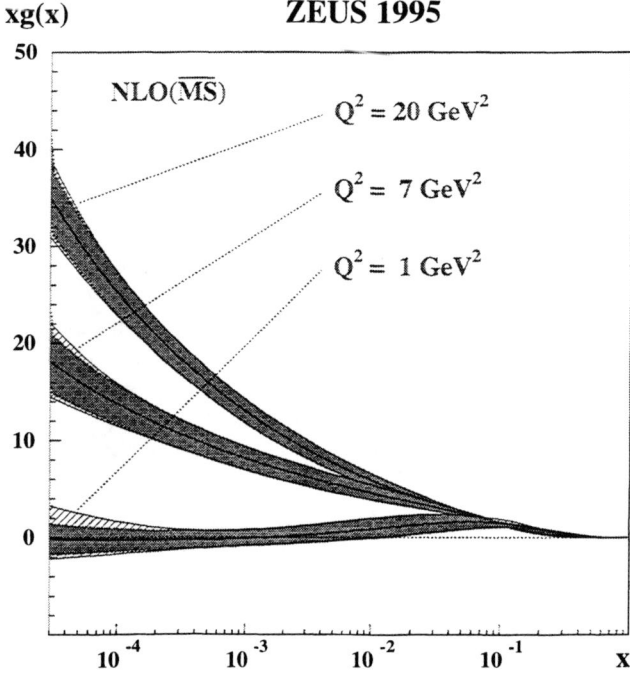

FIGURE 2. The gluon distribution for three different values of Q^2 extracted using the ZEUS NLO QCD analysis. Reprinted with permission from Ref. [21], © 1999 Springer-Verlag.

At high Q^2, ($Q^2 \gg 10$ GeV^2) the rise in the gluon structure function at small x is very well understood [16] in the framework of perturbative QCD (pQCD). An asymptotic expression for the rise is the double logarithmic formula where the gluon distribution grows as

$$G(x,Q^2) \sim \exp\left[\sqrt{\ln\left(\ln\left(\frac{Q^2}{\Lambda^2_{QCD}}\right)\right)\ln(1/x)}\right], \qquad (1)$$

This double logarithmic behavior was tested at HERA. It is claimed that the value of $\alpha_S(Q^2)$ extracted from the fit provides a precise determination of the coupling at $M_Z \sim 91$ GeV and is in agreement with other world data [17]. More detailed NLO QCD fits with different sets of parton distributions [18] have been shown to describe the HERA data for a wide range of x_{Bj} and Q^2. At high Q^2, the deeply inelastic scattering data is a testament to the striking success of perturbative QCD.

In the very high Q^2 regime - $Q^2 \gg 10$ GeV^2, one is not probing the region of extremely small x_{Bj}: for $Q^2=10$ GeV^2, the smallest x_{Bj} available is $\sim 10^{-4}$. In the region of $Q^2 = 1\text{-}10$ GeV^2, at correspondingly smaller x_{Bj}, the situation from the usual pQCD standpoint is less clear [19],[20]. The HERA ZEUS and H1 QCD fits agree with the data but with the price being that one extracts an anomalously small value of the gluon distribution, and one obtains more sea quarks than glue at small x [21],[22]. The anomalously small gluon distribution is seen in Fig. 2 for $Q^2=1$ GeV^2 where, at small x, the distribution is consistent with zero. Several groups have argued that one obtains results that run contrary to our intuition because the standard pQCD approach is breaking down[5]. The Tel Aviv group of Gotsman et al., for instance, claims that there is no pQCD fit that can simultaneously explain the inclusive F_2 data and the large amount of data on the energy dependence of J/ψ photo-production [24]. For a recent discussion of unitarity and long distance effects in J/ψ photo--production, see Ref. [25].

The argument is that screening effects due to large parton densities are important in this regime and have to be taken into account. The physics is still weak coupling though; one still has $\alpha_S \ll 1$ in the $Q^2=1\text{-}10$ GeV^2 regime. Phenomenological models that take these effects into account, and match into the usual pQCD formalism at high Q^2, have been successful in fitting both the inclusive and the diffractive HERA Data [26],[27]][28],[29].

There are therefore tantalizing hints from the HERA data that one is beginning to see the effects of large parton densities in the proton. We will argue below that standard pQCD breaks down when the parton densities become very large. Even though the coupling is weak, the physics will be non--perturbative due to the high field strengths generated by the large number of partons. This is a novel regime of the theory. We will further argue that an eA collider is much better suited to explore this regime even though its x-Q^2 range will be somewhat less extensive than that achieved at HERA.

QCD Is A Colored Glass Condensate At High Energies

In the infinite momentum frame (IMF), the number of partons per unit transverse area, for a fixed resolution of the external probe $Q^2 \gg \Lambda^2_{QCD}$, grows rapidly with the

---

[5] For a summary of recent discussions, see Ref.[23].

energy - or with decreasing x_{Bj}. In this high parton density regime, the corresponding QCD field strength squared[6] becomes $F_{\mu\nu}^2 \sim 1/\alpha_S$: since $\alpha_S(Q^2) \ll 1$, the color field strengths in this regime are large [30]. The non-linearities inherent in the theory become manifest, radically altering the properties of distributions in high energy collisions. For instance, the gluon distribution that was growing slowly now saturates - and grows very slowly - at most logarithmically with decreasing x_{Bj}.

In this high parton density--large field strengths - regime, the saturated gluons, when viewed in the IMF, form a novel state of matter which we will henceforth call a color glass condensate (CGC) [31]. Why a glass, and why a condensate? At small x_{Bj}, most of the partons are rapidly fluctuating gluons that interact weakly with each other. They are however strongly coupled to the large x_{Bj} "hard" parton color charges, that act as random, static, sources of color charge. This is exactly analogous to a glassy system--in particular, one can show that there is a formal analogy to spin glass condensed matter systems [32]. In the latter case, one has a disordered state of spins coupled, say, to random magnetic impurities - in the "quenched" limit, these impurities are static--or long lived.

Further, since the occupation number of the gluons is large, they form a condensate. Being bosons, arbitrary numbers of gluons can pile up in a momentum state. In the classical Bose gas, for instance, Bose-Einstein condensation leads to a dramatic overpopulation of the zero momentum state. In our case, since the gluons are interacting, and have both attractive and repulsive interactions, they "pile" up in a narrow band of states, peaked at a typical momentum we shall call the "saturation" momentum [15],[33],[34],[35]. The saturation scale at a particular x is the density per unit area of all the parton sources at higher x's. One has

$$Q_s^2(x) = \frac{1}{\pi R^2} \frac{dN}{d\eta}, \qquad (2)$$

Here $\eta = \ln(1/x)$ is the rapidity. As the energy increases, or x_{Bj} decreases, this "bulk" scale of the condensate grows and one can have $Q_s \gg \Lambda_{QCD}^2$. The distinction between fields and sources is of course arbitrary-as one decreases x, what were formerly fields turn into sources - thereby increasing the density of sources. This transformation is nothing but the "block spin" renormalization group transformation of Wilson, and the equations describing the evolution to small x_{Bj} are Wilsonian renormalizaton group equations [33],[36]. There has been significant theoretical progress recently in understanding the asymptotic behavior of these renormalization group equations [37].

Thus, because a large scale Q_s is generated at small x, one can predict the behavior of this non--perturbative condensate using weak coupling QCD techniques. An interesting question we don't have the answer to yet is how the coupling constant behaves in this regime - is there a fixed point of the theory at high energies? It would be therefore be absolutely remarkable, and of fundamental interest, if it could be demonstrated empirically that QCD at very high energies is a non-trivial glassy condensate of gluons.

---

[6] $F_{\mu\nu}^2$ is frame independent

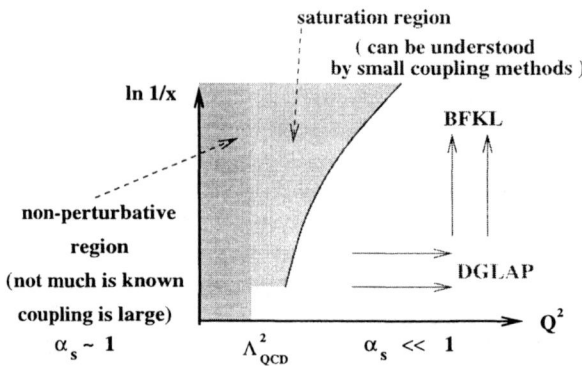

Figure 3. Schematic diagram of the ln(1/x) - Q^2 plane conveying a rough idea of the different regimes of applicability of the different evolution equations. Figure courtesy of Y. Kovchegov.

In Fig. 3 is plotted a (very) schematic diagram of scattering in the $\eta=\ln(1/x)$ versus Q^2 plane. If x_{Bj} is not too small, and Q^2 is large, the Dokshitzer--Gribov-Lipatov-Altarelli-Parisi QCD evolution equations [38] work very well. For a fixed Q^2, the Balitsky-Fadin-Kuraev-Lipatov (BFKL) equations describe the x evolution of distributions in a limited kinematic range. Both of these are linear evolution equations and do not fully take into account the non-linearities of the theory. Indeed, with regard to the latter, it is not clear there is a physical kinematical region available for linear evolution, where these equations apply, before high parton density effects set in. The line in the η-Q^2 plane represents the scale $Q_s(x)$ that separates the extensively studied regime of the well known QCD evolution equations from the saturation regime of the CGC.

We discussed in the previous sub-section how the HERA data may be showing hints of screening effects that may be the precursor to the saturation regime. We will argue below that deep inelastic scattering off large nuclei at high energies may be sufficient to probe this novel regime of the theory.

Probing The Colored Glass Condensate In Ea DIS

At a high energy eA collider, with energies $\sqrt{s}$=60-100 GeV, one will access (roughly) $x_{Bj}=10^{-4}$-10^{-3} for, respectively, Q^2=1-10 GeV^2 -see Fig. 1. These values of x_{Bj} and Q^2 are in the ballpark (even if more limited in range) than those at HERA. However, an eA collider has a tremendous advantage--the parton density in a nucleus, as experienced by a probe at a fixed energy, is much higher than what it would experience in a proton at the same energy. Since the parton density grows as $A^{1/3}$, this effect is more pronounced for the largest nuclei. To probe a comparable parton density in a nucleon, the probe would have to be at much higher energies than presently available.

The physics behind this effect is subtle and is a result of quantum coherence. In DIS at small x_{Bj}, in the target rest frame, the virtual photon splits into a quark--anti-quark pair, that subsequently interacts with the nucleus. If $x_{Bj} \ll 1/(2m_N R_A) \sim 0.01$, the $q\bar{q}$

pair interacts coherently with partons along the entire length of the nucleus. Furthermore, equally importantly, if the transverse separation of the pair $\sim 1/Q$ is smaller than the nucleon size ($Q^2 > \Lambda_{QCD}^2$), the probe will experience, coherently, random p_t kicks from partons in different nucleons along its trajectory. While $<p_t> \sim 0$, fluctuations will be large: $<p_t^2> \sim A^{1/3}$. In the IMF, this effect is interpreted as the $q\bar{q}$ - pair experiencing large fluctuations of color charge in the nuclear "pancake". It is clear, from both viewpoints, that a large scale, proportional to the parton density per unit area, is generated in large nuclei due to quantum mechanical coherence at small x_{Bj} and large Q^2. This scale is none other than the saturation scale $Q_s(x)$ discussed previously.

In a nucleus, one defines

$$Q_s^2 = \frac{1}{\pi R^2} \frac{dN}{dy} \equiv \frac{A^{1/3}}{x^\delta} fm^{-2}, \qquad (3)$$

Here δ is the power of the rise in the gluon distribution in a *nucleon* at the typical $Q^2 \sim Q_s^2$ of interest[7]. At HERA, for Q^2 of a few GeV^2, a reasonable estimate is $\delta \sim 0.3$. Now if we ask at what x_{Bj} in a proton will the probe see the same parton density as in a nucleus, Eq. 3 suggests,

$$x_{proton} = \frac{x_{nucleus}}{\left(\frac{1}{A^{\frac{1}{3}}}\right)^{1/\delta}} \qquad (4)$$

Since the nucleus is dilute, and if, being conservative, we assume that one can't tag on impact parameter-we take the effective $A^{1/3}=4$, then for $\delta \sim 0$, we find $x_{proton} \sim x_{nucleus}/100$. Thus, one would obtain the same parton density in a nucleus at $x_{Bj} \sim 10^{-4}$ and $Q^2 \sim$ a few GeV^2, as would be attained in a *nucleon* at $x_{Bj} \sim 10^{-6}$ and similar Q^2! Put differently, it would take an electron-proton collider with an order of magnitude larger energy than HERA to achieve the same parton density as would be achieved by eRHIC. It is now believed that impact parameter tagging is feasible by counting knock-out neutrons [39],[40] - if so, the gain in parton density in Ea relative to ep would be even more spectacular.

The small x_{Bj} regime has been studied previously in fixed target DIS off nuclei at CERN and Fermilab. In these experiments, the center of mass energy was a factor of 3-5 less than the proposed collider. This corresponds to a factor of 10-25 smaller in Q^2 for the same x_{Bj}. It was therefore difficult to interpret the experimental results at small x_{Bj} in the framework of perturbative QCD. Remarkable phenomena, such as shadowing, were observed-the relation of the experimentally measured shadowing to the physics of parton saturation presented here is at present unclear and deserves to be explored further. It could not be explored at the fixed target experiments because the kinematics corresponded to an intrinsically non-perturbative regime of the theory that

---

[7] This scale must be determined self-consistently.

is not amenable to a weak coupling perturbative QCD based analysis upon which the parton saturation picture rests.

In the following, we will discuss both inclusive and semi-nclusive signatures of the CGC. The latter, in particular, are striking. In this regard as well, the collider environment holds a significant edge since semi-inclusive observables proved very difficult to measure in fixed target eA DIS.

Signatures Of The New Physics Of The CGC

A number of inclusive and semi--inclusive experimental observables exist that will be sensitive to the new physics in the regime of high parton densities. All the inclusive and semi--inclusive observables that were studied at HERA can be studied with eRHIC--with a ZEUS/H1 type detector design [40]. However, due to the remarkable versatility of RHIC, and due to likely improvements in detector design, several new observables can be measured in the small x_{Bj} region for the first time. We will first discuss inclusive variables and the signatures of new physics in these. We will then discuss semi-inclusive observables.

Inclusive signatures of the CGC

An obvious inclusive observable is the structure function $F_2(x_{Bj}, Q^2)$ and its logarithmic derivatives with respect to x_{Bj} and Q^2. ERHIC should have sufficient statistical precision for one to extract the logarithmic derivatives of F_2 (and of its logarithmic derivatives!). Whether the systematic errors at small x_{Bj} will affect the results is not clear at the moment.

The logarithmic derivative $dF_2/d\ln(Q^2)$, at fixed x_{Bj}, and large Q^2, as a function of Q^2, is the gluon distribution. QCD fits implementing the DGLAP evolution equations should describe its behavior at large Q^2. At smaller Q^2, one should see a significant deviation from linear QCD fits--in principle, if the Q^2 range is wide enough, one should see a turnover in the distribution. The Q^2 at which the turnover takes place should be systematically larger for smaller x's and for larger nuclei. Predictions for this quantity, as a function of Q^2, for fixed W^2 and for fixed x, in a phenomenological model, are shown in Figs.4 and 5 respectively.

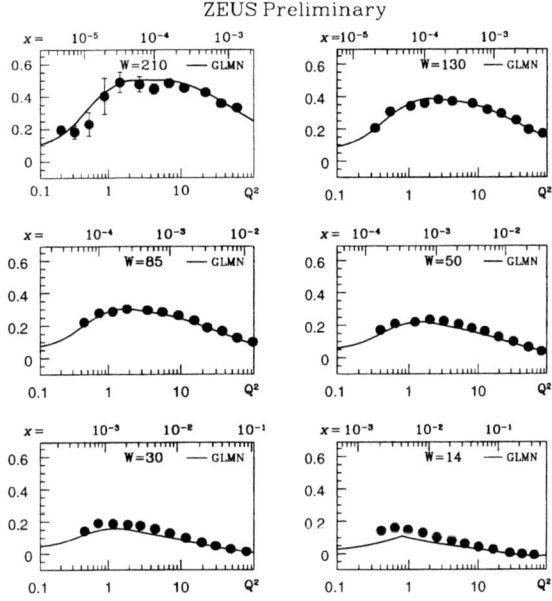

FIGURE 4. The slope $dF_2/d\ln(Q^2)$ versus Q^2 for fixed W^2. Figure from Ref [24].

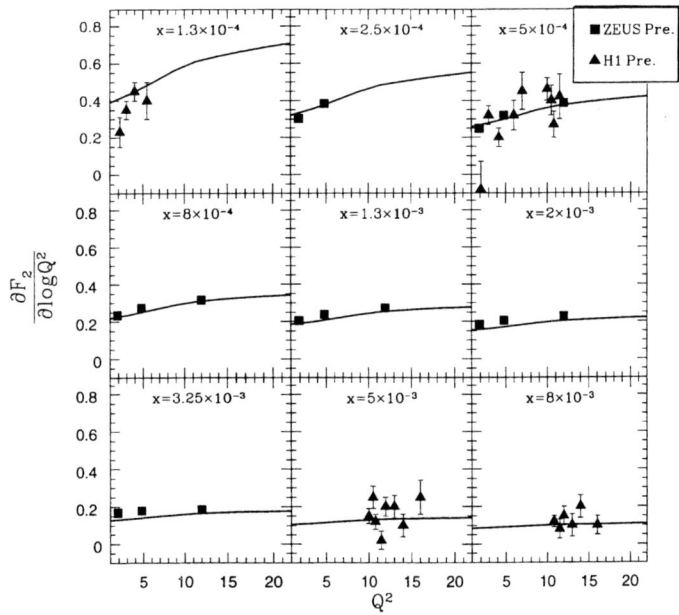

FIGURE 5. The slope $dF_2/d\ln(Q^2)$ versus Q^2 for fixed x

A remarkable feature of eRHIC will be that one can extract the longitudinal structure function $F_L(x_{Bj},Q^2) = F_2 - 2\, x_{Bj}\, F_1$ at small x_{Bj} for the first time. An independent extraction of F_L requires that the energy of the colliding beams be varied significantly. At RHIC, this is feasible. In the parton model, $F_L = 0$ - thus F_L is very sensitive to scaling violations. In particular, it provides an independent measure of the gluon distribution-a fact that makes this quantity very important to measure in its own right [41].

It has been suggested by several authors that $F_2 = F_L + F_T$ is not very sensitive to higher twist saturation effects which may be prominent in both F_L and F_T but may cancel in the sum[42]. An independent measurement of F_L, and thereby of F_T, will confirm this claim. The ratio of F_L/F_T has a very particular behavior in screening/saturation models. In Figs. 6 and 7 is shown the prediction from a particular model for this ratio [43]. The ratio F_L/F_T, for a fixed x_{Bj}, has a maximum at a particular Q^2; this maximum grows with the nuclear size (Fig. 6). As x_{Bj} decreases, the position of the maximum, for each nucleus, increases (Fig. 7). The maximum at which the turnover $Q_{eff}(x,A)$ occurs is related to the saturation scale Q_s. The precise relation is not known currently independently of particular models. To study the A dependence, it might be more useful to look at F_L and F_T separately.

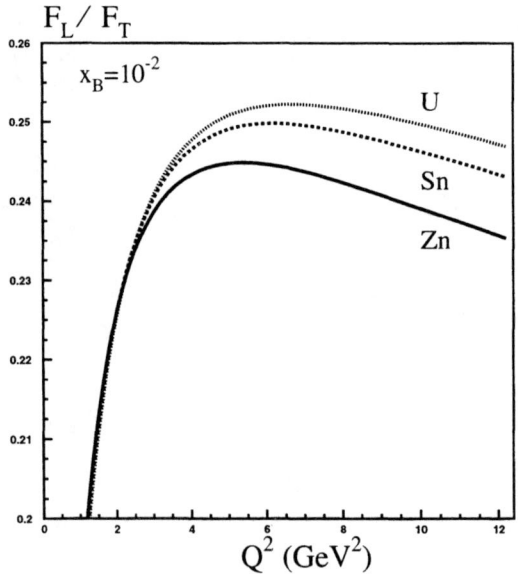

Figure 6. The ratio F_L/F_T as predicted in Ref. [43]. This ratio is plotted as a function of Q^2 for different nuclei and for fixed x_{Bj}.

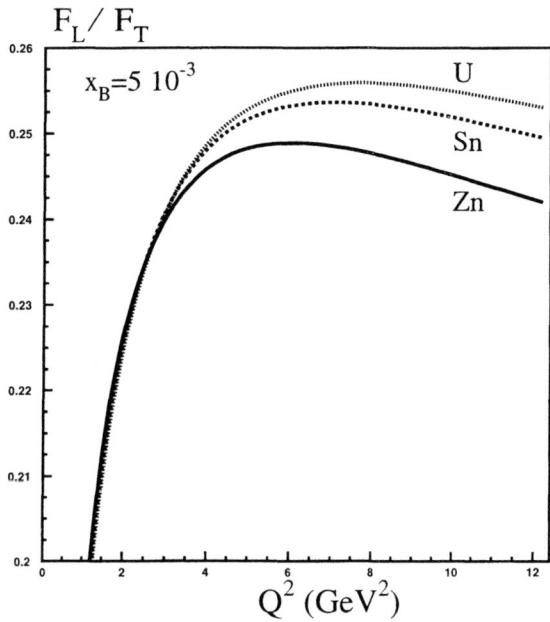

Figure 7. The ratio F_L/F_T as predicted in Ref. [43]. This ratio is plotted as a function of Q^2 for different nuclei and for a different x_{Bj} than Fig. 6.

A very important inclusive observable is nuclear shadowing. Quark shadowing is defined through the measured ratio of the nuclear structure function F_2^A to A times the nucleon structure function F_2^N: $s_{quark} = F_2^A / AF_2^n$. Gluon shadowing is similarly defined to be $S_{gluon} = G_A / AG_N$. Quark shadowing was observed in the fixed target experiments (NMC,E665···) and gluon shadowing, indirectly, through logarithmic derivatives of F_2. However, the gluon shadowing data at the smallest x's are also at very low Q^2-- where the application of perturbative QCD is unreliable.

There are model calculations (see Fig. 8) that suggest that gluon shadowing is very large at small x_{Bj}'s and fairly large Q^2 [28]. eRHIC can help confirm if this is the case. In addition, because of the extended kinematic range of eRHIC, we can determine whether shadowing is entirely a leading twist phenomenon, or if there are large higher twist perturbative corrections. Some saturation models, for instance, predict that perturbative shadowing will become large as one goes to smaller x_{Bj}'s [46]. Isolating perturbative contributions to shadowing from non-perturbative ones will be an interesting experimental and theoretical challenge. Another interesting question is whether shadowing saturates at a particular value of x_{Bj}, for fixed Q^2 and A. How does this value vary with x_{Bj} and Q^2? Does the ratio of quark shadowing to gluon shadowing saturate?

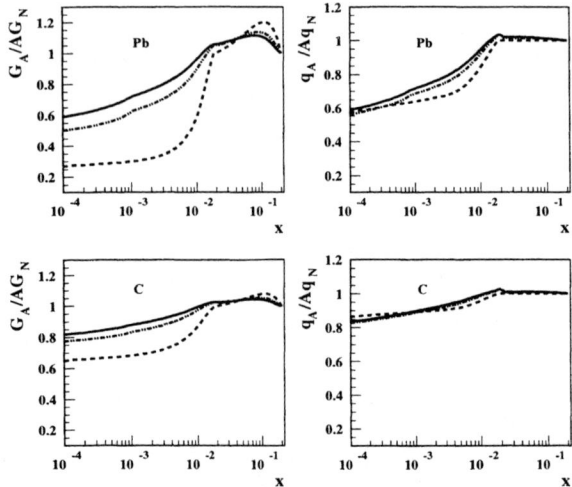

Figure 8. Gluon shadowing $G_A(x,Q^2)/A\, G_N(x,Q^2)$ and quark shadowing q_A/Aq_N versus x_{Bj} for lead (Pb) and Carbon (C). The different curves correspond to $Q = 2$ GeV (dotted), $Q=5$ GeV (dashed) and $Q=10$ GeV (solid). |Figure from Ref. [28], with permission from IOP Publishing Limited.

Finally, it is well known that there is a close relation between shadowing and diffraction. See Fig. 9. Whether this relation persists at high parton densities is not known. In an interesting recent exercise, it has been shown that diffractive *nucleon* data at HERA could be used to predict the shadowing of quark distributions observed by NMC [47],[55]. Significant deviations from the simple relation between shadowing and diffraction, may again suggest the presence of strong non--linearities. At eRHIC the validity of this relation can be explored directly--different nuclear targets are available, and the diffractive structure function may also be measured independently.

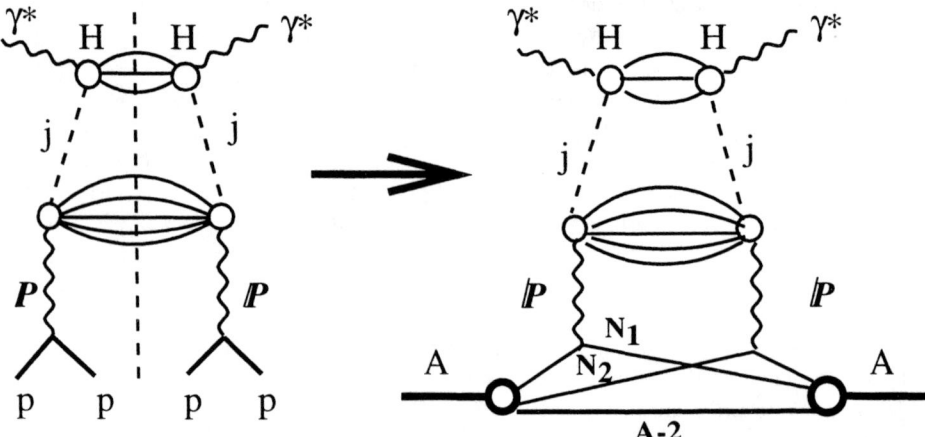

Figure 9. Diagrams demonstrating the relation between gluon induced hard diffraction on protons and the leading twist contribution to nuclear shadowing in DIS. Figure from Ref. [28], with permission from IOP Publishing Limited.

Semi--inclusive signatures of the CGC

The discussion of the relation between shadowing and diffraction provides a smooth segue into the topic of semi--inclusive signatures of the Colored Glass Condensate. While the novel physics of saturation and the formation of a CGC should be visible in inclusive quantities, their most dramatic manifestation will be in semi--inclusive measurements.

The most striking of these semi--inclusive measurements is hard diffraction. Hard diffraction is the phenomenon wherein the virtual photon emitted by the electron fragments into a final state X, with an invariant mass $M_X^2 \gg \Lambda_{QCD}^2$, while the proton emerges unscathed in the interaction. A large rapidity gap--a region in rapidity essentially devoid of particles—is produced between the fragmentation region of the electron and that of the proton. In pQCD, the probability of a gap is exponentially suppressed as a function of the gap size. At HERA though, gaps of several units in rapidity are unsuppressed; one finds that roughly 10% of the cross—section corresponds to hard diffractive events with invariant masses $M_X > 3$ GeV. The remarkable nature of this result is transparent in the proton rest frame; a 50 TeV electron slams into the proton and, 10% of the time, the proton is unaffected, even though the interaction causes the virtual photon to fragment into a hard final state.

The interesting question in diffraction is to study the nature of the color singlet object (the ``Pomeron") within the proton that interacts with the virtual photon since it addresses, in a novel fashion, the central mystery of QCD--the nature of confining interactions within hadrons. In hard diffraction, the mass of the final state is large and one can reasonably ask questions about the quark and gluon content of the Pomeron. A diffractive structure function $F_{2,A}^{D(4)}$ can be defined [48],[49],[50], in a fashion analogous to F_2, as

$$\frac{d_4\sigma_{eA \to eXA}}{dx_{Bj}dQ^2 dx\rho dt} = A \bullet \frac{4\pi\alpha_{em}^2}{xQ^4}$$

$$\left\{1-y+\frac{2}{2\left[1+R_A^{D(4)}(\beta,Q^2,xp,t)\right]}\right\} F_{2,A}^{D(4)}(\beta,Q^2,xp,t), \qquad (5)$$

where, y=Q^2s x_{Bj}, and analogously to F_2, one has $R_A^{D(4)} = F_L^{D(4)} / F_T^{D(4)}$ Also,

$$Q^2 = -q^2 > 0; x_{Bj} = \frac{Q^2}{2P \bullet q}; xp = \frac{q \bullet (P-P')}{q \bullet P}; t = (P-P')^2, \qquad (6)$$

and $\beta = x_{Bj}/x_P$. Here P is the initial nuclear momentum, and P' is the net momentum of the fragments Y in the proton fragmentation region. Similarly, M_X is the net momentum of the fragments X in the electron fragmentation region. An illustration of the hard diffractive event is shown in Fig. 10. Unlike F_2 however, $F_2^{D(4)}$ is not truly universal - it cannot be applied, for instance, to predict diffractive cross-sections in p-A scattering; it can be applied only in other lepton--nucleus scattering studies [50].

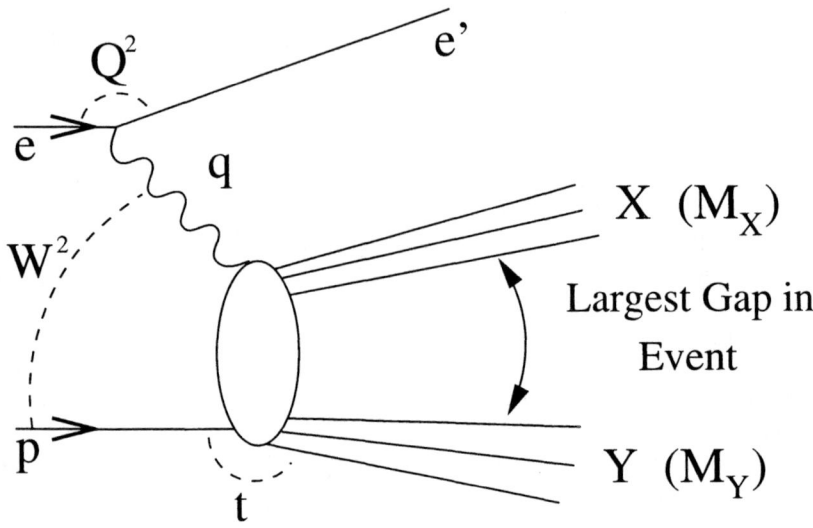

Figure 10. The diagram of a process with a rapidity gap between the systems X and Y. The projectile nucleus is denoted here as p. Figure from Ref. [3].

It is more convenient in practice to measure the structure function $F_{2,A}^{D(3)} = \int F_{2,a}^{D(4)} dt$, where $|t_{min}| < |t| < |t_{max}|$, where $|t_{min}|$ is the minimal momentum transfer to the nucleus, and $|t_{max}|$ is the maximal momentum transfer to the nucleus that still ensures that the particles in the nuclear fragmentation region Y are undetected. An interesting quantity to measure is the ratio

$$R_{A1,A2}(\beta, Q^2, x_P) = \frac{F_{2,A1}^{D(3)}(\beta, Q^2, x_P)}{F_{2,A2}^{D(3)}(\beta, Q^2, x_P)}. \tag{7}$$

For eA at HERA, it was argued that this ratio could be measured with high systematic and statistical accuracy[3] - the situation for eRHIC should be at least comparable, if not better. If $R_{A1,A2} = 1$, one can conclude that the structure of the Pomeron is universal, and one has an A-independent Pomeron flux. If $R_{A1,A2} = f(A1,A2)$, then albeit a universal Pomeron structure, the flux is A--dependent. Finally, if Pomeron structure is A-dependent, some models argue that $R_{A1,A2} = F_{2,A1}/F_{2,A2}$.

We discussed previously that the ratio of $R_D = \sigma_{diffractive}/\sigma_{total}$ at HERA is ~ 10% for $M_X > 3$ GeV. The systematics of hard diffraction at HERA can be understood in saturation models [26]. For eA collisions at eRHIC energies, saturation models predict that the ratio R_D^A can be much higher-on the order of 30% for the largest nuclei [51],[52]. The appearance of a large rapidity gap in 30% of all eA scattering events would be a striking confirmation of the saturation picture. Much theoretical and experimental work needs to be done to flesh out the details of predictions for hard diffractive events at eRHIC. Recent estimates suggest, for instance, that F_L^D/F_T^D, like F_L/F_T may have a peak as a function of Q^2, whose position, likewise, increases with decreasing x_{Bj} and increasing A[43].

An important semi--inclusive observable in eA DIS at high energies is coherent (or diffractive) and inclusive vector meson production. As discussed by Brodsky et al. [44] (see also Ref. [45]), the forward vector meson diffractive leptoproduction cross--section off nuclei

$$\frac{d\sigma}{dt}\bigg|_{t=0}(\gamma^*A \to VA) \propto \alpha_s^2(Q^2)[G_A(x,Q^2)]^2, \qquad (8)$$

for large Q^2. Here V denotes the vector meson. This quantity is clearly very sensitive to the gluon structure function. The ratio of this quantity in nuclei to that in nucleons is therefore (like the ratio of the longitudinal structure function) a probe of gluon shadowing.

In the color dipole picture, the amplitude for diffractive leptoproduction can be written as a convolution of the $q\bar{q}$ component of the γ^* wavefunction times the $q\bar{q}$-nucleus cross--section times the vector meson wavefunction. In the saturation picture, a semi--hard scale is introduced via the $q\bar{q}$-A cross-section-whether this scale is larger or smaller than the scale associated with the size of the vector meson strongly affects the energy and Q^2 dependence of the vector meson cross-sections at small x_{Bj}. Recently, Caldwell and Soares have studied vector meson production at HERA in the Golec-Biernat-Wusthoff model of saturation [26]. They find that the model provides a good description of the cross-section for photo-and electro-production of J/Ψ in a wide Q^2 and energy range. For ρ meson production, the change in the energy dependence as a function of Q^2 is well described by the model but the normalization of the cross-section is not. One possible explanation of this discrepancy is the lack of knowledge about the ρ wavefunction.

We should mention here that it is very important to measure inclusive and diffractive open charm and jets since they provide useful and independent measures of the gluon distribution (and gluon shadowing) at small x_{Bj}. These will complement information on the gluon structure functions obtained from the $\ln(Q^2)$ derivative of F_2, from F_L, and from diffractive leptoproduction of vector mesons.

We may conclude from the above that it is in the measurement of semi--inclusive observables that a future collider environment has a marked superiority over previous fixed target electron-nucleus experiments. Rapidity gaps in eA collisions will be measured for the first time. Coherent and incoherent vector meson production can be studied in great detail in a wide kinematic range with much greater accuracy than previously.

At small x_{Bj}, the high parton densities produce large color fluctuations which are subsequently reflected in large multiplicity fluctuations. One expects for instance the following phenomena: a) a broader rapidity distribution in larger nuclei relative to lighter nuclei and protons, b) Rapidity correlations over several units of rapidity—an anomalous multiplicity in one rapidity interval in an event would be accompanied by an anomalous multiplicity in rapidity intervals several units away [53],[54] and c) a correlation between the central multiplicity with the multiplicity of neutrons in a forward neutron detector [55].

Precision Measurements Of Nuclear Observables At Small x_{Bj}

In much of our discussion, we have focused on the potential of an Ea collider to discover a novel state of saturated gluonic matter - the colored glass condensate. This search, if successful, could revolutionize our understanding of QCD at high energies by providing answers to questions about the nature of confinement in high energy scattering, the origins of multi-article production, the asymptotic behavior of cross-sections, etc.

However, even in the absence of the promise of radically new physics, there is a compelling case to be made for an eA collider. The gluon distribution in a nucleus is ill-understood. Understanding its behavior as a function of x_{Bj} and Q^2, and of its shadowing is of intrinsic interest. At eRHIC, as discussed above, measurements of F_L, and of semi-inclusive quantities promise that the gluon distribution in a nucleus could be independently extracted with high precision. The nuclear gluon distribution extracted with an eA collider can be compared with the distribution extracted from AA and pA collisions.

The A dependence (as a function of x_{Bj} and Q^2) of vector meson production at small x_{Bj}] is also of intrinsic interest, as well of use in interpretations of pA and AA collisions - a particular example being that of J/ψ suppression.

Hard diffraction off nuclei has not been previously measured. At eRHIC, nuclear diffractive structure functions can be measured for the first time. The relation of these to F_2 will, as discussed previously, provide new insight into the relation between diffraction and shadowing.

At intermediate x's, an eA collider provides a laboratory to study the propagation of fast partons through nuclear matter. Color transparency and color opacity, which we have not discussed here, can be studied more extensively than previously [56]. Jet quenching, often cited as a signature of the quark gluon plasma in nuclear collisions[57], can be investigated in the cold nuclear environment of an eA collider [58] and compared to results from AA collisions.

Connections To pA and AA Physics

We will very briefly discuss here the relation of the physics of an eA collider to pA and AA physics at current and future collider facilities.

Relation of eA to pA

In pA scattering at RHIC[8], one also has the opportunity to study the gluon distribution in nuclei. Gluon fusion to jets, vector mesons, open charm and beauty can be measured. Hard and soft diffraction - the size and distribution of energy gaps with energy and nuclear size can also be studied. Scaling violations in Drell-Yan scattering can be measured for the first time.

---

[8] For a recent discussion, see the proceedings of the pA workshop at BNL, Oct. 28th--29th, Eds. S. Aronson and J. C. Peng, at the website www.bnl.gov/rhic/townmeeting/agenda_b.htm

Some of the differences between pA and eA are as follows. In pA scattering, for instance in the signature Drell-Yan process, it is very hard to reliably extract distributions in the region below the Ψ' tail - namely, one requires $Q^2 > 16$ GeV^2. In the x region of interest, one expects saturation effects to be important at lower Q^2 of 1-10 GeV^2. For $Q^2=16$ GeV^2, one might have to go to significantly smaller x's to see large saturation effects. Secondly, the survival probability of large rapidity gaps is smaller in pA relative to eA. This is because the gap is destroyed due to secondary interactions between "spectator" partons in the proton and the "Pomeron" from the nucleus. This does not occur in eA scattering because of course there are no spectator partons in the electron. Thus one expects that diffractive vector meson and jet production in pA should be qualitatively different than what one will see in eA.

Relation of eA to AA

A large variety of models combining hard and soft physics are used to study nuclear collisions at RHIC and LHC energies [59]. Many of these model predictions depend sensitively on the nuclear gluon density--for a recent parametrization of nuclear gluon densities, see Ref. [60]. Data from an eA collider will be essential in further refining these parametrizations.

In the classical approach discussed previously, the relation between the parton distributions in the nuclear wavefunction and the multiplicity of produced gluons simplifies--the initial multiplicity of produced gluons is given in terms of the saturation scale Q_s by the simple relation.

$$\frac{1}{\pi r^2}\frac{dN}{d\eta} = c_N \frac{N_c^2 - 1}{N_c}\frac{1}{4\pi^2 \alpha_s}Q_s^2 \qquad (9)$$

The coefficient c_N can be estimated numerically in classical lattice simulations of nuclear collisions [62] and is determined to be $c_N \sim 1.3$. A similar analysis is used to determine the initial energy of the produced glue [63]. This distribution is only the initial parton distribution--the subsequent possible evolution to a quark gluon plasma [64] is controlled again only by the scale Q_s.

It is therefore conceivable that high energy heavy ion collisions, despite their complexity, may provide insight into the parton distributions in the nuclear wavefunction. An eA collider will confirm and deepen our understanding of what we may learn from heavy ion collisions.

Summary

In this talk, we discussed the physics case for an eA collider. We emphasized the novel physics that might be studied at small x. The interesting physics at intermediate x's has been discussed elsewhere [3].

Plans for an electron--ion collider include, as a major part of the program, the possibility of doing polarized electron--polarized proton/light ion scattering. A discussion of the combined case for high energy electron nucleus and polarized electron--polarized proton scattering will be published separately [65].

ACKNOWLEDGEMENT

I would like to thank the organizers of the EPIC meeting at MIT for inviting me to present the physics case for eRHIC. Many thanks to Abhay Deshpande, Witek Krasny, Yuri Kovchegov, Genya Levin, Larry McLerran, and Mark Strikman for discussions, and for kindly allowing me to use figures from their papers. In particular, I would like to thank Mark Strikman for his detailed comments on the manuscript. I would also like to thank all the participants of the eA meeting at BNL in June/July 2000 for the stimulating discussions that have informed this report. This work was supported at BNL by the Department of Energy under DOE contract number DOE-FG02-93-ER-40764, by RIKEN--BNL, and by an LDRD grant from Brookhaven Science Associates.

REFERENCES

1. Proceedings of the 2nd eRHIC workshop, Yale, April 6th-8th, 2000, BNL-52592.
2. Proceedings of the BNL summer meeting, June 26th-July 14th, 2000, BNL-52606
3. M. Arneodo et al., in Proceedings of *Future Physics at HERA*, DESY, September 25th--26th, 1995, hep-ph/9610423.
4. Proceedings of the EPIC meeting, MIT, Boston, September 14th-16th, 2000.
5. R.Venugopalan, Pramana **55**, 73 (2000) [hep-ph/0005096].
6. F.E.Low, Phys. Rev. D **12**, 163 (1975); S.~Nussinov, Phys. Rev. D **14**, 246 (1976).
7. A.Donnachie and P.V. Landshoff, Phys. Lett. **B296**, 227 (1992) [hep-ph/9209205].
8. M.Arneodo, Phys. Rept. **240** (1994) 301.
9. E.A.Kuraev, L.N.Lipatov and V.S.Fadin, Sov. Phys. JETP **45**, 199 (1977); I.I.Balitsky and L.N.Lipatov, Sov. J. Nucl. Phys. **28**}, 822 (1978).
10. V.S.Fadin and L.N.Lipatov, Phys. Lett. **B429**, 127 (1998) [hep-ph/9802290].
11. G.Camici and M.Ciafaloni, Phys. Lett. **B412**, 396 (1997) [hep-ph/9707390].
12. G.P.Salam, Acta Phys. Polon. **B30**, 3679 (1999) [hep-ph/9910492].
13. L.V.Gribov, E.M.Levin and M.G.Ryskin, Phys. Rept. **100**, 1 (1983).
14. A.H.Mueller and J.Qiu, Nucl. Phys. **B268**, 427 (1986).
15. L.McLerran and R.Venugopalan, Phys. Rev. D **49**, 2233 (1994); *ibid*}, 3352 (1994); Phys. Rev. D **50**, 2225 (1994); Phys. Rev. D **59**, 094002 (1999).
16. A.De Rujula, S.L.Glashow, H.D.Politzer, S.B.Treiman, F.Wilczek and A.Zee, Phys. Rev. D **10**, 1649 (1974); D.J.Gross, Phys. Rev. Lett. **32**, 1071 (1974).
17. R.D.Ball and S.Forte, Phys. Lett. **B358**, 365 (1995) [hep-ph/9506233].
18. H.L.Lai *et al.* [CTEQ Collaboration], Eur. Phys. J. **C12**, 375 (2000) [hep-ph/9903282]; A.D.Martin, R.G.Roberts, W.J.Stirling and R.S.Thorne, Nucl. Phys. Proc. Suppl. **79**, 105 (1999) [hep-ph/9906231]; M.~luck, E.Reya and A.Vogt, Eur. Phys. J. **C5**, 461 (1998) [hep-ph/9806404].
19. L.Mankiewicz, A.Saalfeld and T.Weigl, hep-ph/9706330.
20. G.Altarelli, R.D.Ball and S.Forte, hep-ph/0011270.
21. J.Breitweg *et al.* [ZEUS Collaboration], Eur. Phys. J. **C7**, 609 (1999) [hep-ex/9807010].
22. C.Adloff *et al.* [H1 Collaboration], Z. Phys. **C76**, 613 (1997) [hep-ex/9708016].
23. M.F.McDermott, hep-ph/0008260.
24. E.Gotsman, E.Levin, M.Lublinsky, U.Maor, E.Naftali and K.Tuchin, hep-ph/0010198.
25. L.Frankfurt, M.McDermott and M.Strikman, hep-ph/0009086.
26. K.Golec-Biernat and M.Wusthoff, Phys. Rev. D **59**, 014017 (1999) [hep-ph/9807513]; *ibid*, **60**, 114023 (1999) [hep-ph/9903358].
27. E.Gotsman, E.Levin, U.Maor and E.Naftali, Nucl. Phys. **B539**, 535 (1999) [hep-ph/9808257].
28. L.Frankfurt, V.Guzey and M.Strikman, J. Phys. G **G27**,R23 (2001) [hep-ph/0010248].

29. A.M.Stasto, K.Golec-Biernat and J.Kwiecinski, hep-ph/0007192.
30. A.H.Mueller, Lectures given at *International Summer School on Particle Production Spanning MeV and TeV Energies*, Nijmegen, Netherlands, 8-20 Aug 1999, hep-ph/9911289.
31. E.Iancu, A.Leonidov and L.McLerran, hep-ph/0011241.
32. G.Parisi and N.Sourlas, Phys. Rev. Lett. **43**, 744 (1979).
33. J.Jalilian-Marian, A.Kovner, L.McLerran and H.Weigert, Phys. Rev. D **55**, 5414 (1997) [hep-ph/9606337].
34. Y.V.Kovchegov, Phys. Rev. D **54**, 5463 (1996) [hep-ph/9605446].
35. Y.V.Kovchegov and A.H.Mueller, Nucl. Phys. **B529**, 451 (1998) [hep-ph/9802440].
36. J. Jalilian-Marian, A. Kovner and H. Weigert, Phys. Rev. **D59** 014015 (1999).
37. I. Balitsky, Phys. Rev. **D60** 014020 (1999); Phys. Rev. Lett. **81** 2024 (1998); Y.V.Kovchegov, Phys. Rev. **D61**, 074018 (2000); E.Levin and K.Tuchin, hep-ph/9908317; E.Iancu, A.Leonidov and L.McLerran, hep-ph/0011241.
38. V.N.Gribov and L.N.Lipatov, Sov. J. Nucl. Phys. 15, 78 (1972); G.Altarelli and G.Parisi, Nucl. Phys. **B126**, 298 (1977); Yu.L.Dokshitzer, Sov. Phys. JETP. **73**, 1216 (1977).
39. M.Strikman, M.G.Tverskoii and M.B.Zhalov, Phys. Lett. **B459**, 37 (1999) [nucl-th/9806099].
40. W. Krasny, contribution in Ref. 1.
41. A. Zee, F. Wilczek, and S. B. Treiman, Phys. Rev. D **10**, 2881 (1974); W. Bardeen, A. J. Buras, D. W. Duke, and T. Muta, Phys. Rev. D **18**, 3998 (1978).
42. J.Bartels, K.Golec-Biernat and K.Peters, Eur. Phys.J. **C17**, 121 (2000) [hep-ph/0003042].
43. E.Gotsman, E.Levin, U.Maor, L.McLerran and K.Tuchin, hep-ph/0008280; hep-ph/0007258.
44. S.J.Brodsky, L.Frankfurt, J.F.Gunion, A.H.Mueller and M.Strikman, Phys. Rev. **D 50**, 3134 (1994) [hep-ph/9402283].
45. B.Z.Kopeliovich, J.Nemchick, N.N.Nikolaev and B.G.Zakharov, Phys. Lett. **B324**, 469 (1994) [hep-ph/9311237].
46. J.Jalilian-Marian and X.Wang, Phys. Rev. D **60**, 054016 (1999) [hep-ph/9902411].
47. A.Capella, A.Kaidalov, C.Merino, D.Pertermann and J.Tran Thanh Van, Eur. Phys. J. **C5**, 111 (1998) [hep-ph/9707466].
48. A.Berera and D.E.Soper, Phys. Rev. D **53**, 6162 (1996) [hep-ph/9509239].
49. L.Trentadue and G.Veneziano, Phys. Lett. **B323**, 201 (1994).
50. J.C.Collins, Phys. Rev. D **57**, 3051 (1998) [hep-ph/9709499].
51. E.Levin and U.Maor, hep-ph/0009217.
52. L. Frankfurt and M. Strikman, Phys. Lett. **B382** (1996) 6.
53. V.A.Abramovsky, V.N.Gribov and O.V.Kancheli, Yad. Fiz. **18**, 595 (1973).
54. Y.V.Kovchegov, E.Levin and L.McLerran, hep-ph/9912367.
55. L.Frankfurt and M.Strikman, Eur. Phys. J. **A5**, 293 (1999) [hep-ph/9812322].
56. L.Frankfurt, V.Guzey, W.Koepf, M.Sargsian and M.Strikman, hep-ph/9608492.
57. X.Wang, M.Gyulassy and M.Plumer, Phys. Rev. D **51**, 3436 (1995) [hep-ph/9408344].
58. R.Baier, Y.L.Dokshitzer, S.Peigne and D.Schiff, Phys. Lett. **B345**, 277 (1995) [hep-ph/9411409]; M.Luo, J.W.Qiu and G.Sterman, Phys. ev. D **50**, 1951 (1994). E.Levin, Phys. Lett. **B380**, 399 (1996) [hep-ph/9508414].
59. N.Armesto and C.Pajares, Int. J. Mod. Phys. **15**, 2019 (2000) [hep-ph/0002163].
60. K.J.Eskola, V.J.Kolhinen and C.A.Salgado, Eur. Phys. J. **C9**, 61 (1999) [hep-ph/9807297].
61. A.H.Mueller, Nucl. Phys. **B572**, 227 (2000) [hep-ph/9906322].
62. A.Krasnitz and R.Venugopalan, hep-ph/0007108, *Phys. Rev. Lett* in press.
63. A.Krasnitz and R.Venugopalan, Phys. Rev. Lett. **84**, 4309 (2000) [hep-ph/9909203].
64. A.Dumitru and M.Gyulassy, Phys. Lett. **B494**, 215 (2000) [hep-ph/0006257]; J.Bjoraker and R.Venugopalan, Phys. Rev. **C 63**, 024609 (2001) [hep-ph/0008294]; R.Baier, A.H.Mueller, D.Schiff and D.T.Son, hep-ph/0009237.
65. A.Deshpande and R.Venugopalan, in preparation.

Luminosity Limitations for Electron-Ion Collider[*]

V. A. Lebedev

Thomas Jefferson National Accelerator Facility
12000 Jefferson Avenue, Newport News, VA 23606
Email – lebedev@jlab.org

Abstract. The major limitations on reaching the maximum luminosity for an electron ion collider are discussed in application to the ring-ring and linac-ring colliders. It is shown that with intensive electron cooling the luminosity of 10^{33} cm^{-2}s^{-1} is feasible for both schemes for the center-of-mass collider energy above approximately 15 GeV. Each scheme has its own pros and cons. The ring-ring collider is better supported by the current accelerator technology while the linac-ring collider suggests unique features for spin manipulations of the electron beam. The article addresses a general approach to a choice of collider scheme and parameters leaving details for other conference publications dedicated to particular aspects of the ring-ring and linac-ring colliders.

INTRODUCTION

Currently HERA at DESY and CEBAF at Jefferson Lab are two major players in the study of structure of the proton and light ions. HERA operates with polarized electrons and unpolarized protons, while CEBAF operates with polarized electrons and polarized protons or nuclei. HERA[1] is an electron-proton collider with very high center-of-mass energy (320 GeV) and moderate luminosity ($1.7 \cdot 10^{31}$ cm^{-2}s^{-1}), while CEBAF[2,3] is a fixed target machine with moderate energy and practically unlimited luminosity. The energy of the electron beam at CEBAF can presently be varied from 0.6 to 6 GeV and the machine is expected to be upgraded to 12 GeV[4] within about 5 years, boosting the center-of-mass energy of electron-proton collisions to 4.8 GeV. CEBAF luminosity for operation into the 4π CLAS detector of Hall B is mainly limited by the detector to about 10^{34} cm^{-2}s^{-1}. The luminosity is a few orders of magnitude higher for two other halls, which use spectrometers for particle detection.

For further progress in the study of nucleon spin structure a machine in the intermediate range of energies is required. The energy range of 15 to 50 GeV is currently considered to be interesting, with a request for the luminosity to be 10^{33} cm^{-2}s^{-1} or above. For physics of interest the effective luminosity of the collider is proportional to the square of each beam polarization, $L_{effective} = L_o p_e^2 p_i^2$, and therefore achieving polarization above 70% for both electron and ion beams is of primary importance. Naive model of electron-quark collisions would require a ratio of

---

[*] Work supported by the US DOE under contract #DE-AC05-84ER40150

the electron to proton energy to be about 1 to 6. We will not impose such a constraint but, as one will see later, an optimization of machine parameters yields close energy ratio.

Two general concepts for the collider have been suggested. The first is the classical ring-ring collider[5], where ion and electron beams are stored in independent storage rings of the same circumference. The second concept is the linac-ring collider[6], colliding protons in a storage ring against electrons from an energy recovery linac. Below we will try to analyze advantages and disadvantages of each scheme, as well as major factors limiting the collider luminosity. Reaching 10^{33} cm^{-2}s^{-1} luminosity with minimum cost will be our major optimization criterion. Because the machine luminosity generally grows with energy, we will also address the question at which energy each of the considered schemes can achieve the required luminosity. Although the parameter list and the luminosity optimization considered below are carried out for the electron-proton collider all results are also applicable to the electron-ion collider.

1. LUMINOSITY LIMITATIONS

There are three major limitations on the beam brightness. The first is the space charge effect in the ion beam. It causes nonlinear dependence of particle betatron tune on amplitude (Laslett tune shift) and, consequently, the loss of the particle motion stability in the case of large tune shift. This limits the phase density of the ion beam and is one of the major limitations at low ion energy. The second one is the beam-beam effect at the interaction point (IP) which limits the density of beams for both collider schemes. The third one is intrabeam scattering (IBS) in the ion beam. Although the intensity of IBS decreases with increasing energy, it is still an important limitation even at highest energy considered for the collider. In the estimates below we presume that the horizontal and vertical beam emittances, as well as the horizontal and vertical beta-functions at the IP, are equal for each of the beams (round beams). However, we do not necessarily consider the beam sizes for each of two beams to be equal. In this case the luminosity is determined by the following formula,

$$L = \frac{N_e N_i f_0}{2\pi(\sigma_i^{*2} + \sigma_e^{*2})}, \qquad (1)$$

where N_e and N_i are the number of electrons and ions per bunch, σ_e^* and σ_i^* are the electron and ion beam sizes, and f_0 is the bunch frequency.

1.1 Laslett Tune Shift Limit

We will start our consideration from the Laslett tune shift. Its value is determined by the following formula

$$\Delta v_i = \frac{Z^2 e^2 N_i R}{2\sqrt{2\pi}\, m_i c^2 \gamma_i^3 \beta_i^2 \sigma_{si} \varepsilon_i}, \qquad (2)$$

where Ze is the ion charge, $m_i = A m_p$ is the ion mass, R is the storage ring mean radius, γ_i and β_i are the ion relativistic factors, ε_i is the ion beam emittance, and σ_{si} is the rms longitudinal beam size. Substituting the ratio N_i/ε_i from the above equation into the luminosity formula and choosing the beta-function at the IP to be equal to the bunch length, $\beta^* = \sigma_{si}$, one obtains the following remarkably simple formula for the luminosity limit due to the Laslett tune shift:

$$L_{\Delta v} = \sqrt{\frac{2}{\pi}} \frac{\eta}{Z} \frac{I_e B_d}{e^2 \left(1 + \sigma_e^{*2}/\sigma_i^{*2}\right)} \gamma_i^2 \beta_i \Delta v_i \longrightarrow$$

$$\left(\frac{\eta}{0.2}\right)\left(\frac{I_e}{1\,A}\right)\left(\frac{B_d}{4\,T}\right)\left(\frac{\Delta v_i}{0.05}\right)\left(\frac{E_i/A}{15\,GeV/nucleon}\right)^2 \frac{1.06 \cdot 10^{33}\,cm^{-2}s^{-1}}{Z\left(1 + \sigma_e^{*2}/\sigma_i^{*2}\right)}.$$

(3)

Here B_d is the magnetic field of ring dipoles and η is the fraction of the machine circumference covered by the dipoles, I_e is the electron beam current, and we expressed the ring radius through the bending magnetic field, the beam momentum and the dipole occupation factor η.

The values of parameters in Eq. (3) with the exception of the electron beam current and ion beam energy are well determined and cannot be significantly increased. In particular, the electron cooling sets the maximum value of Laslett tune shift, Δv, to about 0.05 determined by the distance to the nearest non-linear resonance. Actually, the cooling cools the beam to the equilibrium between cooling and heating coming from non-linear resonances and intrabeam scattering. If machine parameters are chosen so that the last does not dominate, the beam is cooled as far as it is not heated by the resonance. In absence of cooling higher order resonances play more significant role and Δv is smaller. In this case the emittance is rather determined by the ion source emittance and the emittance growth in the course of acceleration. It is currently expected that $\Delta v_i = 0.01$ can be achieved without cooling.

The luminosity limit is proportional to the occupation factor and magnetic field of the dipoles. Therefore superconducting dipoles are strongly favored even in the case of comparatively small energy (~10-20 GeV), when conventional wisdom prefers a normal conducting synchrotron. The request for two IPs suggests that the ion ring should be a racetrack. Then, the interaction regions and other systems such as RF system, beam injection, electron cooling and Siberian snakes for suppression of depolarizing spin resonances should be located in the straight sections. The length of the straight sections does not depend much on the ring energy and expected to be slightly above 100 m. Using a strong magnetic field in the dipoles decreases the ring circumference and increases the luminosity limit. But at a proton energy of about 20 GeV the bending radius of SC dipoles is quite small (~10-20 m) which complicates the dipole design for very high magnetic field. The choice of superferric dipoles with about 4 T field is a reasonable compromise. Further increase of the magnetic field makes dipoles more complicated but does not bring significant gain in the machine circumference because it starts to be limited by the length of the straight sections. Taking all of the above into account we can estimate $\eta = 0.2$ for 4 T dipoles. For an energy of 20 GeV/nucleon, this yields a ring circumference of about 500 m.

The luminosity limit of Eq. (3) does not depend on the ion beam current and is proportional to the electron beam current. The recent commissioning of B-factories[7,8] suggests that an electron beam current of 2 A can be achieved in a storage ring with parameters required for the collider. That determines that to reach 10^{33} cm^{-2}s^{-1} luminosity one needs a proton energy above 15 GeV for the ring-ring scheme. For the linac-ring scheme the major limitation for electron beam current comes from the electron injector[6]. It sets the current maximum to about 0.2-0.3 A. In this case the energy of proton beam needs to be above about 30 GeV to achieve 10^{33} cm^{-2}s^{-1} luminosity, taking into account the factor of almost 2 which can be obtained with an electron beam size significantly smaller than the ion beam size, $\sigma_e^{*2} \ll \sigma_i^{*2}$. As one will see below such a choice is not limited by other constraints.

Note that in the absence of ion beam cooling Δv is about five times smaller. For both schemes this requires an additional increase of ion energy by factor of $\sqrt{5}$ to achieve 10^{33} cm^{-2}s^{-1} luminosity. One can also see from Eq. (3) that the luminosity per nucleon LA is proportional to A/Z and therefore practically does not depend on Z for fully stripped ions. As will be seen below intrabeam scattering puts more severe request for the ion beam energy increase in the absence of cooling.

1.2 Luminosity Limit due to Beam-beam Effects

Beam-beam effects in colliders are well known. They shift the betatron tunes of the beams and set the following two luminosity limits,

$$L_i = \frac{I_i E_i \xi_i}{e^3 Z \beta_i^*} \frac{2}{1+\sigma_i^{*2}/\sigma_e^{*2}} = $$
$$\left(\frac{I_i}{1A}\right)\left(\frac{E_i}{15\,GeV}\right)\left(\frac{\xi_i}{0.02}\right)\left(\frac{7\,cm}{\beta_i^*}\right)\frac{2}{1+\sigma_i^{*2}/\sigma_e^{*2}} 1.85 \cdot 10^{33}\,cm^{-2}s^{-1}, \quad (4)$$

$$L_e = \frac{I_e E_e \xi_e}{e^3 Z \beta_e^*} \frac{2}{1+\sigma_e^{*2}/\sigma_i^{*2}} = $$
$$\left(\frac{I_e}{1A}\right)\left(\frac{E_e}{5\,GeV}\right)\left(\frac{\xi_e}{0.035}\right)\left(\frac{7\,cm}{\beta_e^*}\right)\frac{2}{1+\sigma_e^{*2}/\sigma_i^{*2}} 1.08 \cdot 10^{33}\,cm^{-2}s^{-1}, \quad (5)$$

which corresponds to the tune shift limitations in the ion and electron beams. Here $E_i = A m_p c^2 \gamma_i$ and $E_e = m_e c^2 \gamma_e$ are the energies of the ion and electron beams, and the parameters $\xi_i = e^2 Z N_e \beta_i^*/(4\pi E_i \sigma_e^{*2})$ and $\xi_e = e^2 Z N_i \beta_e^*/(4\pi E_e \sigma_i^{*2})$ correspond to the linear tune shifts in the ion and electron beams. There is not much freedom in choice of the parameters in the above equations. The achievable value of betatron tune shift for the ion beam depends on the beam cooling. We currently believe that with effective electron cooling (10 to 100 s damping time) ξ_i can be close to 0.02 per IP while without cooling it should be an order of magnitude lower. The beta-function at the IP is determined by the ion bunch length and is limited by the beam separation, the chromaticity of the final focus and its aperture. All these considerations limit β_i to the range of 6 to 10 cm for ion beam energy from 15 to 30 GeV. The luminosity limit described by Eq. (4) does not really depend on the collider scheme (both schemes have similar ion rings) and for the ion energy of 15-30 GeV sets the required value of the proton current to the range of 1-2 A.

In the case of the luminosity limit due to electron beam tune shift described by Eq. (5) the electron beam parameters depend on the collider scheme. Considerable experience acquired on the electron-positron colliders suggests that $\xi_e = 0.035$ can be achieved. For the linac-ring scheme with single IP when one can accept quite significant electron beam emittance growth after the collision the tune shift can be as large as one[6] and for practical machine parameters the beam-beam effects do not limit ξ_e. With two interaction points the tune shift will be limited by the emittance growth after the first collision and $\xi_e \leq 0.2$.

For the linac-ring scheme there is another limitation on the product $\xi_e \xi_i$ which usually puts more severe limitations on possible collider parameters. In this case the interaction of electron beam with the ion beam transfers an electromagnetic excitation from the ion beam head to its tail and thus acts similar to the transverse impedance of the ion ring causing the ion beam kink instability above the threshold[9,10]. Taking into account an estimate of Ref. [6] and more accurate simulation results of Ref. [11] the threshold can be parametrized in the following form,

$$\xi_e \xi_p \leq \frac{v_{si}}{5\pi} \frac{\beta_e^*}{\sigma_{si}} , \qquad (6)$$

where v_{si} is the dimensionless synchrotron tune. Taking into account that the synchrotron tune for a high energy proton synchrotron hardly can be made more than 0.01 and ξ_p is desired to be about 0.005-0.01 that limits ξ_p to be less than about 0.1.

1.3 Intrabeam Scattering

Another important limitation on the beam brightness is determined by intrabeam scattering (IBS). For the above collider parameters the longitudinal energy spread in the beam frame is significantly smaller than the transverse one. That allows one to get comparatively simple formulas to describe IBS. In this case IBS transfers the energy from the transverse degrees of freedom to the longitudinal one and the growth rate can be approximated by the following formula[12],

$$\frac{d}{dt}\left(\theta_\parallel^2\right) \equiv \frac{d}{dt}\left(\frac{p_\parallel^2}{p}\right) = \frac{(Ze)^4 N_i}{4\sqrt{2} A^2 m_p^2 c^3 \gamma_i^3 \beta_i^3 \sigma_{si}} \left\langle \frac{\Xi_\parallel(\theta_x,\theta_y)}{\sqrt{\theta_x^2+\theta_y^2}} \frac{L_C}{\sigma_x \sigma_y} \right\rangle_s . \qquad (7)$$

Here

$$\Xi_\parallel(x,y) \approx 1 + \frac{\sqrt{2}}{\pi} \ln\left(\frac{x^2+y^2}{2xy}\right) - 0.055 \left(\frac{x^2-y^2}{x^2+y^2}\right)^2 , \qquad (8)$$

averaging is performed along the beam orbit, $\sigma_x = \sqrt{\varepsilon_x \beta_x + D^2 \theta_\parallel^2}$, $\sigma_y = \sqrt{\varepsilon_y \beta_y}$, $\theta_x = \sqrt{\varepsilon_x/\beta_x}$ and $\theta_y = \sqrt{\varepsilon_y/\beta_y}$ are the beam sizes and angular spreads along the ring, and

$$L_C = \ln\left(\sqrt{\frac{\pi}{2} \frac{\sigma_x \sigma_y \sigma_{si} \gamma_i}{N_i}} \left(\frac{A m_p c^2 \gamma_i^2 \beta_i^2 (\theta_x^2 + \theta_y^2)}{2 e^2}\right)^3\right) \qquad (9)$$

is the Coulomb logarithm.

In the smooth focusing approximation, for equal horizontal and vertical emittances, $\varepsilon_x = \varepsilon_y$, equal betatron tunes, $v_x = v_y$, and small contribution of energy spread into the beam size, $D(\Delta p/p) \ll \sigma_x$, the momentum spread growth rate can be written in the following form

$$\Lambda_{\parallel} \equiv \frac{1}{\theta_{\parallel}^2} \frac{d}{dt}\left(\theta_{\parallel}^2\right) = \frac{Z^4}{8A^2} \frac{e^4 N_i L_C}{m_p^2 c^3 \gamma_i^3 \beta_i^3 \sigma_{si} \varepsilon_x^{3/2} \theta_{\parallel}^2} \sqrt{\frac{R}{v_x}} \quad (10)$$

The heating of the longitudinal degree of freedom, consequently, causes cooling for both transverse degrees of freedom; but there is another mechanism, which additionally heats the horizontal degree of freedom. At regions with non-zero dispersion, changes in longitudinal momentum change the particles reference orbits, which additionally excites the horizontal betatron motion,

$$\frac{d\varepsilon_x}{dt} = \left\langle A_x \frac{d\theta_{\parallel}^2}{dt} \right\rangle_s \quad (11)$$

where

$$A_x = \frac{D^2 + (D'\beta_x + \alpha_x D)^2}{\beta_x} \quad (12)$$

Finally, one can write for the emittance growth rates

$$\begin{bmatrix} \Lambda_x \\ \Lambda_y \end{bmatrix} \equiv \frac{1}{\varepsilon_{x,y}} \frac{d\varepsilon_{x,y}}{dt} = \frac{(Ze)^4 N_i}{8\sqrt{2} A^2 m_p^2 c^3 \gamma_i^3 \beta_i^3 \sigma_{si} \varepsilon_{x,y}}$$

$$\left\langle \frac{1}{\sqrt{\theta_x^2 + \theta_y^2}} \frac{L_C}{\sigma_x \sigma_y} \begin{bmatrix} 2A_x \Xi_{\parallel}(\theta_x, \theta_y) - \frac{\beta_x}{\gamma_i^2} \Xi_{\perp}(\theta_x, \theta_y) \\ -\beta_y \Xi_{\perp}(\theta_y, \theta_x) \end{bmatrix} \right\rangle_s \quad (13)$$

where

$$\Xi_{\perp}(x, y) \approx 1 + \frac{2\sqrt{2}}{\pi} \ln\left(\frac{\sqrt{3x^2 + y^2}}{2y^2} x\right) + \frac{0.5429 \ln(y/x)}{\sqrt{1 + \ln^2(y/x)}} \quad (14)$$

The energy conservation requires $\Xi_{\perp}(x,y) + \Xi_{\perp}(y,x) = 2\Xi_{\parallel}(x,y)$ which for considered approximation is fulfilled with accuracy better than 1%.

In the smooth focusing approximation for equal emittances and betatron tunes we obtain for the horizontal emittance growth rate,

$$\Lambda_x = \frac{Z^4}{16 A^2} \frac{e^4 N_i L_C}{m_p^2 c^3 \gamma_i^3 \beta_i^3 \sigma_{si} \varepsilon_x^{5/2}} \sqrt{\frac{R}{v_x}} \left(\frac{2}{v_x^2} - \frac{1}{\gamma_i^2}\right) \quad (15)$$

Note that for considered collider the length of the straight sections is close to the length of arcs and therefore Eq. (15) underestimates this growth rate by about factor of two.

1.3 Choice of Basic Parameters

The limitations considered above allow one to choose the basic machine parameters for both collider schemes. As a major goal we will consider achieving the luminosity of 10^{33} cm^{-2}s^{-1} for the electron-proton collider with the minimum machine energy and, consequently, the minimal cost. In this case, the Laslett tune shift (see Eq. (3)) and the electron beam current limitations set the minimum energy of the proton ring. They are 15 GeV for the ring-ring collider and 30 GeV for the linac-ring collider.

The distance between bunches is determined by the beam separation after collision and was chosen to be 3 m for both schemes. The limitation coming from the beam-beam effects determines the proton beam current and the energy of the electron beam. Main parameters of the considered colliders are shown in Table 1.

Although the energies of electron beams for the linac-ring and the ring-ring colliders are equal they are set by different limitations. For the ring-ring collider the energy is determined by the beam-beam tune shift of electrons in the field of protons. For the linac-ring collider the energy is determined by the kink instability threshold; and parameters are chosen so that the beam is at the instability threshold of Eq. (6). The threshold of the kink instability is proportional to the synchrotron tune and therefore achieving highest possible tune is desirable. Expressing the tune through the bunch length and the energy spread we obtain the following formula:

$$\nu_s = \frac{\gamma_i}{1-\gamma_i^2} \frac{q\sigma_{si}}{2\pi R} \frac{eV_0}{Am_p c^2 \theta_\parallel}, \qquad (16)$$

TABLE 1. Parameters of Ring-ring and Linac-ring Colliders with Luminosity of 10^{33} cm^{-2}s^{-1}

	Ring-ring		Linac-ring	
	Electrons	Protons	Electrons	Protons
Center of mass energy, GeV	13.8		19.3	
Kinetic beam energy, GeV	3	15	3	30
Circumference, m	420	420	-	650
Betatron tunes	≈11	≈12	-	≈16
Critical kinetic energy, GeV	-	7.2	-	9.7
Beam current, A	1.7	1.8	0.27	2.7
Beta-functions at IP, cm	7	7	7	7
Rms normalized beam emittance, mm·mrad	780	2.26	130	1.4
Rms beam size at IP, μm	96	96	40	55
Laslett tune shift	-	0.05	-	0.05
Beam-beam tune shift per IP	0.032	0.006	0.15	0.0029
Synchrotron tune	-	0.0075	-	0.007
Rms momentum spread	6·10^{-4}	8·10^{-4}	<5·10^{-3}	8·10^{-4}
Rms bunch length, cm	≤3	7	≤0.3	7
Longitudinal intrabeam scattering lifetime, $\Lambda_\parallel^{-1}$, s	-	110	-	80
Transverse intrabeam scattering lifetime, Λ_x^{-1}, s	-	370	-	150

where V_0 is the total voltage of the ion ring RF system, and q is the ring's harmonic number. Further decreasing of the momentum spread, $\theta_\parallel$, is limited by the intrabeam scattering and the single bunch longitudinal instability. Therefore the only free parameter we have is the RF voltage. Its increase requires increasing the momentum compaction factor and, consequently, decreasing the betatron tune, which leads to a fast growth of the transverse IBS growth rate. As result a compromised value $v_x \approx 16$ has been chosen.

Note that the threshold of the kink instability was computed in the linear theory, which overestimates the instability threshold. We currently work on a more detailed non-linear simulation of the instability and we believe that this more accurate theory should increase the instability threshold by a factor of two. Consequently, the electron beam energy could be decreased to about 1.5 GeV, but in the case of two IPs this additional decrease of the electron beam energy is limited by deterioration of the electron beam emittance after the first collision.

If fully striped ions are used in the machine optimized for protons the Laslett tune shift is going to be the major limitation for luminosity. Actually, in this case $A/Z \approx 2$, and the energy per nucleon is about two times lower. Substituting this into Eq. (3) yields that the luminosity per nucleon, LA, is also two times smaller than for the proton case. Usually one would like to keep the same bunch length and the beam size at the IP and, consequently, the same beam emittance. Then, Eq. (2) yields that the number of ions should be $4Z$ times less than the number of protons and the ion beam current should be 4 times less than the proton beam current. That determines that the tune shift in the electron beam is one forth of the tune shift for the electron-proton case, and the tune shift of the ion beam is the same as for the proton beam. The IBS increments grow by approximately $Z/2$ times and, as one will see in the next section, they can be compensated by increased strength of the electron cooling.

2. ELECTRON COOLING

The intrabeam scattering in the ion beam is so strong that it is impossible to reach the required luminosity without strong cooling of the ions. Electron cooling is the only cooling method, which works for the ion density discussed above. For the collider parameters the velocity spread in the ion beam is sufficiently large and cooling can be considered non-magnetized. Then the cooling force in the beam frame is[13]:

$$\mathbf{F}'(\mathbf{v}') = \frac{4\pi n'_e Z^2 e^4}{m_e} \int \frac{\mathbf{v}' - \mathbf{v}'_e}{|\mathbf{v}' - \mathbf{v}'_e|^3} f_e(\mathbf{v}'_e) d\mathbf{v}'^3_e \quad , \tag{17}$$

where $f_e(\mathbf{v}'_e)$ is the electron distribution function. The longitudinal velocity spread for the electron beam accelerated in an electrostatic accelerator is usually much smaller than its transverse energy spread. In our case of very high energy electron cooling, it is expected that the electron beam will be accelerated by a low frequency linear accelerator with energy recovery, and therefore the longitudinal velocity spread is expected to be significantly higher. For the estimate we assume that the longitudinal and transverse energy spreads of electrons are equal. Then, the friction force can be approximated by the following formula:

$$\mathbf{F}'(\mathbf{v}') \approx \frac{4\pi n'_e Z^2 e^4 L_{Ce}}{m_e} \frac{\mathbf{v}'}{\left(v'^2 + 2.4 v'^2_e\right)^{3/2}} \quad , \tag{18}$$

where $\overline{v'^2_e} = \overline{v'^2_{ex}} = \overline{v'^2_{ey}} = \overline{v'^2_{ez}}$, and L_{Ce} is the Coulomb logarithm.

To find the damping decrement one needs to perform averaging of the force over betatron motion. For the sake of this estimate we consider that the ion longitudinal velocity spread is much smaller than the transverse one, $\overline{v'^2_x}, \overline{v'^2_y} \gg \overline{v'^2_z}$, and we put that $v'_z = \beta_i c \theta_\parallel$, $v'_x = \sqrt{2} \beta_i \gamma_i c \theta_x \cos\varphi_x$ and $v'_y = \sqrt{2} \beta_i \gamma_i c \theta_y \cos\varphi_y$, where a factor of $\sqrt{2}$ takes into account that the amplitude is $\sqrt{2}$ times larger than the rms value. After performing averaging and returning back to the lab frame we obtain the following estimate for the damping decrements:

$$\tau_\parallel^{-1} \approx \frac{16\pi n_e Z^2 e^4 L_{Ce}}{3 m_e m_i \gamma_i^5 \beta_i^3 c^3} \frac{\eta_c}{\left(2\theta_x^2 + 0.66\left(\theta_\parallel^2/\gamma_i^2\right) + 1.6\theta_e^2\right)\sqrt{\left(\theta_\parallel^2/\gamma_i^2\right) + 2.4\theta_e^2}} ,$$

$$\tau_\perp^{-1} \approx \frac{16\pi n_e Z^2 e^4 L_{Ce}}{3 m_e m_i \gamma_i^5 \beta_i^3 c^3} \frac{\eta_c}{\left(2\theta_x^2 + 1.2\left(\theta_\parallel^2/\gamma_i^2\right) + 2.8\theta_e^2\right)^{3/2}} .$$

(19)

Here the decrements are defined as $\tau_\perp^{-1} \equiv 1/\varepsilon \cdot d\varepsilon/dt$ and $\tau_\parallel^{-1} \equiv 1/\Delta p^2 \cdot d\Delta p^2/dt$, θ_e is the angular spread in the electron beam, η_c is a fraction of the ring orbit used for cooling, and the following two approximate equations have been used to perform averaging:

$$\int_0^\pi\int_0^\pi \frac{dxdy}{\pi^2} \frac{1}{\left(\cos^2 x + \cos^2 y + \Delta^2\right)^{3/2}} \approx \frac{2}{3} \frac{1}{\Delta\left(1+\frac{2}{3}\Delta^2\right)} ,$$

$$\int_0^\pi\int_0^\pi \frac{dxdy}{\pi^2} \frac{\cos^2 x}{\left(\cos^2 x + \cos^2 y + \Delta^2\right)^{3/2}} \approx \frac{1}{1.55\Delta\left(1+\left(\frac{2}{1.55}\right)^{2/3}\Delta^2\right)^{3/2}} ,$$

(20)

Assuming that electron beam size is three times larger than the rms ion beam size in the cooling section, $r_{eb} = 3\sqrt{\varepsilon_i \beta_c}$ we can finally rewrite Eq. (19) in the following form

$$\tau_\parallel^{-1} \approx \frac{8 Z^2 e^3 I_0 L_{Ce}}{27 m_e m_i \gamma_i^5 \beta_i^4 c^4 \varepsilon_i^2} \frac{\eta_c}{\left(1+\left(0.33\frac{\theta_\parallel^2}{\gamma_i^2}+0.8\theta_e^2\right)\frac{\beta_c}{\varepsilon_i}\right)\sqrt{\left(\theta_\parallel^2/\gamma_i^2\right)+2.4\theta_e^2}} ,$$

$$\tau_\perp^{-1} \approx \frac{4\sqrt{2} Z^2 e^3 I_0 L_{Ce}}{27 m_e m_i \gamma_i^5 \beta_i^4 c^4 \varepsilon_i^{5/2}} \frac{\eta_c \sqrt{\beta_c}}{\left(1+\left(0.6\frac{\theta_\parallel^2}{\gamma_i^2}+1.4\theta_e^2\right)\frac{\beta_c}{\varepsilon_i}\right)^{3/2}} .$$

(21)

which will be used for the cooling time estimates considered below. Here $I_0 = \pi r_e^2 n_e \beta_i c$ is the electron beam current.

Table 2 presents parameters of the required cooling devices for the ring-ring and the linac-ring colliders considered in the previous section. The following issues have determined the choice of the parameters. The cooling decrements are proportional to the cooling length and one would like to choose this length as long as possible. To keep the ion beam inside the electron beam along the entire cooling length, the ion beam beta-functions should be larger, or equal to the cooling length. However an increase of beta-functions increases electron beam size and it decreases the electron density, which compensates the growth of the decrements due to the decrease of the angular spread. As one can see from Eq. (21) the effect of beta-function increase on damping decrements depends on the actual values of parameters. For the considered here parameters, an increase of beta-functions decreases the decrements. Also note that an increase of beta-functions raises the requirements on the angular spread of electrons due to smaller angular spread in the ion beam. Therefore, we choose the beta-functions to be equal to the cooling length for both cases. The cathode temperature and size set the minimum for the angular spread in the electron beam. We believe that the electron beam acceleration and transport needs to be done (and can be done) with a sufficiently small additional emittance growth. Then, the transverse temperature of electrons is equal to the cathode temperature and the longitudinal temperature was chosen to be equal to the transverse one. A smaller longitudinal temperature does not bring any significant gain, while the larger one affects both decrements.

TABLE 2. Electron Cooling Parameters for Ring-ring and Linac-ring Colliders		
	Ring-ring	Linac-ring
Kinetic energy of electrons, MeV	8.2	16.3
Peak electron beam current, A	3	10
Mean electron beam current in energy recovery linac and 40 cm bunch, A	0.4	1.3
Electron beam radius, cm	0.6	0.34
Electron density in the beam frame, cm^{-3}	$3.2 \cdot 10^7$	$1.7 \cdot 10^8$
Beta-function of proton beam at the center of cooling section, m	30	30
Rms proton beam size at the center of cooling section, cm	0.2	0.11
Cooling section length, m	30	30
Rms angular spread in the proton beam,	$6.7 \cdot 10^{-5}$	$3.7 \cdot 10^{-5}$
Rms transverse velocity spread of ions in the beam frame, cm/s	$3.4 \cdot 10^7$	$3.7 \cdot 10^7$
Rms longitudinal velocity spread of ions in the beam frame, cm/s	$2.4 \cdot 10^7$	$2.1 \cdot 10^7$
Effective electron temperature in the beam frame, eV	0.2	0.2
Rms velocity spread of electrons in the beam frame, cm/s	$1.9 \cdot 10^7$	$1.9 \cdot 10^7$
Relative rms energy spread of electrons	$6.2 \cdot 10^{-4}$	$6.2 \cdot 10^{-4}$
Relative rms angular spread of electrons	$3.7 \cdot 10^{-5}$	$1.9 \cdot 10^{-5}$
Longitudinal emittance damping time, s	110	70
Transverse emittance damping time, s	240	150

As one can see from Table 2, the electron velocity spread is already comparable to the ion velocity spread and its further increase would cause fast decrease of the damping decrements. There are two major issues limiting the angular spread for electrons: uncontrolled transverse dipole fields and the space charge effects. The first issue is important only inside the cooling section where achieving the beam angle variation below 10-20 mrad is not going to be a challenging problem. This issue is worse for the linac-ring case where the angular perturbations have to be 2 times lower. The space charge effects are a major concern at low energy and solution of this problem will be another challenge. If an energy recovery linac is chosen for the acceleration of electrons, there is another effect, which can significantly affect the transverse velocities. This is a time dependent focusing of the accelerating cavities, which focuses differently particle in the head and in the tail of the bunch.

If for the ring-ring case the electrostatic acceleration to 8 MeV is still feasible, this is not a viable option for the linac-ring scheme with 16 MeV, where the energy recovery linac is the only choice. In that case to maintain effective cooling for all particles in the ion bunch the electron bunch should have a bunch length of at least 5σ, corresponding to about 40 cm. In this case achieving $6 \cdot 10^{-4}$ energy spread will pose a challenge. One of possible solutions can be an addition of higher harmonics to correct the accelerating profile. In particular adding the third and the fifth harmonics allows one to reach the accelerating gradient uniformity within $\pm 10^{-4}$ band for 45 cm bunch and 3 m bunch spacing.

The parameters of the electron coolers required for both projects are well beyond current state-of-the-art technology for the beam current, and the angular and momentum spread in the electron beam. The required electron beam current for the linac-ring scheme is three times higher and the angular spread is two times lower making this choice significantly more difficult, if possible at all. Also note that the electron beam parameters were calculated for the most optimistic case. In reality, we may have difficulties to achieve the desired velocity spread in the electron beam. That will require higher electron beam current making the project more complicated.

3. NO COOLING SCENARIO

Taking into account that the electron cooling is still a significant pending problem we also consider a collider with no cooling at the top energy. However some cooling can be used at low energy to shape the beam. In this case the IBS is the major obstacle in achieving high brightness in the ion beam; usually the longitudinal IBS is more limiting. Combining Eqs. (1) and (10) and taking into account that the bunch length and the beta-function at the IP are equal one obtains the following luminosity limitation:

$$L_{IBS} = \frac{4A^2}{\pi Z^4} \frac{m_p^2 c^3 \gamma_i^3 \beta_i^3 \theta_\parallel^2 I_e \Lambda_\parallel}{e^5 L_C \left(1 + \sigma_e^{*2} / \sigma_p^{*2}\right)} \sqrt{\frac{\varepsilon_x v_x}{R}} \quad . \tag{22}$$

For high energy collider without cooling the emittance is determined by the injector, and the normalized emittance, $\varepsilon_{xn} = \varepsilon_x \gamma_i$, is conserved. The ratio R/γ_i is determined by the magnetic field of dipoles. Thus, we can write the following scaling for the luminosity limit: $L_{IBS} \propto \gamma_i^2 \theta_\parallel^2 I_e \Lambda_\parallel \sqrt{\varepsilon_{xn} v_x}$. As one can see, the only effective free parameter to compensate a longer IBS time is the ion beam energy. Table 3 depicts the tentative parameters of the ring-ring collider with 10^{33} cm^{-2}s^{-1} luminosity and minimum center of mass energy.

TABLE 3. Tentative Parameters of Ring-ring Collider with Luminosity of 10^{33} cm^{-2}s^{-1} and without Cooling

	Electrons	Protons
Center of mass energy, GeV	42.6	
Kinetic beam energy, GeV	3	150
Circumference, m	1600	1600
Beam current, A	1.7	0.8
Beta-functions at IP, cm	5	10
Rms normalized beam emittance, mm·mrad	110	6.6
Rms beam size at IP, μm	30	64
Laslett tune shift	-	$2\cdot10^{-4}$
Beam-beam tune shift per IP	0.023	0.002
Rms momentum spread	$6\cdot10^{-4}$	$8\cdot10^{-4}$
Rms bunch length, cm	≤ 3	10
Longitudinal intrabeam scattering lifetime, $\Lambda_\parallel^{-1}$, hour	-	17
Transverse intrabeam scattering lifetime, Λ_x^{-1}, hour	-	23

In the case of the linac-ring collider the ion ring energy should be increased by another factor two or three. As one can see, although achieving the desired luminosity is still feasible with high-energy ion ring, its energy becomes disproportionally high.

4. DISCUSSION

Achieving the luminosity of 10^{33} cm^{-2}s^{-1} for the electron-proton collider is a challenging problem. Two possible collider schemes have been proposed. They are the ring-ring and the linac-ring colliders. Both of them require electron cooling of the proton beam at the collision energy. The electron beam parameters for the cooler are far beyond of the current electron cooling technology, and the development of cooling is one of the highest priorities of the project R&D. In the comparison of two collider schemes considered above, the electron beam current for the linac-ring collider has been limited to 200-300 mA. Even with this optimistic choice supported by estimates, but staying far away from the current technology level, the luminosity of 10^{33} cm^{-2}s^{-1} requires two times higher energy for the ion ring and significantly more complicated electron cooling. Although, at the present time the linac-ring collider does not look competitive to the ring-ring if a new machine with ultimate luminosity is built, the linac-ring collider can be a good choice for already existing high energy synchrotrons. In the case of RHIC collider it does not look reasonable to build a full circumference electron ring, but in a ring of smaller circumference the beam-beam effects are much worse. A linac with energy recovery may be a much better choice.

If a collider optimized for the electron-proton collisions is used for the electron-ion collisions with fully stripped ions the ultimate luminosity for the electron-ion mode is expected to be about half of luminosity for the proton-electron case with approximately the same requirements for the electron cooling.

ACKNOWLEDGMENTS

The author is grateful to I. Ben-Zvi, Ya. Derbenev, A. Hutton, L. Merminga, R. Li, G. Krafft, S. Nagaitsev and Yu. Shatunov for useful and stimulating discussions. I also would like to thank A. Bogacz and M. Tiefenback for help in the editing of the article.

REFERENCES

1. Barber, D, P., "Electron and Positron Polarization at HERA; Past and Future," *These proceedings*.
2. Douglas, D. R., York R. C., and Kewisch J., *Proc. of the 1989 Particle Accelerator Conference*, 1989, pp.557-559.
3. Bowling, B., et al., *Proc. of the 1991 Particle Accelerator Conference*, 1991, pp. 446-448
4. "The Science Driving the 12 GeV Upgrade of CEBAF" edited by L. Cardman, Jefferson Lab, 2000.
5. Koop, I, A., et al., "Conceptual Design Study of the Electron-Proton Storage Ring Collider with Polarized Beams," *These proceedings*.
6. Merminga L., et al., "An Energy Recovery Electron Linac-On Ring Collider," *These proceedings*.
7. Funakoshi Y., et al., "KEKB Performance," *Proceedings of the 2000 European Part. Accel. Conf.*, 2000. pp. 38-44.
8. Seeman J.T., et al., "Status Report of PEP-II performance," *Proceedings of the 2000 European Part. Accel. Conf.*, 2000. pp. 28-32.
9. Li R. and Bisognano J., "A Strong-Strong Simulation on the Beam-Beam Effect in a Linac/Ring B-Factory," *Proceedings of the 1993 Part. Accel. Conf.*, 1993. pp. 3473-3475.
10. Perevedentsev E.A., Valishev A.A., "Characteristics and possible cures of the Head-Tail Instability of Colliding Bunches," *Proceedings of the 2000 European Part. Accel. Conf.*, 2000. pp. 1223-1225.
11. Li R., Lebedev V.A., Bisognano J. J., "Analysis and Simulation of Beam-Beam Kink Instability in a Linac-Ring Electron-Ion Collider," *To be submitted to Proceedings of the 2001 Part. Accel. Conf.*
12. Lebedev V.A., et al., *NIM-A* **391**, 176-187 (1997).
13. Budker G. I. and Skrinsky A. N., *Usp.Fiz.Nauk* **124**, 4, p.561(1978).

Overview of European Plans for future Lepton/Photon Scattering Facilities

Dietrich von Harrach

Institut für Kernphysik der Universität Mainz, D 55099 Mainz, FRG

Abstract. The status of the European discussions for future electron scattering facilities is reported. It appears that plans for a European electron scattering facility (ELFE) at around 25 GeV are most advanced. Both CERN and DESY have presented options for such a machine. The European electron scattering community has made large progress in formulating a coherent physics program for such a facility. This program is centered on deeply virtual exclusive processes and is complemented by a rich program for semi inclusive processes, spin-observables and nuclear effects. Other options at higher energies, centering around the future use of the HERA complex or the planned TESLA project, have also been worked out in detail.

I INTRODUCTION

The understanding of hadronic structure is one of the remaining unsolved problems of physics on the microscopic scale. Although it appears that QCD is the right framework for strong interaction physics, it has become a commonplace to state that we do not have a solid understanding of the origin of the (luminous) mass of the universe, which is dominated by baryons. We know that hadrons contain an infinite number of quarks and gluons, but up to now we do not really know how masses and other static or low resolution properties of hadrons are to be calculated starting off the fundamental QCD-Lagrangian. There are beautiful symmetries and phenomenological models, like the SU3 constituent quarkmodel, but we do not know why they work and what are their limits.

There are competing experimental approaches to the hadronic structure problem. One of the approaches are hard scattering processes measuring structure functions and related quantities . Another approach is hadron spectroscopy which is trying to access properties of the hadronic wave function by the observation of systematics and symmetries of the excitation spectrum and the decay widths. A specific example is the use of antiprotons. Reactions with antiprotons have added considerably to the knowledge of the meson sector but have failed to produce a clear and undisputable signal for gluonballs or hybrids since there is no unique way to convert spectroscopic information into information about the wave function.

The use of leptons interacting via photon or electroweak boson exchanges has the large attraction of combining a weakly interacting probe of well controlled energy momentum transfer with the complex multi particle system containing charged constituents. The well proven perturbative methods allow us to define a hierarchy of operators which, at least in principle, can be extracted separately. The command of perturbative methods finally allows us to overcome one of the limitations of the electroweak probe, that is the access of gluon distributions. All this can be complemented by the addition of the spin degrees of freedom. While most of the past theoretical work has concentrated on structurefunctions, which were studied experimentally in exclusive and semiinclusive processes, it has recently been discovered that exclusive processes open new ways to access hadronic structure. There is an enormous theoretical activity in this field which has come up with a number of predictions and relations between hitherto unrelated aspects of hadronic structure. Only few data for the new observables are available up to now. However they fall short in their kinematical range or they do not have sufficient kinematical resolution to isolate individual exclusive channels. It appears that the properties of a facility like ELFE are particularly well adapted to study these processes.

The cost of an electron accelerator with a high luminosity and a high duty factor for fixed target experiments presently sets a limit to the accessible energy range of about 30 GeV. It appears that instantaneous luminosities in excess of a few times $10^{36} cm^{-2} s^{-1}$ cannot be handled easily in a large acceptance coincidence experiment due to the large total hadronic cros ssection of quasireal photons. This defines a natural lower limit to the cross section of order $10^{-40} cm^{-2}$. The cross sections for exclusive processes mentioned above fall off with inverse powers of the four momentum transfer of typically 6-8. They reach the $10^{-40} cm^{-2}$ barrier at about $Q^2 \approx 20\,\text{GeV}^2$ in the most favourable x region between $0.1 < x < 0.3$. The advantage in going to higher collision energies s is the extension to lower x, if comparable luminosities can be achieved. With present day techniques, colliders seem to be limited to luminosities of about $10^{33} cm^{-2} s^{-1}$ and will not be able to gain in Q^2 range. With an increase in s by a factor 10 however the small x-region $0.01 < x < 0.1$ will be accessible with reasonable Q^2.

For the study of exclusive processes the separation from other channels, containing more particles in the final state, proceeds preferentially by a kinematical fit, requiring momentum conservation. Anticoincidence may help in some cases but will require full coverage of the solid angle.

The scale for the required kinematical resolution is set by the mass of the π-meson. Roughly a resolution of 25-50 MeV in all (n-1) momenta is needed to guarantee a missing mass resolution of a fraction of the pion mass. This results in a separation from a channel containing one additional neutral pion. Such a channel clearly corresponds to different quantum numbers of the baryonic system.

At ELFE the energy spread of the beam is about 0.1% at an energy of 25 GeV. At higher energies recirculating machines will increase the relative energy spread due to synchrotronradiation and thus fail the resolution criterion. The construction of large acceptance and high resolution detectors appears possible [2] in this energy range.

The measurement of exclusive processes in a collider scenario at higher s (e.g. ENC 7.5 GeV e + 35 GeV p corresponding to s=32 GeV compared to s=7 GeV for ELFE) respecting the resolution requirements formulated above appears possible, if powerful forward/backward spectrometer magnets are integrated into the beamline design of the collision zone. The difficulty to measure the recoil proton in DVCS in a fixed target experiment, however, is exchanged for the difficulty to detect a low energy electron. In fact at low $x_{Bj} \approx 0.1$ $x_{crit} = 0.1 \cdot \frac{E_e}{E_p} = 0.02$ one has to deal with electron energies of order 100 MeV.

The real obstacle for the use of a collider for the investigation of deeply virtual exclusive processes remains its limited luminosity. Even if the luminosity of $10^{33} cm^{-2} s^{-1}$ can be obtained in double polarization experiments, it will fall short compared to fixed target experiments with DNP targets, except for the small x region where there is no alternative to a collider.

The only existing ep collider is HERA. It was conceived at an energy s=300 GeV to be able to access physics beyond the standard model. The bulk of its data are concentrated in the small x region down to $x \approx 10^{-5}$. There are exciting predictions about the behaviour of nuclei at such low values of x which would make the use of HERA as an eN collider attractive. The upcoming heavy ion program at the CERN LHC may require information from an eN collider. A similar relation has been formulated for the RHIC collider and the ERHIC project discussed together with the EPIC project at this conference.

The exploitation of the spin degree of freedom for the analysis of hadronic structure, in which it appears to play a decisive role, is still in its infancy. Recent results from HERMES demonstrate progress in the field of semiinclusive and (almost) exclusive spin observables like azimuthal asymmetries. More is expected from the COMPASS experiment which has started commissioning the detector this summer. The limitations of HERMES in energy and of COMPASS in luminosity have led to a discussion of future prospects. There have been intense discussions about a future use of HERA with polarized protons and deuterons in addition to the already existing polarized electron and positron beams. Such a program will have a potential for discoveries in the low x region. For example it could reveal information about the polarized structure function of a photon. It would also be able to access the polarized gluon structurefunction at low x. The risks of such a program are, however, considerable since it requires effective luminosities $L_{eff} = P_e^2 \cdot P_p^2 \cdot L$ of $\approx 10^{31} cm^{-2} s^{-1}$ to provide enough sensitivity.

II EXISTING FACILITIES

A MAMI, ELSA and GRAAL

In Europe presently exists a wide range of facilities for the investigation of hadrons with the electroweak probe. At the low energy end with presently 880

MeV MAMI microtron at Mainz is pursuing a program to measure formfactors and polarizabilities of nucleons and mesons and the properties of the low lying excited states of the nucleon and their decays. The current program, making extensive use of polarized electrons, is concentrating on the neutron electric formfactor, on the strangeness contribution to the electric formfactor via parity violaton, on the pion polarizability and on the GDH spin sumrule. The ELSA stretcher ring at Bonn presently is setting up a crystal barrel detector which is concentrating on the neutral final states in the decay of nucleon resonances. First encouraging results have been presented at photon beam energies of 2 GeV. The second experimental facility at ELSA is measuring the spindependent photon cross sections complementing the Mainz GDH program on the higher energies 800 MeV $<$ E $<$ 3 GeV. At the European syncrotron radiation facility ESRF in Grenoble, France, experiments with laser backscattered photons are performed. Polarized photons up to 1.5 GeV are used to study strangeness and vector meson production.

The MAMI facility at MAINZ is now realizing an upgrade program to extend the energy range to 1.5 GeV. A fourth acceleration stage with a double-sided microtron DSM is being built. It will preserve the cw-operation with polarized currents up to 100 μA.

B HERMES

The HERMES facilty at DESY is pursuing its program of semi-inclusive and exclusive measurements with the circulating polarized electron or positron beam of 27.5 GeV of HERA on a polarized storage cell gas target. This experiment is presently unique in this combination and has made several discoveries in the azimuthal hadron distributions. It is presently the only experiment which allows to test predictions for deeply virtual exclusive processes and serves as a starting point for discussions for the planned ELFE facility, operating in the same energy range. Since its missing mass resolution is of order 150 MeV it cannot separate truely exclusive channels and needs some extra assumptions on the behaviour of the more complex background processes containing additional mesons.

The luminosity of HERMES is limited by two facts:
(i)the density of polarized storage cell targets is presenly not exceeding $10^{14} cm^{-2}$ due to limitations in the atom beam sources. Together with the electron/positron beam current of HERA of 50 mA the luminosity is limited to $2\,10^{31}\ cm^{-2}s^{-1}$. With the typical polarization values of $P_e \approx 0.6$ and $P_p \approx 0.86$ the effective undiluted luminosity is $5\,10^{30} cm^{-2}s^{-1}$.
(ii)The second limitation, which does not become effective with polarized targets, comes through the lifetime of the electron beam. Since HERMES runs concurrently with the two ep collider experiments, it is supposed to contribute not more than 20% to the lifetime. This limits the allowed luminosity for protons to $10^{32} cm^{-2}s^{-1}$. If allowed to use up the residual beam, HERMES has realized unpolarized luminosities for protons or nucleons (D_2) of $0.8\,10^{34} cm^{-2}s^{-1}$ for measuring times of order 1 h.

Such an operation appears possible if HERMES can be the only user of the HERA electron ring. The overall duty factor for such a mode of operation will be about 50%. If this operation would be possible for a few years a considerable upgrade of the detector, offering better resolution and higher data rates, would be necessary.

C COMPASS

CERN has a long tradition using 100-300 GeV muon beams for hadron structure experiments. The uniqueness of this energy together with the high natural polarization and the long radiation length of muon beams has led to a new experiment COMPASS, which is essentially geared to the measurement of the gluon polarization via charm and "high" $p_\perp$ hadron pairs from quasireal photon interactions. Measurements with transverse target polarization to measure h_1, making use of the Collins effect at transverse target polarization, are also foreseen. The new experiment is being setup now with a completely new tracking and particle identification system. It makes use of an extremely large polarized DNP target of about 1.50 m length and hadron acceptance of ±180 mrad. There have been considerable delays in the fabrication of the superconducting target magnet. Data taking with a reduced setup is foreseen to start in 2001.

The COMPASS experiment has pushed the limits of the muon beam experiments in several ways. It uses the optimum energy-dutycycle combination (160 GeV muons, 400 GeV protons, 3 s out of 14s) and operates at the thermal limit of the production target. The phase space of the muon beam has been optimized for the use of a long target with minimum diameter to maximize luminosity and minimize secondary interactions of hadrons. Large $p_\perp$ hadrons are favoured by this setup.

A further upgrade of muon beam experiments is not obvious.

The energy range of leptons of a few 100 GeV corresponding to cm energies of s=15-40 GeV (100-800 GeV) appears well adapted to the study of hadronic structure, which we believe is governed by some natural nonperturbative length scales of order of 0.1-1 fm. The transition from the asymptotic behaviour at resolution scales 1/Q smaller than this to the nonperturbative scales can be covered over a wide x-range down to x$\approx 10^{-3}$. The x-region below, at least for nucleons, appears to be governed by an asymtotic behaviour studied in detail by HERA. For nuclei with large radii qualitatively new phenomena are predicted in this region and a higher cm-energy appears to be necessary for their study.

We will discuss two attempts to overcome the limitations of the muon beam experiments which go by the names ENC (electron nucleon collider) and TESLA-N (fixed target experiments with a single TESLA beam).

D HERA

Two experiments H1 and ZEUS are continuing their program. DESY is presently realizing an upgrade program enhancing the luminosityfrom the present maximum of $2\ 10^{32} cm^{-2} s^{-1}$ (about 100 pb^{-1} per year) by a factor of five essentially by low-β insertions. Spin rotators are installed for all experiments. The proton beam energy has been increased to 920 GeV. This will considerably enhance the program searching for deviations from the standard electroweak predictions. This program is planned to extend through the next five years. The future of the HERA complex will depend on the decisions on the TESLA proposal expected in 2001. There are plans to reuse the electron injector syncrotron PETRA as a synchrotron radiation facility.The proton ring will be left untouched for an ep option using the TESLA electron beam. Finally the electron ring has been discussed as a stretcher ring for the ELFE at DESY option.

III PROJECTS

A The ENC Discussion

The idea to build a dedicated electron nucleon collider came out of discussions for the future of the GSI laboratory. It is the major German national laboratory for nuclear physics with a main emphasis on heavy ion physics. The idea originally grew out of the problem to measure electromagnetic radii of instable socalled halo nuclei with one or more extremely loosely bound nucleons. Two working groups were set up in 1996 to study its physics potential and its realizability [13]. The physic potential was seen in semi-inclusive experiments especially in connection to the spin and flavour structure of the nucleon and the study of nuclear effects on the parton structure and the hadronisation process. As a particular strength the accessibility of the target fragmentation region and the large charm signal from secondary vertices was identified. The ideal energy range was seen around s=30 GeV. A luminosity around $10^{33} cm^{-2} s^1$ and of course double polaristion was required.

The energy was felt not to be optimal for the investigation of high gluon densities at low x. Due to the charge to mass ratio the cm-energy for e-N collisions with Lead or Uranium is only about 19 GeV which means that at $Q^2 = 1\,GeV^2$ the lower limit of the Bjorken x is $x \approx 0.003$ and $x \approx 0.03$ at $Q^2 = 10\,GeV^2$, which is relevant for the charm threshold. Although shadowing at high virtualities was discovered at such energies by the EMC, detailed investigations of the scale dependence are not possible , since they would require a lever arm in Q^2 of at least $3 - 10\,GeV^2$.

At the time of the ENC discussion the deeply virtual exclusive processes were not yet intensely discussed. As I have pointed out in the introduction, a collider at $s \approx 30\,GeV$ with a good (proton)forward spectrometer could provide data for the low x region $0.005 <$ x < 0.05 e.g. for the evaluation of sumrules.

For the dicussions of similar projects in other parts of the world it might be interesting to know why this project was not pursued after the Seeheim meeting [12] in 1997. The project failed to get support from GSI, since it was felt not to be related sufficiently to the history and the main projects of the laboratory to withstand the fierce opposition from the DESY laboratory, which could not see any virtue in a project of a smaller edition of the HERA collider 300 km south of Hamburg, possibly draining resources from the TESLA project.

B The ELFE project

The ELFE project has been originally conceived and promoted by the French nuclear physics community after the shutdown of the Saclay ALS linac in 1989. A number of international workshops were organized to identify the physics program and the machine parameters. A project for a 15 GeV cw machine of the CEBAF type emerged with a wide program without a clear "mission statement". The project received criticism for its conceptional weakness and never got unanimous support even from the hadron structure community. Especially the comparatively low beam energy of 15 GeV was considered unsuitable for the study of QCD in the transition from the scaling region to scales of chiral symmetry breaking and confinement. This and the high cost of the stand-alone concept led to stagnation during the early nineties.

Later the idea of using the HERA electron ring as a stretcher for the planned pulsed TESLA electron beam was born. This scheme would provide a cw-beam of about 30 GeV with a 30 μA beam current. The feasability was studied in some detail and the study entered as an appendix in the TESLA project report [4] [5]. Figure 1 shows the principle. The electron beam is taken out of the south linac after being accelerated to 27 GeV and then fed back to the HERA ring. This is similar to the operation of a planned SASE free electron laser, which also will be fed by bunches from the TESLA south arm after acceleration to about 50 GeV.

An update, pointing out some technical details and a cost estimate has been worked out recently [3] and has been submitted to be included in the TESLA technical design report.

Expected performances of	ELFE@DESY
Energy range	15-27 GeV
Maximum current	30 μA
Duty-factor	88 %
Bunch spacing (433.33 MHz)	2.3 ns
Horizontal emittance (90% of particles)	4 mmμrad at 15 GeV
	12 mmμrad at 25 GeV
Energy spread (FWHM)	$1.2\,10^{-3}$ ($\sigma = 0.51\,10^{-3}$) at 15 GeV
	$2.2\,10^{-3}$ ($\sigma = 0.93\,10^{-3}$) at 25 GeV

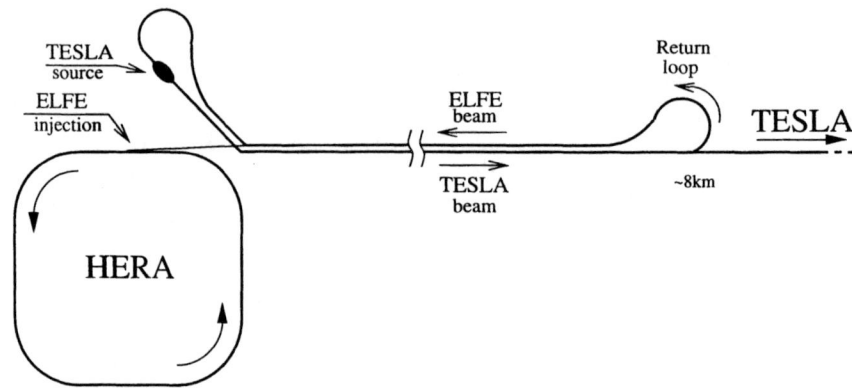

FIGURE 1. Scheme of the ELFE at DESY option with the HERA electron ring used as a stretcher for the 10 Hz TESLA pulses

Two facts are important: Obviously the ELFE@DESY is linked to the realisation of the TESLA project or at least a part of it. The TESLA project will be reviewed together with other large science projects in the summer of 2001.

The use of the HERA electron ring as a stretcher will require some modifications of its lattice which are not compatible with an operation in $\vec{e}\vec{p}$ or eA mode.

In the end of 2000 LEP ceased operation after a dramatic discussion about the significance of a number of events compatible with a Higgs particle of a mass of about 114 GeV. It was proposed to reuse the cavities and the radiofrequency equipment to built a 25 GeV cw electron accelerator. The project was studied by a CERN group of accelerator physicists. A design report was published [1]. The machine shown in figure 2 is of the recirculating linac design with 7 passages having an extension of roughly 2.0 x 0.5 km².

ELFE at CERN performance parameters	
Top energy	25 GeV
Beam current on target	100 μA
Beam power on target	2.5 MW
Injection energy	0.8 GeV
Number of passes	7
Energy gain per pass	3.5 GeV
Relative r.m.s. momentum spread at 25 GeV	$\leq 10^{-3}$
Emittance at 25 GeV	≤ 30 nm
Bunch repetition time on target	2.8 ns

As shown in figure 3 it would fit on the CERN ground in the north area close to the present muon beam experiments.

The physics case of ELFE was presented at CERN in November 2000. It appears that there are competing plans for the use of LEP cavities for a neutrino factory.

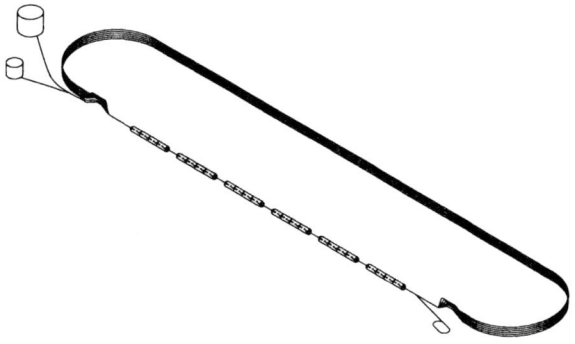

FIGURE 2. The ELFE at CERN design is a recirculating linac with 7 passages

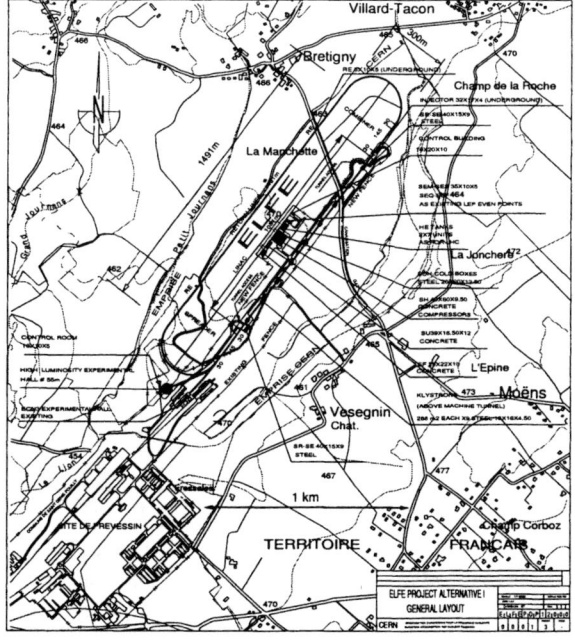

FIGURE 3. Alternative 1 for the location of ELFE in the CERN north area

1 The ELFE physics program

It is obvious that the two large high energy laboratories in Europe would have the means to realize an electron machine on their premises at a cost of about 125 MEuro(DESY) or 250 MEuro(CERN), taking advantage of the existing infrastructure and existing hardware. It is also obvious that an ELFE program would not represent the mainstream in high energy physics. In contrast the necessity to deepen the understanding of QCD in the nonperturbative regime is violently contested in the high energy community. Being financially (over)committed in the LHC and the TESLA project, additional money would be needed in both laboratories to include an extra nuclear physics program.

The nuclear physics community in Europe, represented by NUPECC, has lobbied primarilary for a radioactive beam facility. The fact that the initial ELFE physics program was not supported unanimously, even by the hadron structure community, weakened the case.

The GSI as one of the major nuclear physics laboratories is presently trying to include the sub-nuclear QCD aspect by favouring the spectroscopic approach. A combination of a radioactive beam facility and an antiproton storage ring similar to the super LEAR project is proposed.

Presently there are clear signs that the ELFE case is considerably strengthened. The realisation that the deeply virtual exclusive processes are important for the understanding of hadronic structure, and that an ELFE project is ideally suited to access them came in several steps in the last few years culminating in two workshops in April (Valencia) and September (Hamburg) 2000. A document has been produced [2], which spells out the physics case and the experimental consequences. It is obvious that ELFE now has a clear mission which is supported by a large and growing community of experimental and theoretical physicists. This goes with the clear understanding that not all interesting hadron structure aspects can be covered by a single machine. The higher energy option at $s \approx 30\,\text{GeV}$ is important for the smaller x region and a clear connection to the scaling region and even higher energies are required for the high gluon field strength physics in eA collisions.

C TESLA-N

The TESLA project [4] comprises two linear accelerators each of about 15 km length colliding electrons and positrons at an initial cm energy of s=250+250 GeV with an upgrade option of s=500+500 GeV. The pulse structure of 5 Hz for the e^+e^--collisions leaves 6.2 10^6 electron buckets per second with a spacing of 770 ps unused. The basic idea of TESLA-N is to fill the unused buckets of the north(positron) arm with polarized electrons and use them after magnetic separation from the main beam for a fixed target experiment. This operation corresponds to relatively low duty factor of 0.5 % which is essentially limited by the available cooling power. The luminosity for coincidence experiments is limited to about

$0.75 \; 10^{35} \text{cm}^{-2}\text{s}^{-1}$ if the fraction multiple hadronic events per bunch crossing the target has to be kept below 3%. In a recent report [6] the physics potential of such a setting has been explored. It was shown that for double polarization experiments one could exceed the sensitivities of the COMPASS experiment by a factor of 100 and in addition be less restricted in the detection of the hadronic final state. Taking the dilution of a DNP ammonia target, the electron polarization of $P_e = 0.8$ and the proton polarization of $P_p = 0.9$ into account the effective luminosity will be of the order $1.2 \; 10^{33} \text{cm}^{-2}s^{-1}$. This number is hard to beat with a collider, even if it is operating at full polarization.

The additional investments for TESLA-N would consist in the construction of a microtron for the injection of polarized electrons into the north arm, a separator stage and an underground experimental hall plus a beam dump.

A three stage forward dipole spectrometer has been designed to cover the Q^2 region from 0.5 GeV^2 to 100 GeV^2. The beam energy can be lowered from 250(500) GeV by either selecting a different phase of the r.f. or by separating the TESLA-N beam earlier from the TESLA positron beam.

It is conceivable that a part of the ELFE program could be realized with TESLA-N running at a lower energy but with an increased duty factor. The extra investment into additional cooling for the first 25-50 GeV to run at 5% duty factor are comparable to the cost estimate for ELFE@DESY. The first stage of the TESLA-N could incorporate the design of the ELFE detector to reach a missing mass resolution of 25-50 MeV.

It may happen that the realisation of TESLA will happen in several steps. A 50 GeV electron stage could serve for the construction of a free electron laser. TESLA-N at a lower energy or ELFE@DESY could already be implemented at this stage and be adapted later to run with the full system. If the ELFE physics could be covered by TESLA-N then there would be room to continue collider operation in HERA. Two projects have been intensely discussed over the last five years: HERA-$\vec{p}$ for the measurement of polarized structure functions and HERA-N for the investigation of eA collisions.

D Double Polarized HERA

The study of polarized structure functions at very high collision energies requires the injection of polarized protons, deuterons and ^{3}He into the HERA proton ring and means to preserve the polarization through the accelleration and during storage. A solution [7] has been worked out for this. The physics scope of HERA-$\vec{p}$ has been formulated in a series of three workshop proceedings [8]. The polarized gluon distribution of the nucleon could be measured with different methods (scale breaking of g_1, 1+two jet events) over a large x-region. A unique measurement would be the determination of the polarized structure function of the photon. This measurement would only need a moderate integrated luminosity of 50 pb^{-1} while the rest of the program would typically need 500 pb^{-1} which could be obtained by running several years (≈ 3) under optimum conditions with 70% polarization for both beams and a luminosity $10^{32} \text{cm}^{-2}\text{s}^{-1}$.

E   HERA-N

The discussions around a new phase of QCD at extremely high gluon densities have led to the project of colliding heavy ions like lead or uranium with electrons in HERA. This requires a heavy ion injector for the HERA proton ring. Obviously the energy of the ion beam per nucleon is lower by the charge to mass ratio of approximately 0.38 while the luminosity per nucleon is predicted to be similar to the ep luminosity.

The physics case has been studied in two workshops [9]. It was found that a qualitatively new behaviour of the structure functions and their scale dependence could be found already at moderate integrated luminosities of 10 pb^{-1} per nucleon. Also the expected attenuation of jets in the nuclear medium could be studied. Obviously these questions are related to the description of heavy ion collisions at LHC or RHIC. Depending on the invariant mass M of the typical processes, structure function information is needed around $x \approx \frac{M}{\sqrt{s_{NN}}}$, which at LHC means $x \approx 2 \cdot 10^{-4} (M = 1\, GeV) - 2 \cdot 10^{-2} (M = 100\, GeV)$. This would be well adapted to the kinematical range of HERA-N $x_{min} = \frac{Q^2}{s} \approx 2.5 \cdot 10^{-4} (Q^2 = 10\, GeV^2)$.

If the European heavy ion community would see a necessity to have eA data then HERA could be complemented with a heavy ion injector, similar to the SPS, in relatively short time.

F   The CERN Neutrino Factory

There is a world wide excitement about the mass and mixing scenarios in the neutrino sector. Muon collection and cooling facilities have been first discussed in the connection with muon colliders but they are now seen as the ideal sources for well focussed neutrino beams. The muon and neutrino beams of typically 30-50 GeV which are produced can also be used for the investigation of the hadron structure. Neutrinos are particularly interesting due to their highly selective couplings to spin and flavour of quarks. New parity violating spin structure functions would be accessible and of course a high accuracy study of the strange and charm sector. A working group [10] has been formed to study these options in the context of the CERN neutrino factory proposal [11].

It is not obvious whether the neutrinos or the intermediate muon beams can also be used to study exclusive reactions, which are now a main topic in the ELFE discussion. The neutrino beams have a wide energy spectrum which does not allow to employ missing mass techniques to ensure exclusiveness of charged current neutrino reactions.

IV SUMMARY

In this talk I have described a large variety of existing experimental facilities in Europe investigating the hadron structure problem utilizing lepton beams. There is a strong but diverse community which, in close contact with the international community, is seeking to define the next steps in the development of their science.

A particular situation in Europe is given by the fact that the relevant experimental facilities are in the hands of large laboratories which see their mission primarily at the high energy frontier. Hadron structure physics experiments are nevertheless accepted and supported albeit within limits defined by the primary goals of the large laboratories.

A dedicated investment into hadron structure physics of several 100MEuros therefore is necessary to develop at least one out of the list of options which have been worked out so far.

The definition of a clear and well focussed program is necessary to gather the largest part of the existing community behind one project. The progress which has been made in this direction is considerable.

I want to thank all my friends hidden in the "et al." who have helped by their initative to promote the discussions for the development of hadron structure physics.

REFERENCES

[1] K. Aulenbacher et al., CERN Report 99-10, December 1999

[2] M. Anselmino et al., NuPECC Report (*in press*);
http://www-daphnia.cea.fr/Sphn/Elfe/Report/ELFE_phys.pdf

[3] E. de Sanctis et al., ELFE (*contribution to the TESLA TDR, unpublished*) December 2000

[4] R. Brinkmann, DESY Internal Report DESY-TESLA-95-14.

[5] R.Brinkmann et al. Nucl. Phys. A622 (1997) 187c

[6] M. Anselmino et al., DESY 00-160, hep-ph/0011299

[7] SPIN Collaboration and DESY Polarization Team, UMHE 99-05

[8] Proceedings of Workshop "Future Physics at HERA 1995/1996 eds. G. Ingelmann, A. DeRoeck and R. Klanner, DESY (1996)
Physics with Polarized Protons at HERA, eds. A. DeRoeck and T. Gehrmann, DESY-Proc. 1998-01
"Polarized Protons at High Energy - Accelerator Challenges and Physics Opprortunities", eds. D. Barber, A. DeRoeck, V. Hughes and F. Willeke, DESY-Proc. 1999-,May 1999

[9] Procedings of Workshop "Future Physics at HERA 1995/1996", eds. G. Ingelmann, A. DeRoeck and R. Klanner, DESY (1996), www.desy.de/~heraws96
Workshop on "Physics with HERA as eA collider", 25-26 May 1999, DESY Hamburg, www.desy.de/heraea

[10] Working Group "Neutrino DIS", convener: M. Mangano, , mlm.home.cern.ch/mlm/mucoll/nudis.html

[11] "Prospective Study of Muon Storage Rings at CERN", eds. B. Autin, A. Blondel, J. Ellis, CERN Report 99-02 (1999)
"High Intesity Proton Source and Neutrino Factory", CERN/SPS/779 (2000)
[12] Slide- Report of the Joint DESY/GSI/NuPECC Workshop, Seeheim, March 3/4, 1997, GSI REPORT 97-04
[13] "Conceptual Design Study of the GSI Electron-Nucleon Collider", Budker Institute of Nuclear Physics, GSI-Report 97-07 (1997)

SPIN AND FLAVOR STRUCTURE
OF THE NUCLEON

Semi-inclusive Studies of the Quark Spin and Flavor Structure of the Proton

U. Stösslein and E.R. Kinney
University of Colorado, Boulder, CO 80309-0390, USA

Abstract. Initial semi-inclusive studies using a polarized electron-proton collider facility, 5 GeV on 50 GeV, are performed. The kinematic range $0.001 < x < 0.7$ and $1 < Q^2 < 100$ GeV^2 is explored based on an assumed integrated luminosity of 1 fb^{-1}, and assuming 70% polarization for each of the beams. The access of the flavor separated quark contributions to the nucleon spin seems to be feasible with good precision over a wide kinematic range, in particular a first determination of the poorly known strange quark polarization at $x < 0.05$ would be possible with such a machine.

Precise determination of the quark spin and flavor structure of the nucleon remains one of the central goals in the study of the quark-gluon dynamics. While the structure is becoming well determined in the range of intermediate Bjorken x where valence quarks predominate, there is still large uncertainty in the structure of the virtual sea at low x as well as in the valence structure at high x. Inclusive lepton scattering alone is unable to resolve the details of this structure as the charge weighted sum of all the quark distributions is measured, rather than individual flavor. Significant progress can be made by using semi-inclusive scattering in which hadrons produced in a photon-quark reaction are detected in coincidence with the scattered lepton. Knowledge of the identity of these hadrons and their kinematic correlation with the momentum and energy of the virtual photon allow one to separate the contributions from the different quark flavors. Combined with the use of polarized targets and beams, one learns the spin contribution of the individual flavors as well. The spin contribution of the strange quarks is especially important as knowledge of their role in nucleon structure has been one of the most highly sought pieces of information in recent years. This relatively tiny component of the nucleon wavefunction appears to have a much more significant role than expected in determining the properties of the nucleon, such as spin.

Fixed target experiments suffer from the fact that the so-called current hadrons are produced at forward angles in the lab frame simply due to the Lorentz boost of the beam, which is difficult to instrument adequately, especially as one tries to increase the luminosity to gain significant statistical accuracy. In addition, almost all of the fragments of the remaining target nucleon are lost at small angles and energies. Correlation of these target fragments with the directly produced hadrons would likely enhance the power of the semi-inclusive technique, as the dilution due to hadron production in the fragmentation process would be lessened.

A polarized ion-electron collider such as the proposed EIC would be an ideal facility for semi-inclusive studies. The collider kinematics would open up the final state into a large solid angle in the lab, allowing much more complete identification of the hadronic final state, both in the current and target kinematic regions of fragmentation phase space, and at the same time allow high luminosity operation.

Initial studies of spin-flavor structure have focused on semi-inclusive asymmetries measured on the proton only; deuteron beams will allow study of the neutron, which typically increases the sensitive to the down valence quarks. The simulation used events generated by the standard deep inelastic generator LEPTO [1] for collisions of 5 GeV electrons on 50 GeV protons, here parton distribution functions were taken from [2]. Hadronization is performed using the LUND string model, implemented in JETSET. Inclusive and semi-inclusive cross sections as well as the flavor tagging probabilities (co-called purities) were determined from the reaction products where a specific leading hadron ($z = E_{hadron}/E_{photon}$, $0 < z < 1$, and $x_F = 2p_{h\|}/W$, $0 < x_F < 1$) is detected and identified (perfectly) in coincidence with the scattered lepton. Lepton and leading hadron are required to have polar angles greater than $5°$ from the axis of the colliding beams, and momenta greater than 1 GeV; no other assumptions are made about the detector capabilities. For the sake of concreteness an integrated luminosity of 1 fb^{-1} has been used to determine the size of the simulation sample. For the statistical precision of the polarized sample for each of the beams a polarization of 70% was taken into account.

The inclusive and semi-inclusive asymmetries were combined from the simulated flavor tagging probabilities and polarized parton distributions [5]. The asymmetries have been analyzed using the purity method developed at the SMC [3] and HERMES [4] experiments. This method relates the set of measured asymmetries to the polarization of flavor separated quark distributions via a matrix of purities derived from knowledge of the unpolarized quark distributions and the fragmentation functions which describe the hadronization of a particular flavor quark into a specific type of hadron. We assume here that these functions are known. A leading order formalism in perturbative QCD is used throughout.

Two different analyses have been performed. In the first, more conservative case, the four hadron asymmetries from both charge states of pions and kaons are used are used to derive quark polarizations for up and down distributions, the polarization of the up and down (light anti-quark) sea, and the polarization of the strange sea quarks, see in Ref. [6] for further details. The results are shown in Fig. 1, in which only statistical uncertainties are displayed by the error bars.

In the second analysis, it is assumed that the up and down quark distributions are known sufficiently well, that one make take them as given, and directly determine the strange quark distribution from any of the specific hadron asymmetries, as all of them depend on the strange quark distribution. A sample of the results on the polarized strange distribution extracted from the K^- asymmetry is shown in Fig. 2 in which the asymmetry is also displayed. As in the previous figure, only statistical

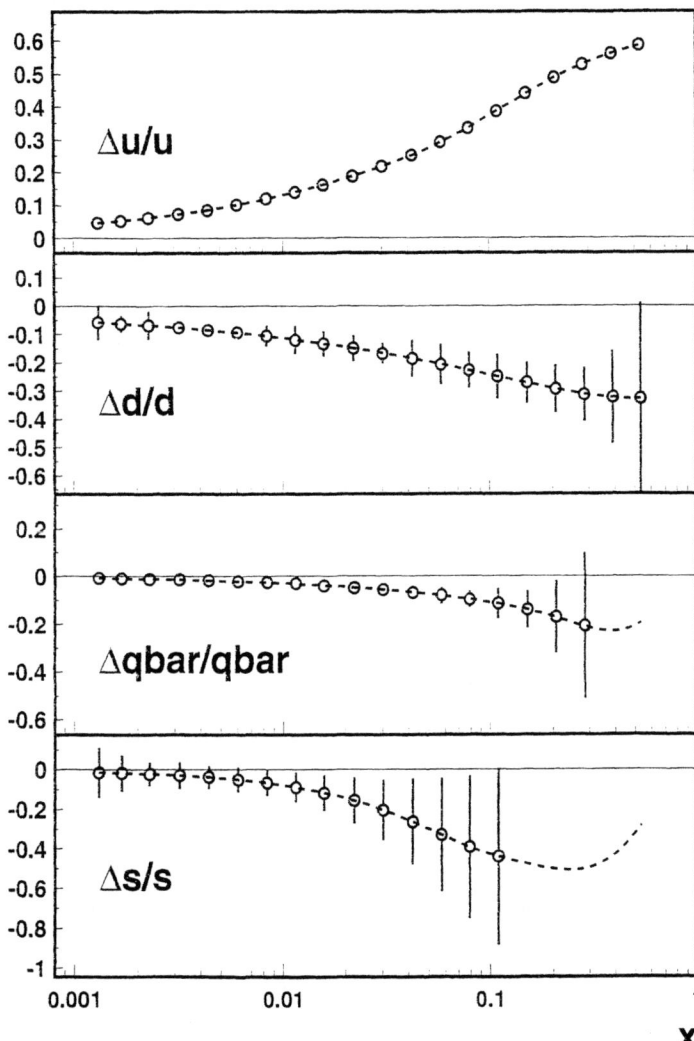

FIGURE 1. Expected statistical precision of the quark polarizations for up, down, the light anti-quarks and the strange quarks, using polarized [5] and unpolarized [2] parton distribution functions. Here the four charged pion and Kaon asymmetries were chosen as input. The measured average Q^2 values per x bin are not shown here and are in the range of 1.1 GeV^2 at lowest x to 40 GeV^2 at highest x.

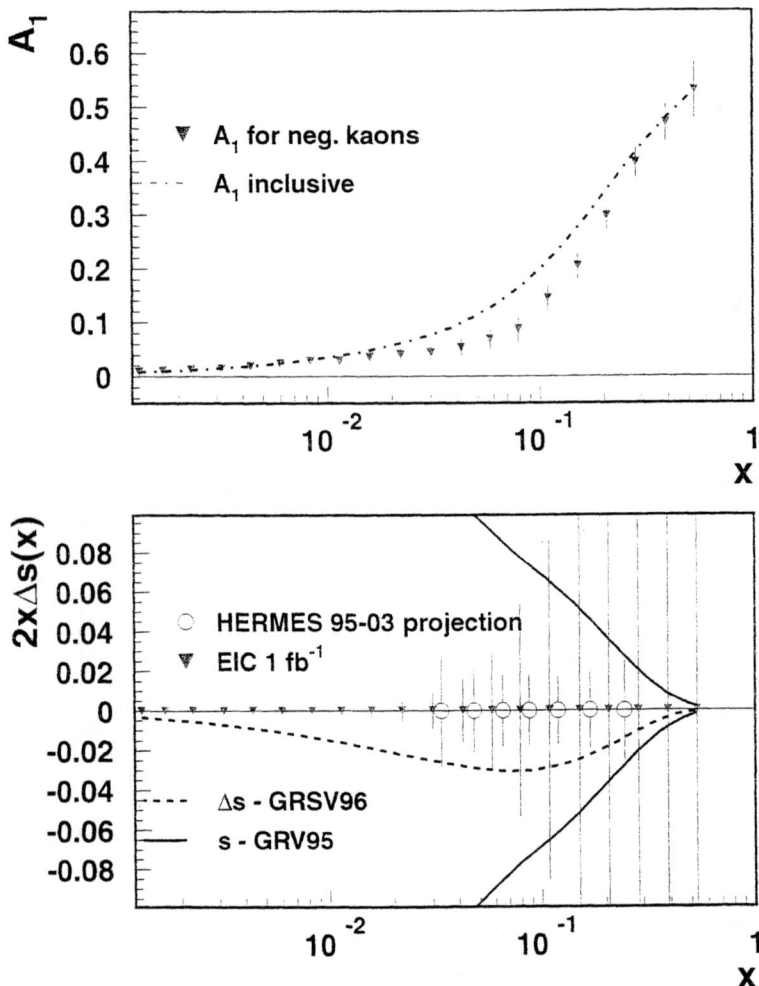

FIGURE 2. The upper plot shows the simulated K^- asymmetry ($p > 1$ GeV) at measured Q^2 values (not shown). The lower plot shows the expected statistical precision of the strange quark distribution for an EIC simulation in comparison to the projected result of a HERMES analysis [7]. The poitivity constraint given by the unpolarized strange quark distribution [2] is also plotted.

uncertainties are indicated. The results are compared with the precision expected from currently planned measurements are the HERMES experiment.

Both of the figures show that the use of standard analysis techniques on semi-inclusive data will yield a much more precise determination of the nucleon spin-flavor structure, especially when combined with the results expected from RHIC, CERN and DESY in the next five years. If one assumes that advances are made in the detailed understanding and description of the fragmentation process, and in making such analyses consistently in next-to-leading order, then it is likely that even more precision will be obtained as well as a deeper understanding of the nucleon structure.

REFERENCES

1. Information and code via WWW page http://www3.tsl.uu.se/thep/lepto/ .
2. M. Glück, E. Reya, and A. Vogt, Z. Phys. **C 67** (1995) 433.
3. SMC, B. Adeva *et al.*, Phys. Lett. **B369** (1996) 93; Phys. Lett. **B420** (1998) 180.
4. HERMES Collab., K. Ackerstaff et al, Phys. Lett. **B464** (1999) 123.
5. M. Glück, E. Reya, M. Stratmann, and W. Vogelsang, Phys. Rev. **D 53** (1996) 4775.
6. H. Kobayashi, PhD thesis, Tokyo Institute of Technology (2000).
7. HERMES Collaboration, "The HERMES Physics Program and Plans for 2001–2006," HERMES Report No. 00-003 (2000).

Establishing Evidences of Factorization in Semi-Inclusive Electron Scattering

Xiaodong Jiang

*Department of Physics and Astronomy, Rutgers University,
Piscataway, New Jersey 08855, USA*

Abstract. The issue of factorization in semi-inclusive deep-inelastic electron scattering is discussed. Existing evidences of factorization are briefly reviewed. We argue that by measuring yield ratios in $(e, e'\pi^+)$ and $(e, e'\pi^-)$ reactions on hydrogen and deuterium, factorization can be demonstrated independent of the knowledge of fragmentation functions. The implications for the design of an e-A collider are discussed.

THE ASSUMPTION OF INDEPENDENT FACTORIZATION

In the naive quark-parton model, at high enough energy transfer, the so-called Independent Factorization is assumed between the processes of quark scattering and quark fragmentation in semi-inclusive lepton scattering. This assumption implies:

$$Y^h(x, z) \propto \sum_i e_i^2 \left[q_i(x) D_{q_i}^h(z) + \bar{q}_i(x) D_{\bar{q}_i}^h(z) \right], \tag{1}$$

where $Y^h(x, z)$ is the semi-inclusive yield of hadron h, $z = E^\pi/\nu$ is the fraction of the energy transfer carried by the hadron h, e_i is the quark charge, $q_i(x)$ and $\bar{q}_i(x)$ are the quark and anti-quark distributions of flavor i. The fragmentation functions $D_{q_i}^h(z)$ represent the probability that a quark of flavor i fragments into a hadron h. For simplicity, we drop the notation of Q^2 on quark distributions and fragmentation functions.

The Independent Factorization assumption is a rather strong statement about the characteristic of quark confinement in nucleon. It defines an energy scale above which the fragmentation of the struck quark no longer entangled with its environment. The exact energy scale of the onset of the factorization could only be established from experimental data. Factorization assumption was widely accepted in high energy muon scattering at 100 GeV energy scale. However, in electron scattering, since high quality semi-inclusive data was limited to below 30 GeV, evidence of factorization became an issue of concern. Although factorization is assumed in recent analysis of HERMES data, the validity of this assumption demands a fair scrutiny.

EXISTING EVIDENCES OF FACTORIZATION

Three types of method have been suggested to provide evidences of factorization in semi-inclusive electron scattering:

1. Measure d_v/u_v through $(e, e'\pi^+)$ and $(e, e'\pi^-)$ reaction to compare with data from neutrino scattering.

2. Extract fragmentation functions and demonstrate that they are universal functions.

3. Measure ratio of $(\bar{d} - \bar{u})/(u - d)$ to demonstrate it is independent of z.

By measuring the yield ratios of $(e, e'\pi^+)$ and $(e, e'\pi^-)$ on hydrogen and deuterium, if factorization applies, one can extract the ratio of up and down valence quark distributions to be compared with data from neutrino scattering. Although preliminary d_v/u_v results of HERMES measurements (with positron energy of 27.5 GeV) were shown to be consistent neutrino data [1], no published results are available at the moment.

It was also shown in HERMES data [2] that the extracted fragmentation functions $(D_u^{\pi^+} + D_u^{\pi^-})$ agree reasonably well with parameterization obtained from e^+e^- annihilation. By presenting the fragmentation functions as universal functions in different processes, one hopes to demonstrate the quark fragmentation does not related to the quark distributions. In real measurements, however, hadrons from the quark fragmentation may not all be covered and sizable acceptance corrections have to be introduced. In addition, certain types of fragmentation model have to be introduced in simulations. Furthermore, the fragmentation functions in e^+e^- annihilation are usually described in fits of 13-parameter of which the systematic uncertainties are unknown.

The third method [3], within the uncertainties of the HERMES data [4], clearly demonstrated a z-independent behavior of the ratio:

$$\frac{\bar{d}(x) - \bar{u}(x)}{u(x) - d(x)} = \frac{J(z)[1 - r(x,z)] - [1 + r(x,z)]}{J(z)[1 - r(x,z)] + [1 + r(x,z)]}, \qquad (2)$$

in which $J(z) = \frac{3}{5}\left(\frac{1+D'(z)}{1-D'(z)}\right)$ with $D'(z) = D_u^{\pi^-}/D_u^{\pi^+}$, and

$$r(x,z) = \frac{Y_p^{\pi^-}(x,z) - Y_n^{\pi^-}(x,z)}{Y_p^{\pi^+}(x,z) - Y_n^{\pi^+}(x,z)}. \qquad (3)$$

It should be noted that the HERMES measurements were limited by systematical uncertainties and the knowledge of fragmentation functions was also required.

ESTABLISHING EVIDENCES OF FACTORIZATION

We point out that evidences of factorization can be established without requiring any knowledge of fragmentation functions [5]. By measuring yield ratios in $(e, e'\pi^+)$ and $(e, e'\pi^-)$ reactions on hydrogen and deuterium, fragmentation functions can be canceled out if one acknowledges isospin symmetry and charge conjugation. Futhermore, these ratios can be demonstrated to be sensitive to the sea quark distributions.

Assuming isospin symmetry and charge conjugation, the number of light quark fragmentation functions is reduced to two type: the favored (D^+) and the unfavored (D^-) fragmentation functions:

$$D^+ \equiv D_u^{\pi^+} = D_d^{\pi^-} = D_{\bar{u}}^{\pi^-} = D_{\bar{d}}^{\pi^+}, \tag{4}$$
$$D^- \equiv D_u^{\pi^-} = D_d^{\pi^+} = D_{\bar{u}}^{\pi^+} = D_{\bar{d}}^{\pi^-}.$$

Neglecting heavy quark contributions, with the Independent Factorization assumption, the yield of π^+ in deep inelastic scattering on proton and neutron can be expressed as:

$$Y_p^{\pi^+}(x, z) = A \left(4u(x)D^+(z) + d(x)D^-(z) + 4\bar{u}(x)D^-(z) + \bar{d}(x)D^+(z) \right), \tag{5}$$
$$Y_n^{\pi^+}(x, z) = A \left(4d(x)D^+(z) + u(x)D^-(z) + 4\bar{d}(x)D^-(z) + \bar{u}(x)D^+(z) \right),$$

where A is a common kinematic factor.

If four independent yields $(Y_p^{\pi^+}, Y_p^{\pi^-}, Y_n^{\pi^+}$ and $Y_n^{\pi^-})$ are measured, ratios can be formed in which the fragmentation functions cancel each other:

$$t_1(x) = \frac{Y_p^{\pi^+}(x, z) + Y_p^{\pi^-}(x, z)}{Y_n^{\pi^+}(x, z) + Y_n^{\pi^-}(x, z)} = \frac{4u(x) + d(x) + 4\bar{u}(x) + \bar{d}(x)}{4d(x) + u(x) + 4\bar{d}(x) + \bar{u}(x)}, \tag{6}$$

$$t_2(x) = \frac{Y_p^{\pi^+}(x, z) - Y_p^{\pi^-}(x, z)}{Y_n^{\pi^+}(x, z) - Y_n^{\pi^-}(x, z)} = \frac{4u(x) - d(x) - 4\bar{u}(x) + \bar{d}(x)}{4d(x) - u(x) - 4\bar{d}(x) + \bar{u}(x)}. \tag{7}$$

Or simply as:

$$r_1(x) = \frac{4 - t_1(x)}{4t_1(x) - 1} = \frac{d(x) + \bar{d}(x)}{u(x) + \bar{u}(x)}, \tag{8}$$

and

$$r_2(x) = \frac{4 + t_2(x)}{4t_2(x) + 1} = \frac{d(x) - \bar{d}(x)}{u(x) - \bar{u}(x)}. \tag{9}$$

A clear signature of factorization will be the apparent scaling of r_1 and r_2 with respect to different values of z.

Precise measurements of r_1 and r_2 at different x and Q^2 will provide strong and independent constraints on the quark distribution functions. While r_1 is the valance quark distribution ratio, r_2 is sensitive to the sea quark distributions. As an example, the values of r_1 and r_2 at Q^2=2.23 GeV^2 are plotted in Fig. 1 from CTEQ5M [6] predictions. The sensitivity of r_1 and r_2 to different values of $\bar{d}/\bar{u}$ ratios are also shown while the value of $\bar{d} + \bar{u}$ is fixed by CTEQ5M.

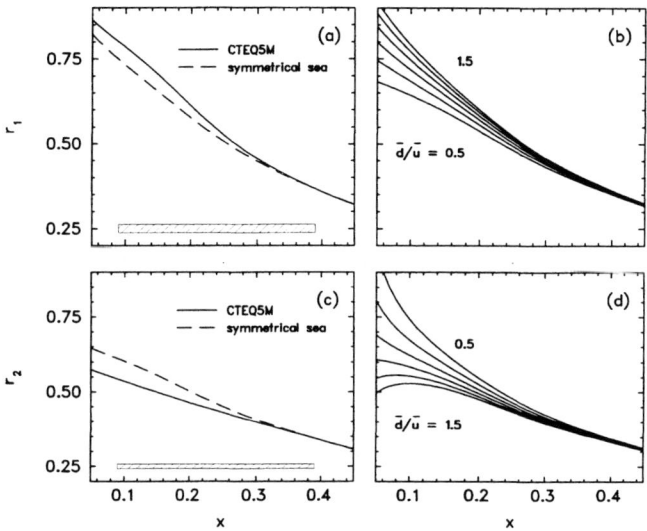

FIGURE 1. The ratio of r_1 and r_2 at Q^2=2.23 GeV^2 from CTEQ5M predictions. The dashed lines represent the value when $\bar{d} - \bar{u} = 0$ is forced while $\bar{d} + \bar{u}$ is fixed by CTEQ5M. The shaded area illustrates the sensitivity corresponding to a measurement of 1.25 % (0.5 %) error on r_1 (r_2). In plots (b) and (d) six different $\bar{d}/\bar{u}$ values uniformly spread from 0.5 to 1.5 are assumed in the CTEQ5M predictions.

Neutron yield can be extracted from the difference of deuteron and hydrogen yield, the nuclear binding effects and corrections due to Fermi motion are expected to be small in the region of $x < 0.4$ [7]. With $Y_D^{\pi^\pm} = Y_n^{\pi^\pm} + Y_p^{\pi^\pm}$, r_1 and r_2 become:

$$r_1 = -1 + \frac{3}{5\rho_1 - 1}, \qquad r_2 = -1 + \frac{5}{3\rho_2 + 1},$$

where

$$\rho_1 = \frac{Y_p^{\pi^+} + Y_p^{\pi^-}}{Y_D^{\pi^+} + Y_D^{\pi^-}}, \qquad \rho_2 = \frac{Y_p^{\pi^+} - Y_p^{\pi^-}}{Y_D^{\pi^+} - Y_D^{\pi^-}},$$

can be measured directly from experiment.

CONSIDERATIONS IN EPIC

Assuming EPIC is a 5 GeV/c (e) + 50 GeV/c (p) collider with luminosity of 10^{33} $cm^{-2}s^{-1}$. With the same detector setting and the same magnet setting, the following arrangements appear to be attractive: 5.0 GeV/c (e) + 25.0 GeV/c (D) and 5.0 GeV/c (e) + 12.5 GeV/c (p). Since the eN systems have the same center of mass energy in both cases, the detector phase space will be the same in ep and eD reaction which would minimize detector related systematic uncertainties. The estimated inclusive $p(e, e')$ cross sections are listed in Table-1 for selected kinematics. An inclusive cross section corresponding to 1.0 nb/GeV^2 in the table will only require a 100 hour run to reach 0.5% statistical accuracy within a bin size of $\Delta x = 0.1$, $\Delta Q^2 = 1.0$ GeV^2. Clearly, statistical accuracy is not a concern in such measurements.

TABLE 1. Typical inclusive cross sections of $p(e, e')$.

E' GeV	Q^2 GeV^2	x	$\frac{d^2\sigma}{dxdQ^2}$ nb/GeV^2
1.0	2.22	0.010	1070
	3.20	0.015	387
	4.40	0.020	163
2.0	4.45	0.028	130
	8.81	0.051	17
	14.10	0.076	4
3.0	13.22	0.100	4
	17.08	0.120	1.8
4.0	17.63	0.187	1.0

The most difficult issue in such a yield ratio measurement is to determine the relative luminosity ratio in ep/eD to better than 1.0% level in a collider experiment. Very forward angle tagging scheme can be implemented as a relative luminosity monitor. Flexibility of the machine to switch between ep mode and eD mode will be necessary.

CONCLUSION

We suggest that evidence of factorization can be established by taking yield ratio measurements in $(e, e'\pi^+)$ and $(e, e'\pi^-)$ reactions on hydrogen and deuterium. A clear signature of factorization will be the apparent scaling of the ratios r_1 and r_2 respect to different values of z. Precise measurements of r_1 and r_2 will provide strong and independent constraints on the valence as well as the sea quark distributions. Although similar measurements can be done with a fixed target at

Jefferson Lab with 6-12 GeV electron beam, the proposed EPIC collider will provide a much higher center-of-mass energy and a much wider kinematic coverage for such measurements.

REFERENCES

1. J. E. Belz, HERMES Collar., Proceedings of the Symposium of Meson and Nucleon Interactions, Vancouver, Canada, August 1997.
2. P. Geiger, Ph.D. thesis, Ruprecht-Karls-University, Heidelberg, 1998
3. J. Levelt *et al.*, *Phys. Lett.* B **263**, 498 (1991),
4. K. Ackerstaff *et al.*, *Phys. Lett.* B **464**, 123 (1999),
5. X. Jiang and R. Ransome, Jefferson Lab proposal P00-115.
6. H. L. Lai, *et al.*, *Phys. Rev.* D **55**, 1280 (1997).
7. W. Melnitchouk, *et al.*, *Phys. Lett.* B **435**, 420 (1998).

(Vector) Meson Production and Duality

R. Ent

Jefferson Lab, 12000 Jefferson Avenue, Newport News, VA 23606

Abstract. At high enough energies, hadronic cross sections, if averaged over an appropriate energy range, must coincide with a perturbative QCD description. One famous example in deep inelastic scattering is termed Bloom-Gilman duality. This quark-hadron duality shows that the nucleon resonance region closely mimics the deep inelastic region where we assume single quark scattering to be dominant. This Bloom-Gilman duality was recently found to work to high precision to far lower momentum transfers, and far smaller regions in invariant mass, than anticipated. Implications for using the spin/flavor selectivity of (polarized) electron-proton scattering and/or (polarized) meson electroproduction to examine such duality in more detail are discussed.

I INTRODUCTION

Three decades ago, Bloom and Gilman observed a fascinating correspondence between the resonance electroproduction and deep inelastic kinematic regimes of inclusive electron-nucleon scattering [1,2]. Specifically, it was observed that the resonance strength could be related to the deep inelastic strength via a scaling variable which allowed the comparison of the lower missing mass squared, W^2, and lower four-momentum squared, Q^2, resonance region data to the higher W^2 and Q^2 deep inelastic data. Further, this behavior was observed over a range in Q^2 and W^2, and it was found that, with changing Q^2, the resonances move along, but always average to, the smooth scaling curve typically associated with deep inelastic scattering. This behavior clearly hinted at a common origin for resonance (hadron) electroproduction and deep inelastic (partonic) scattering, termed quark-hadron, or Bloom-Gilman, duality.

A global kind of quark-hadron duality is well established: low-energy resonance production can be shown to be related to the high-energy behavior of hadron-hadron scattering [3]; the familiar ratio of $e^+e^- \to$ hadrons over $e^+e^- \to$ muons uses duality to relate the hadron production to the sum of the squared charges of the quarks: here duality is guaranteed by unitarity [4]; in Perturbative QCD (PQCD) the high-momentum transfer behavior of nucleon resonances can be related to the high-energy transfer behavior of deep inelastic scattering [4,5]. Poggio, Quin, and Weinberg [6] suggested that inclusive hadronic cross sections at high energies, averaged over an appropriate energy range, had to approximately coincide with a quark-gluon perturbation theory. However, in general, it is not clear why duality should also work in a localized region, and even at relatively low momentum transfers.

II INCLUSIVE SCATTERING

Inclusive deep inelastic scattering on nucleons is a firmly-established tool for the investigation of the quark-parton model. At large enough values of W^2 ($= M^2 + Q^2(1/x - 1)$, with M the proton mass and x the Bjorken scaling variable) and Q^2, QCD provides a rigorous description of the physics that generates the Q^2 behavior of the nucleon structure function $F_2 = \nu W_2$. The well-known logarithmic scaling violations in this structure function, predicted by asymptotic freedom, played a crucial role in establishing QCD as the accepted theory of strong interactions [7,8].

An analysis of the resonance region at smaller W^2 and Q^2 in terms of QCD was presented in Refs. [9,10], where Bloom and Gilman's duality was re-interpreted, and the integrals of the average scaling curves were equated to the $n = 2$ moment of the F_2 structure function. The Q^2 dependence of these moments can be described by ordering the contributing matrix elements according to their twist (= dimension - spin) in powers of $1/Q^2$. It was concluded [9] that the fall of the resonances along a smooth scaling curve with increasing Q^2 was attributed to the fact that there exist only small changes in these lower moments of the F_2 structure function due to higher twist effects. Such effects are inversely proportional to Q^2, and can therefore be large at small Q^2. If not, averages of the F_2 structure function over a sufficient range in x at moderate and high Q^2 are approximately the same.

Recently, high precision data on the F_2 structure function from Jefferson Lab [11] have quantified these earlier observations, and demonstrated that duality works to better than 10% for both the total nucleon resonance region and each of the separate low-lying nucleon resonance regions, for $Q^2 \geq 0.5$ (GeV/c)^2. This is illustrated in Fig. 1, where the nucleon resonance data for various Q^2 is compared to parameterizations of deep inelastic scattering data at constant $Q^2 = 5$ and 10 (GeV/c)^2 [12]. Such behavior shows that the distinction between the nucleon resonance region and the deep inelastic region is spurious; if properly averaged, the nucleon resonance regions closely mimic the deep inelastic region.

Even more surprising, it was found that the nucleon resonance region data *at all* Q^2 oscillate around a single smooth curve, as shown in Fig. 2. This curve coincides with the deep inelastic scaling curve at $Q^2 > 0.5$ (GeV/c)^2, consistent with Bloom-Gilman duality, and resembles neutrino/anti-neutrino xF_3 data or a valence-like sensitivity only [13] below $Q^2 \approx 0.5$ (GeV/c)^2. This is perhaps not too surprising in the quark model where the nucleon resonances act as valence quark transitions, while at low Q^2 not many sea quarks can be "seen" yet. However, it is surprising that all the strongly-interacting nucleon resonances shuffle their strength around as function of Q^2 to closely follow a single scaling curve. Do we see duality down to the lowest values of Q^2?

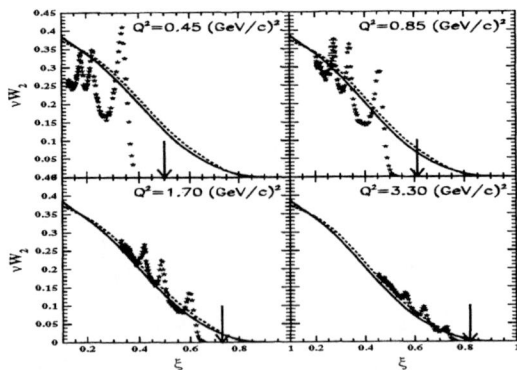

FIGURE 1. Sample hydrogen νW_2 structure function spectra obtained at $Q^2 = 0.45, 0.85, 1.70$, and 3.30 $(\text{GeV/c})^2$ and plotted as a function of the Nachtmann scaling variable ξ [14]. Arrows indicate elastic kinematics. The solid (dashed) line represents the NMC fit of deep inelastic structure function data at $Q^2 = 10$ $(\text{GeV/c})^2$ ($Q^2 = 5$ $(\text{GeV/c})^2$).

Future investigations would fully utilize the spin/flavor selectivity of the electron-proton reaction. E.g., it is well known that, at low Q^2, the $N - \Delta$ quark model transition contributes negatively to the polarized structure function g_1. Does duality only work if one averages over a large enough invariant mass region to provide positive definite results? Do the polarized structure functions exhibit a valence-like sensitivity too, similar as the F_2 unpolarized structure function? These questions remain completely unresolved to date.

III SEMI-INCLUSIVE SCATTERING

Duality in the case of semi-inclusive meson photo- and electroproduction has not been experimentally tested. Here duality would manifest itself with an observed scaling in the meson plus resonance final state [15].

Assuming one is in a kinematics region that mimics single-quark scattering, in analogy with the inclusive scattering case, the question here is whether the remaining part of the process can be described in terms of a process where the struck quark hadronizes into the detected meson. Such a factorization approach is strictly valid at asymptotic energies only, as at low energies there may not be clear separation of target and current fragmentation regions [16]. To what extent factorization applies at lower energy is an open question, although recent data supports factorization at lower energies than previously anticipated [17].

However, as in the inclusive case where the nucleon resonances average at low energies to the scaling curve, the nucleon resonances remaining in the final state after having produced a fast meson may also average to the fragmentation function. If duality holds for semi-inclusive scattering, the overall scale of scattering

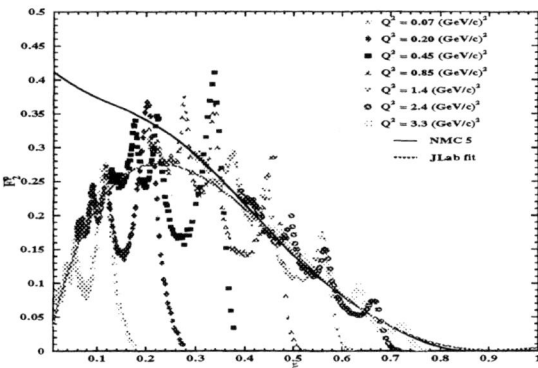

FIGURE 2. Sample hydrogen νW_2 structure function spectra obtained at various Q^2 and plotted as a function of the Nachtmann scaling variable ξ. The data at $Q^2 = 0.07$ and 0.20 (GeV/c)^2 are from older SLAC expriments. The solid line represents the NMC fit of deep inelastic structure function data at $Q^2 = 5$ (GeV/c)^2. The light solid line represents a fit of the various nucleon resonance spectra.

in the low-W' region must mirror that at high W', with W' the invariant mass of the residual system after the meson has been removed. This may come about *if* the various decay channels from resonances with varying W' interfere such that factorization holds. Obviously, this would require a non-trivial interference of these decay channels but this may not be unlikely if one realizes that duality has also been observed in hadronic τ decays [18].

In practice, one can extract the meson yield $\frac{dN^m}{dz}$ over a range of elasticity z (the fraction of total energy transfer ending up in the fast meson) at several values of x and Q^2. This allows the comparison of $\frac{dN^m}{dz}$ in the resonance region to that in the deep inelastic regime, which we obtain from the quark model or from parameterizations of data. Sparse information from both older Cornell data and recent JLab data strongly suggests that of order 10 GeV beam energy will provide the right kinematics region to study the onset of the duality phenomenon in meson electroproduction [19,20].

If so, one could perform a systematic study of duality for current fragments and target fragments with an energy of order 200 GeV. At such energies, a clear separation between current and target fragmentation regions exists [16], enabling a precise investigation of duality in the current fragmentation region with various meson or baryon tags, in addition to an investigation of duality in the target fragmentation region. In order to shed light on the transition from strongly-interacting matter to Perturbative QCD, one must understand the origin of duality. To accomplish this, it must be determined what energy region one has to average over, and what energies one needs to reach, such that hadronic processes equal a perturbative quark-gluon theory.

If one understands duality it may be used as a tool. It can give guidance to cuts typically used to select "hard scattering" regions. E.g., it shows that the $W^2 > 4$ cut to select deep inelastic events is spurious. One can access the very large x region, where, without escape, one encounters the nucleon resonance region. This could provide us with data for parton distribution functions in the strict valence region, and allow investigations of the Q^2 evolution of large-x parton distribution functions. One can utilize duality for a complete spin/flavor and valence-sea decomposition of parton distribution functions. Furthermore, using tags of various mesons one can address questions like: Does one enhance sensitivity to sea quarks if one tags with kaons or ϕ's? What are the vector mesons dual to?

Higher energies also enable one to investigate duality for the heavy quark sector, where calculations are more readily performed [21], and to investigate duality between hadrons and jets [22]: here one can argue that a jet similarly comes about by non-trivial interference between the produced hadrons.

REFERENCES

1. Bloom, E.D. and Gilman, F.J., *Phys. Rev. D* **4**, 2901 (1971).
2. Bloom, E.D., and Gilman, F.J. *Phys. Rev. Lett.* **25**, 1140 (1970).
3. Collins, P.D.B., *An Introduction to Regge Theory and High Energy Physics*, Cambridge University Press, Cambridge, 1977.
4. Close, F.E., *An Introduction to Quarks and Partons*, Academic Press, Great Britain, 1979.
5. Roberts, R. *The Structure of the Proton*, Cambridge University Press, Cambridge, 1990.
6. Poggio, E.C., Quin, H.R., and Weinberg, S., *Phys. Rev. D* **13**, 1958 (1976).
7. Altarelli, G., *Phys. Rep.* **81**, 1 (1982).
8. Buras, A.J., *Rev. Mod. Phys.* **52**, 199 (1980).
9. DeRujula, A., Georgi, H., and Politzer, H.D., *Phys. Lett.* **B64**, 428 (1976).
10. DeRujula, A., Georgi, H., and Politzer, H.D., *Annals Phys.* **103**, 315 (1977).
11. Niculescu, I. et al., *Phys. Rev. Lett.* **85**, 1186 (2000).
12. Arneodo, M. et al., *Phys. Lett.* **B364**, 107 (1995).
13. Niculescu, I. et al., *Phys. Rev. Lett.* **85**, 1182 (2000).
14. Nachtmann, O., *Nucl. Phys.* **B63**, 237 (1975).
15. Afanasev, A., Carlson, C.E., and Wahlquist, C., *hep-ph/0002271* (2000).
16. Sloan, T., Smadja, G. and Voss, R., *Phys. Rep.* **162**, 45 (1988).
17. Ackerstaff, K. et al., *Phys. Rev. Lett.* **81**, 5519 (1998).
18. Shifman, M, *hep-ph/0009131* (2000).
19. Bebek, C.J. et al., *Phys. Rev. Lett.* **34**, 759 (1975); *Phys. Rev. Lett.* **37**, 1525 (1976); *Phys. Rev. D* **15**, 3085 (1977).
20. Ent, R., Mkrtchyan, H., and Niculescu, G., *JLAB Proposal E00-004* (2000), unpublished.
21. Voloshin, M.B., and Shifman, M., *Sov. J. Nucl. Phys.* **47**, 511 (1988); Isgur, N., *Phys. Rev. D* **40**, 101 (1989); *Phys. Lett.* **B448**, 111 (1999).
22. Ochs, W., *hep-ph/9910319* (1999).

Generalized Parton Distributions and Deep Virtual Compton Scattering

D. Hasell, R. Milner, and K. Takase

Laboratory for Nuclear Science, Massachusetts Institute of Technology, Cambridge, MA 02139, USA

Abstract.
A brief description of generalized parton distributions is presented together with a discussion on studying such distributions via deep virtual Compton scattering. The kinematics, estimates of rates, and accuracies achievable for measuring DVCS utilizing a 5 + 50 GeV ep collider are also provided.

I INTRODUCTION

Generalized parton distributions, GPD's, can be used to relate a large number of exclusive processes in electron-nucleon reactions. In certain limits GPD's reduce to the normal parton distributions and nucleon form factors. Thus, they not only provide a powerful technique for QCD calculations but may also lead to better understanding of nucleon properties such as charge and spin structure. To study generalized parton distributions a number of reactions must be measured and included in a global analysis. One possibly interesting reaction could be deep virtual Compton scattering, DVCS.

The following section describes how deep virtual Compton scattering can be used to help measure GPD's. Then there is a brief section on the kinematics of DVCS including some estimates for the rates and accuracies possible in measuring DVCS at a proposed 5 + 50 GeV ep collider.

II DEEP VIRTUAL COMPTON SCATTERING ON THE PROTON

Measurement of deep virtual exclusive processes on the proton over a large kinematic range is highly desirable to investigate the Generalized Parton Distributions of the proton. The most promising of these non-forward hard exclusive processes are deep virtual Compton scattering (DVCS) and longitudinal electro-production of vector or pseudo-scalar mesons at large momentum transfers.

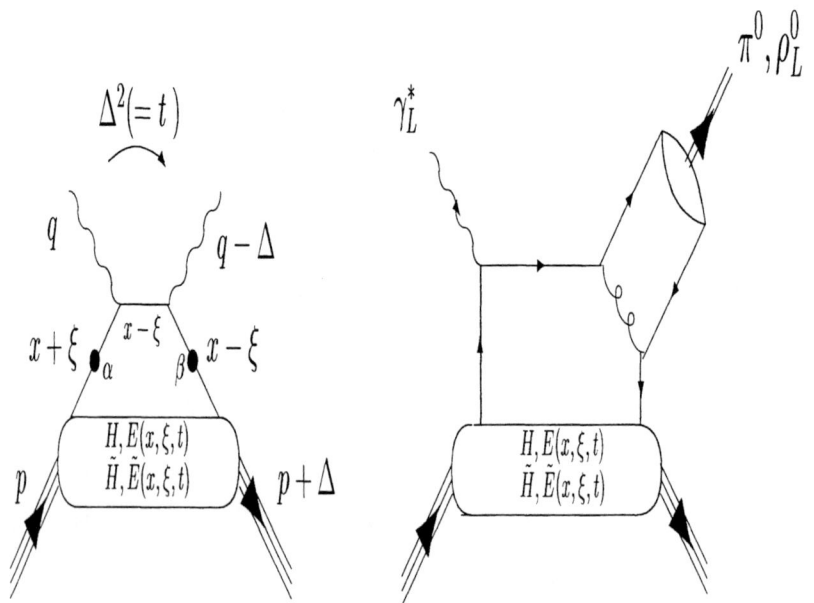

FIGURE 1. Leading order diagrams for DVCS (left) and for longitudinal electro-production of mesons (right).

The leading order QCD diagrams for DVCS and hard meson electro-production are shown in Fig. 1. The hard scale in Fig. 1 is the momentum transfer Q^2, which should be large (at least several $(\text{GeV}/c)^2$) so as to be in the deep inelastic scaling regime for inclusive scattering. It has been proven [2,3] that the leading order DVCS amplitude in the forward direction can be factorized into a hard scattering part (which is exactly calculable in PQCD) and a soft, non-perturbative nucleon structure piece as illustrated on the left side of Fig. 1. The nucleon structure information can be parametrized, at leading order, in terms of four (quark helicity conserving) generalized structure functions. These functions are the GPD's denoted by $H, \tilde{H}, E, \tilde{E}$ and defined above in terms of the three variables x, ξ, and t.

III   DVCS KINEMATICS

For the reaction $ep \to ep\gamma$ the equation for conservation of energy and momentum can be written as $k + N = k' + N' + g$ where the energy-momentum vectors k, N, k', N', and g represent the initial state electron and proton; and the final state electron, proton, and photon.

For an electron-proton collider with zero (head-on) crossing angle and choosing the proton momentum to be along the $+Z$ then the standard DIS variables can be

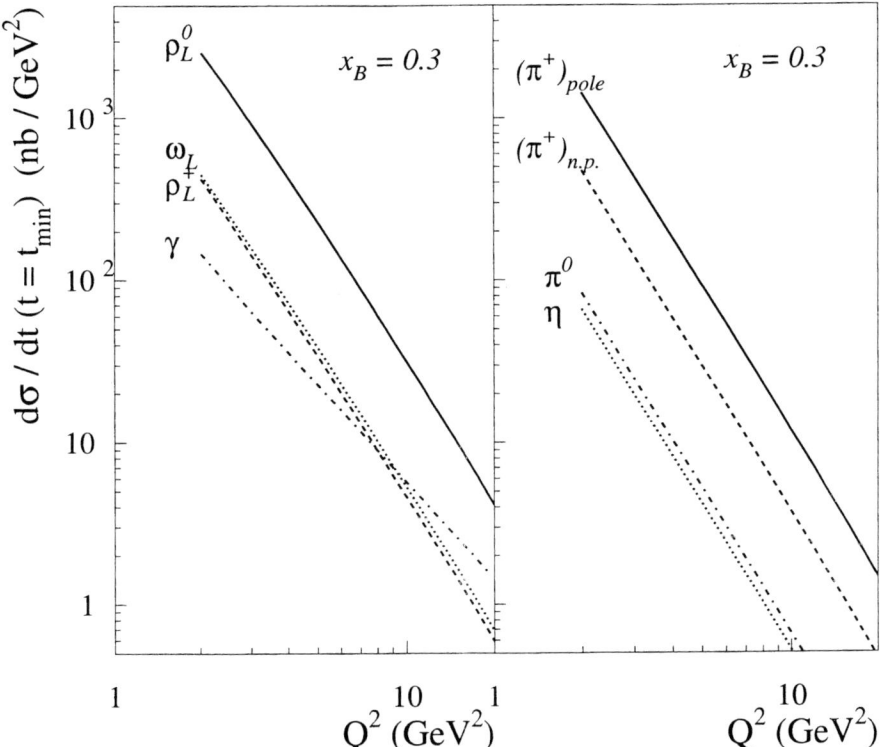

FIGURE 2. Scaling behavior of the L.O. predictions [1] for the forward differential electro-production cross section on the proton, for vector mesons (left panel) and pseudo-scalar mesons (right panel). For the π^+ channel, the pion pole contribution (full line, $(\pi^+)_{pole}$) is shown separately from the $\tilde{H}$ contribution (dashed line, $(\pi^+)_{n.p.}$). Also shown is the scaling behavior of the forward transverse cross section $\frac{d\sigma_T}{dt}$ for the leading order DVCS cross section (dashed-dotted line in left panel).

expressed as:

- the square of the centre of momentum frame total energy:

$$s = (k+N)^2 = m_{0e}^2 + m_{0p}^2 + 2k \cdot N \quad (1)$$
$$\approx 4 E_l E_N \quad (2)$$

- the square of the four-momentum transfer:

$$-Q^2 = q^2 = (k-k')^2 \quad (3)$$
$$= 2m_{0e}^2 - 2k \cdot k' \quad (4)$$
$$\approx -4 E_l E_{l'} \cos^2 \frac{\theta_{l'}}{2} \quad (5)$$

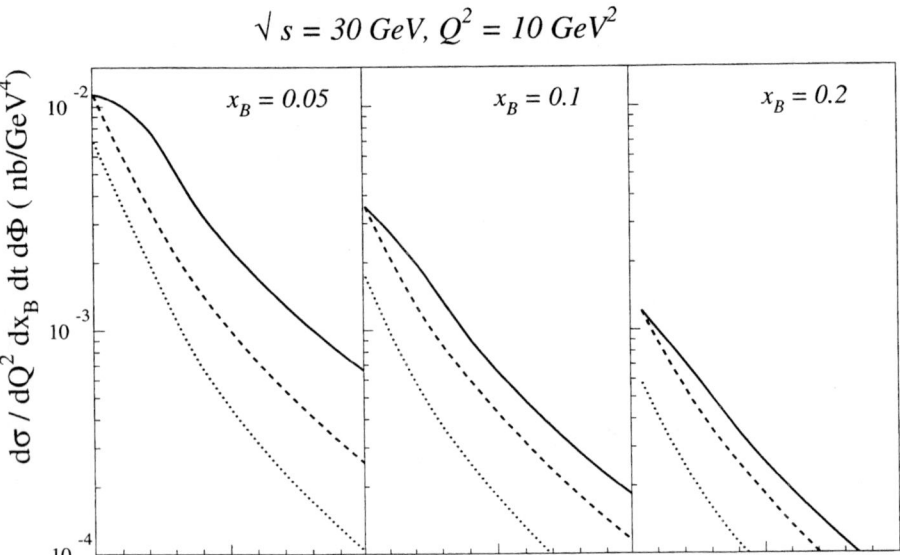

FIGURE 3. Cross section at $Q^2 = 10$ (GeV/c)^2 and three x values for the DVCS process for 5 GeV electrons colliding with 50 GeV protons.

- the energy transfered to the proton (in the proton's rest frame):

$$\nu = \frac{q \cdot N}{m_{0p}} \tag{6}$$

$$\approx \frac{2 E_N E_l}{m_{0p}} \left(1 - \frac{E_{l'}}{E_l} \sin^2 \frac{\theta_{l'}}{2}\right) \tag{7}$$

- the fraction of the available energy transferred to the proton (in the proton's rest frame):

$$y = \frac{m_{0p}\nu}{k \cdot N} = \frac{q \cdot N}{k \cdot N} = 1 - \frac{k' \cdot N}{k \cdot N} \tag{8}$$

$$\approx 1 - \frac{E_{l'}}{E_l}(\sin^2 \frac{\theta_{l'}}{2}) \tag{9}$$

- the Bjorken scaling factor:

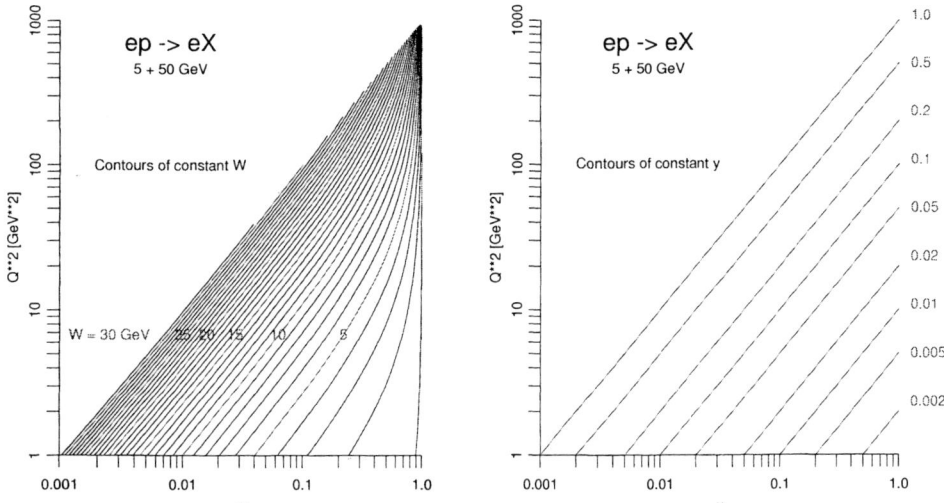

FIGURE 4. Kinematic range (Q^2 versus x) for different values of W and y.

$$x = \frac{Q^2}{2m_{0p}\nu} = \frac{Q^2}{2q \cdot N} \qquad (10)$$

$$\approx \frac{E_l}{E_N} \left(\frac{\frac{E_{l'}}{E_l} \cos^2 \frac{\theta_{l'}}{2}}{1 - \frac{E_{l'}}{E_l} \sin^2 \frac{\theta_{l'}}{2}} \right) \qquad (11)$$

- the mass squared of the resulting hadronic system:

$$W^2 = (N+q)^2 = (N'+g)^2 \qquad (12)$$
$$= m_{0p}^2 - Q^2 + 2m_{0p}\nu = m_{0p}^2 + 2N' \cdot g \qquad (13)$$
$$\approx s - 2E_{l'}\left[E_l + E_N + (E_l - E_N)\cos\theta_{l'}\right] \qquad (14)$$

Given the initial state (ie. s fixed) any pair of Q^2, x, y, or W^2 fixes the reaction kinematics and can be determined by measuring $E_{l'}$ and $\theta_{l'}$.

Figures 4 and 5 illustrate the kinematic range for a proposed $5 + 50$ GeV ep collider.

The errors in determining x and Q^2 due to uncertainties in $E_{l'}$ and $\theta_{l'}$ are shown in figure 6. The lines near the centres of each bin in x and y represent the extent of the error in reconstructing the kinematic invariants in that region due to small uncertainties in the measured quantities.

DVCS kinematics can be considered as the incident (and scattered) lepton providing a source of virtual photons of four-momentum q. Assuming the scattered lepton is well measured q is known and the kinematics reduces to two-body kinematics described by $q + N = W = N' + g$ which implies:

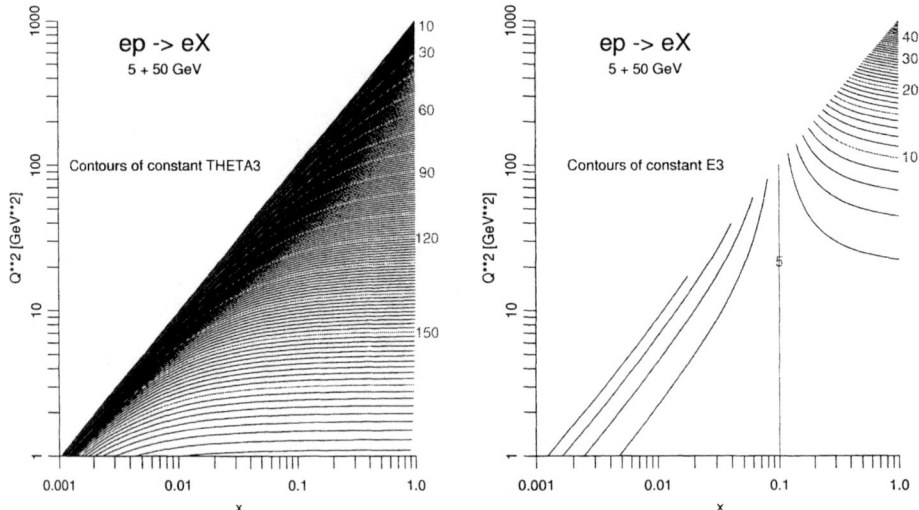

FIGURE 5. Kinematic range (Q^2 versus x) for different values of $\theta_{l'}$ and $E_{l'}$.

$$N' \cdot g = \frac{1}{2}(W^2 - m_{0p}^2) \quad (15)$$

$$W \cdot N' = \frac{1}{2}(W^2 + m_{0p}^2) \quad (16)$$

$$W \cdot g = \frac{1}{2}(W^2 - m_{0p}^2) \quad (17)$$

Assuming W^2 is known from measuring the scattered electron then the above equations can be used to determine the reaction kinematics from any pair of E, θ, or ϕ of either the scattered proton or photon. Solutions from measuring the energy of both the photon and proton or measuring the scattering angles of both are also possible.

For example, assume the scattered photon is described by

$$g \equiv (E_\gamma, E_\gamma \sin\theta_\gamma \cos\phi_\gamma, E_\gamma \sin\theta_\gamma \sin\phi_\gamma, E_\gamma \cos\theta_\gamma) \quad (18)$$

then

$$W \cdot g = \frac{1}{2}(W^2 - m_{0p}^2) = \frac{1}{2}Q^2(\frac{1}{x} - 1) \quad (19)$$

$$= E_\gamma[E_N + E_l - E_{l'} + |\vec{P}_{l'}|\sin\theta_{l'}\sin\theta_\gamma \cos(\phi_{l'} - \phi_\gamma) \quad (20)$$

$$-(|\vec{P}_N| - |\vec{P}_l| - |\vec{P}_{l'}|\cos\theta_{l'})\cos\theta_\gamma] \quad (21)$$

Given any pair of E_γ, θ_γ, or ϕ_γ then the third parameter is fixed by the above relationship hence the photon is completely determined and by conservation of energy-momentum so is the scattered proton. A similar derivation is possible for the proton using $W \cdot N' = \frac{1}{2}(W^2 + m_{0p}^2)$.

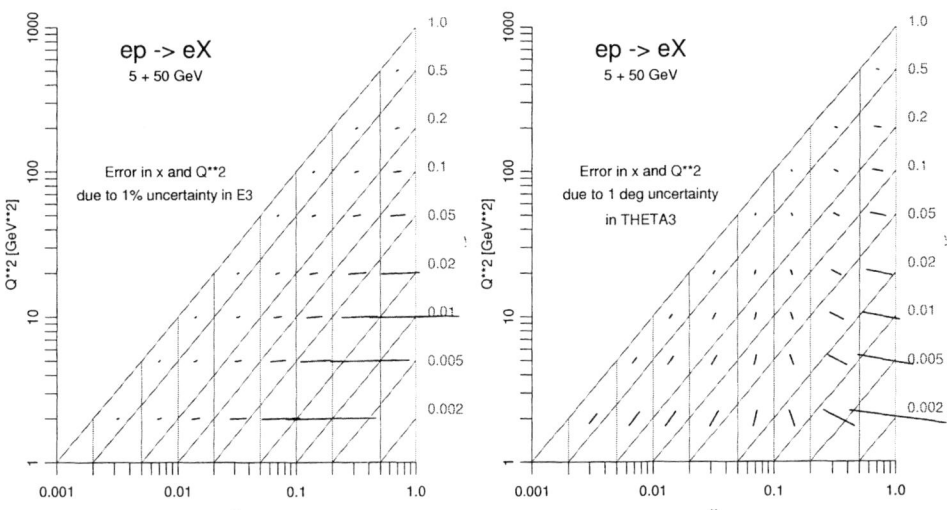

FIGURE 6. Error in reconstructing x and Q^2 due to a 1% uncertainty in $E_{l'}$ or a 1° uncertainty in $\theta_{l'}$

In DVCS it is generally more useful to consider the four-momentum transfered to the proton by the virtual photon represented here by t

$$t = (q - g)^2 = (N' - N)^2 \tag{22}$$

$$= -\frac{Q^2}{x} + 2E_\gamma[E_N - |\vec{P}_N|\cos\theta_\gamma] \tag{23}$$

which is independent of the azimuthal angle. This equation and the previous equation for $W \cdot g$ permit the DVCS kinematics to be determined from knowing any pair of t, E_γ, θ_γ, or ϕ_γ.

The physically accessible range of t is typically where $|t| << Q^2$ since the cross section falls quickly with t. Also this regime allows perturbative QCD calculations to be performed. Some representative kinematics for DVCS is listed in Table III. This shows, for three pairs of x and Q^2, the energy and scattering angle for the scattered electron (ie used to determine x and Q^2) and the kinematic range in energy and scattering angles for the scattered proton and photon for the case where $|t| < 1$ GeV^2. As one expects the scattered proton energy decreases at higher x. However, for this range in t the proton remains within 1.2° of the incident proton direction and can assume any azimuthal angle. The scattered photon energy range increases with x while the mean polar angle decreases. The range in polar angle for the photon also becomes smaller at higher x. The photon azimuthal angle is relatively stable opposite that of the scattered electron.

Figure 7 shows t versus the photon energy E_γ in the restricted range $|t| < 1$ GeV^2. The vertical lines represent contours of constant θ_γ while the curved lines are contours of constant azimuthal angle ϕ_γ. As expected there is a strong correlation

$Q^2 = 10$, $x = 0.05$	$\Rightarrow$	($E_{l'} = 4.5$ GeV, $\theta_{l'} = 141.1°$, $\phi_{l'} = 0°$)	
proton	$46 < E_{N'} < 49$ GeV	$\theta_{N'} < 1.2°$	$0 < \phi_{N'} < 360°$
photon	$1.8 < E_\gamma < 4.5$ GeV	$55° < \theta_\gamma < 95°$	$160 < \phi_\gamma < 200°$
$Q^2 = 10$, $x = 0.1$	$\Rightarrow$	($E_{l'} = 5.0$ GeV, $\theta_{l'} = 143.1°$, $\phi_{l'} = 0°$)	
proton	$42 < E_{N'} < 48$ GeV	$\theta_{N'} < 1.2°$	$0 < \phi_{N'} < 360°$
photon	$2.5 < E_\gamma < 8.0$ GeV	$30° < \theta_\gamma < 53°$	$160 < \phi_\gamma < 200°$
$Q^2 = 10$, $x = 0.2$	$\Rightarrow$	($E_{l'} = 5.25$ GeV, $\theta_{l'} = 144.0°$, $\phi_{l'} = 0°$)	
proton	$35 < E_{N'} < 45$ GeV	$\theta_{N'} < 1.2°$	$0 < \phi_{N'} < 360°$
photon	$5.0 < E_\gamma < 14.0$ GeV	$14° < \theta_\gamma < 25°$	$160 < \phi_\gamma < 200°$

TABLE 1. Ranges of the kinematic variables for DVCS with $|t| < 1$ GeV^2 for x and Q^2 value pairs corresponding to the calculations in figure 3.

between polar angle and energy. Thus, measuring one of these quantities well effectively fixes the other. Hence, it is not practical to try to extract DVCS kinematics by measuring just $E\gamma$ and θ_γ. Better measurements would be $E\gamma$ and ϕ_γ or ϕ_γ and θ_γ. From the spacing of the contours it is clear that near the edges of the energy range uncertainties in either energy or azimuthal angle will lead to large uncertainties in t. If the experiment and/or analysis can be restricted to the central region of the energy range then relatively easy measurements of $E_\gamma \pm 0.1$ GeV and $\phi_\gamma \pm 1°$ would permit t to be calculated to precisions of $\sim \pm 0.02$.

Of course the above discussion is based on an exactly known x and Q^2 measurement. In practice uncertainties in these variables can dramatically affect the determination of t. However, if the analysis is further restricted to regions where x and Q^2 are well measured (see figure 6) then this should not be a problem.

It is also possible to derive information from the scattered proton if the detector and/or machine permit measurements to be made on the scattered proton near the incident beam direction and energy. The equation for t can also be written as

$$t = (N' - N)^2 \tag{24}$$
$$= 2m_{0p}^2 - 2[E_N E_{N'} - |\vec{P}_N||\vec{P}_{N'}|\cos\theta_{N'}] \tag{25}$$

In practice measuring just the scattered proton's polar angle provides a reasonable measure of t (± 0.1 GeV^2) as illustrated in figure 8. provided the angle can be measured to $< 0.1°$. Such measurements could be possible using the lattice of the collider as a spectrometer for the proton. This is only possible for a fraction of the scattered proton's azimuthal range thus the statistics are significantly reduced over techniques measuring the photon over the entire azimuthal range. However, a dedicated experiment making precise measurements of the scattered electron and proton in this restricted regime could be quite compact and significantly cheaper than a general purpose 4π detector.

Figure 9 indicates the expected event rates for DIS and the region accessible for studying DVCS in bins of x and y for a $5 + 50$ GeV ep collider after an integrated luminosity of 2000 pb^{-1}. The DVCS calculations have used the cross sections

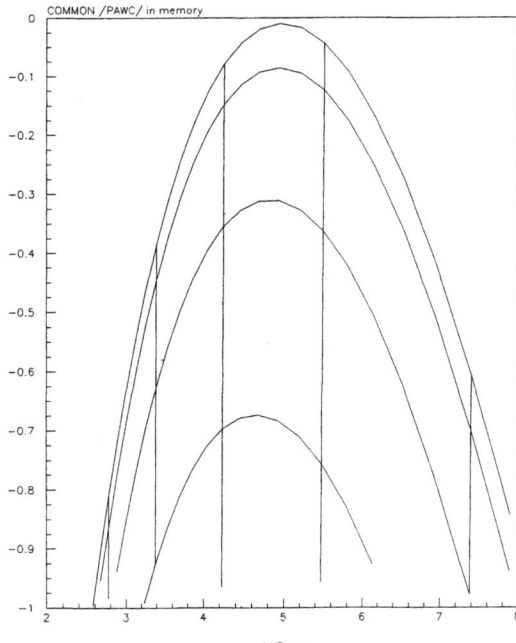

FIGURE 7. t versus the photon energy E_γ in the restricted range $|t| < 1\,\text{GeV}^2$ for $Q^2 = 10\,\text{GeV}^2$ and $x = 0.1$. The vertical lines represent contours of constant $\theta_\gamma = 50°$, $45°$, $40°$, $35°$, and $30°$ going left to right. The curved lines are contours of constant azimuthal angle $\phi_\gamma = 180°$, $175°$, $170°$, and $165°$ going top to bottom.

from figure 3. In the region where x and Q^2 are well determined there are several thousand DVCS events which however have to be resolved from the ten's of millions of DIS events. This should not be a problem provided a central tracking detector shows the scattered electron with a good track and shows no track for the scattered photon. The key elements are to measure the scattered electron's energy and scattering angles together with the scattering angles of the photon. The detector must also verify that the event is clean ie no other particles produced and that the proton (to good approximation) remains intact. Additionally tagging the scattered proton would improve the measurement.

REFERENCES

1. M. V. Polyakov and C. Weiss, Phys. Rev. **D60** (1999) 114017;
   M.V. Polyakov and C. Weiss, Phys. Rev. **D60** (1999) 114017;
   M. Diehl, T. Feldmann, R. Jakob and P. Kroll, Phys. Lett. **B460** (1999) 204;
   M. Vanderhaeghen, P. Guichon and M. Guidal, Phys. Rev. **D60** (1999) 094017.
2. X. Ji, Phys. Rev. Lett. **78** (1997) 610; Phys. Rev. **D55**, (1997) 7114.
3. A.V. Radyushkin, Phys. Lett **B380**, (1966) 417.

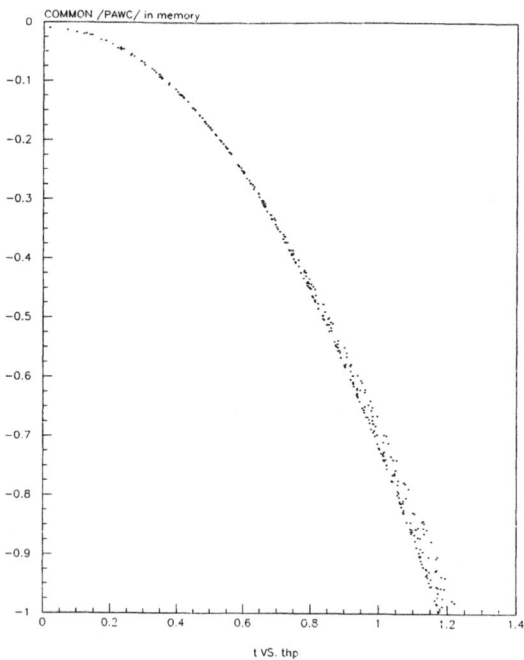

FIGURE 8. t versus the scattered proton polar angle $\theta_{N'}$ for $|t| < 1$ GeV^2 and $Q^2 = 10$ GeV^2 and $x = 0.1$.

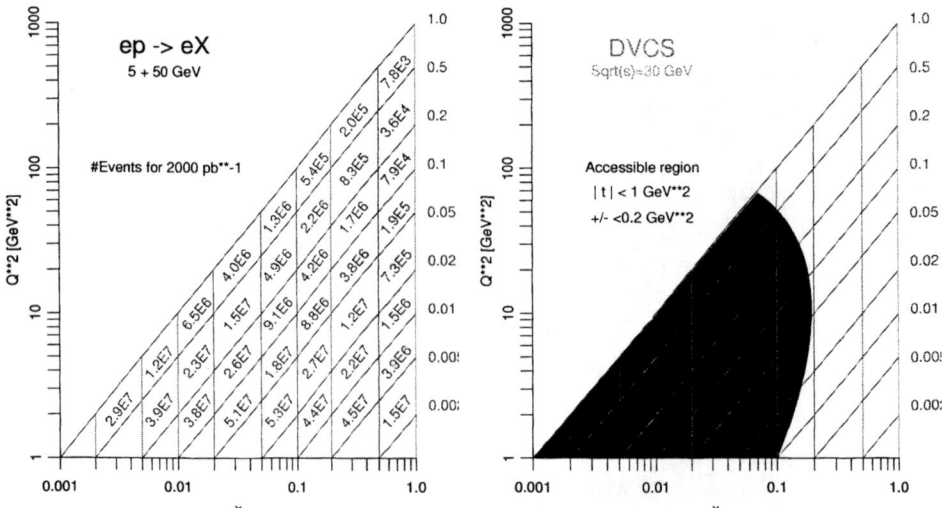

FIGURE 9. Expected event rates for DIS and the accessible region for studying DVCS in bins of x and y for a $5 + 50$ GeV ep collider after an integrated luminosity of 2000 pb^{-1}.

SEMI-EXCLUSIVE PROCESSES

Generalized Parton Distributions and the Dependence of Parton Distributions on the Impact Parameter

Matthias Burkardt

Department of Physics
New Mexico State University
Las Cruces, NM 88003

Abstract. Generalized parton distributions (GPDs) provide a link between form factors, parton distributions and other observables. I discuss the connection between GPDs and parton distributions as a function of the impact parameter. Since this connection involves GPDs in the limit of vanishing skewedness parameter ξ, i.e. when the off-forwardness is purely transverse, I also illustrate how to relate $\xi \neq 0$ data to $\xi = 0$ data, which is important for experimental measurements of these observables.

INTRODUCTION

Deeply virtual Compton scattering experiments provide a useful tool for probing off-forward or generalized parton distributions (GPDs) [1,2]

$$\bar{p}^+ \int \frac{dx^-}{2\pi} \langle p'|\bar{q}(\frac{-x^-}{2})\gamma^+ q(\frac{x^-}{2})|p\rangle e^{ix\bar{p}^+ x^-} = H^q(x,\xi,t)\bar{u}(p')\gamma^+ u(p) \quad (1)$$

$$+ E^q(x,\xi,t)\bar{u}(p')\frac{i\sigma^{+\nu}\Delta_\nu}{2M}u(p),$$

$$\bar{p}^+ \int \frac{dx^-}{2\pi} \langle p'|\bar{q}(\frac{-x^-}{2})\gamma^+\gamma_5 q(\frac{x^-}{2})|p\rangle e^{ix\bar{p}^+ x^-} = \tilde{H}^q(x,\xi,t)\bar{u}(p')\gamma^+\gamma_5 u(p) \quad (2)$$

$$+ \tilde{E}^q(x,\xi,t)\bar{u}(p')\frac{\gamma_5\Delta^+}{2M}u(p),$$

where $x^\pm = x^0 \pm x^3$ and $p^+ = p^0 + p^3$ refer to the usual light-cone components, $\bar{p} = \frac{1}{2}(p+p')$, $\Delta = p - p'$, and $t \equiv \Delta^2$. The skewedness in Eqs. (1,2) is defined as $\xi \equiv \frac{\Delta^+}{\bar{p}^+}$. From the point of view of parton physics in the infinite momentum frame (IMF), GPDs have the physical meaning of the amplitude for the process that a quark is taken out of the nucleon with longitudinal momentum fraction x and then inserted back into the nucleon with a four momentum transfer Δ^μ [3]. GPDs play multiple roles and in a certain sense they interpolate between form factors and conventional parton distributions (PDs) [1,2]: for $\xi = t = 0$ one recovers conventional PDs, i.e. longitudinal momentum distributions in the IMF, while when one integrates $H^q(x, \xi, t)$ over x, one obtains a form factor, i.e. the Fourier transform of a position space density (in the Breit frame!). One of the new physics insights that one can learn from these GPDs is the angular momentum distribution [4]. Others include meson distribution amplitudes. [1]

In this note, we will discuss the limit $\xi \to 0$, but $t \neq 0$, i.e. when the momentum transfer is purely transverse. In this limit, the "E-terms" in Eqs. (1) and (2) drop out and one finds

$$\int \frac{dx^-}{4\pi} \langle p'|\bar{q}(\frac{-x^-}{2})\gamma^+ q(\frac{x^-}{2})|p\rangle e^{ixp^+x^-} = H^q(x, 0, -\Delta_\perp^2) \tag{3}$$

$$\int \frac{dx^-}{4\pi} \langle p'|\bar{q}(\frac{-x^-}{2})\gamma^+ \gamma_5 q(\frac{x^-}{2})|p\rangle e^{i x \bar{p}^+ x^-} = \tilde{H}^q(x, 0, -\Delta_\perp^2), \tag{4}$$

with $p^{+\prime} = p^+$. Eqs. (3) and (4) very much resemble the definitions for ordinary twist-2 PDs, with the only difference being the fact that the $\perp$ momenta of the initial and final state are not the same. The situation here is very analogous to the relation between the forward and off-forward matrix elements of a current, i.e. between a charge and charge form factor. The main difference is of course that the operator entering the 'form factor' in Eq. (4) is not the current operator, but the operator that measures longitudinal momentum distributions. From this analogy, and since charge form factors have the physical interpretation of the Fourier transform of the position space charge distribution, it is natural to expect a similar interpretation also for GPDs.

GPDS FOR $\xi = 0$

In the following we will use a light-front (LF) Fock expansion to represent GPDs for $\xi = 0$ as overlap integrals between LF wave functions $\Psi_N(x, \mathbf{k}_\perp)$ summed ($\sum_N$) over Fock components [6] [2]

$$H^q(x, 0, -\Delta_\perp^2) = \sum_N \sum_j \int [dx]_N \int [d^2 k_\perp]_N \delta(x - x_j) \Psi_N^*(x, \mathbf{k}_\perp') \Psi_N(x, \mathbf{k}_\perp), \tag{5}$$

where $\sum_j$ denotes the sum over all quarks with flavor q in that Fock component and $\mathbf{k}'_{\perp,i} = \mathbf{k}_{\perp,i} - x_i \Delta_\perp$ for $i \neq j$, while $\mathbf{k}'_{\perp,j} = \mathbf{k}_{\perp,j} + (1 - x_j)\Delta_\perp$. Upon switching to the coordinate representation in the $\perp$ direction

$$\Psi_N(x, \mathbf{k}_\perp) = \int \frac{[d\mathbf{b}_\perp]}{(2\pi)^N} e^{-i\mathbf{k}_\perp \cdot \mathbf{b}_\perp} \tilde{\Psi}_N(x, \mathbf{b}_\perp) \tag{6}$$

it is straightforward to see that

---

[1] For a discussion of this connection in the context of QCD_{1+1} see Ref. [5].
[2] Note that this is an exact expression provided one knows the Ψ_N for all Fock components

$$H^q(x, 0, -\Delta_\perp^2) = \int [dx] \int [d\mathbf{b}_\perp] \sum_N \sum_j \delta(x - x_j) e^{i\Delta_\perp \cdot (\mathbf{b}_{\perp,j} - \mathbf{R}_\perp)} \left|\tilde{\Psi}_N(x, \mathbf{b}_\perp)\right|^2. \quad (7)$$

where we have introduced the $\perp$ *center of momentum*

$$\mathbf{R}_\perp \equiv \sum_i x_i \mathbf{b}_{\perp,i}.. \quad (8)$$

Eq. (7) illustrates that GPDs for $\xi = 0$ can be interpreted as Fourier transforms of *impact parameter dependent PDs*

$$H(x, 0, -\Delta_\perp^2) = \int d^2\mathbf{r}_\perp q(x, \mathbf{r}_\perp) e^{-i\Delta_\perp \cdot \mathbf{r}_\perp}$$
$$\tilde{H}(x, 0, -\Delta_\perp^2) = \int d^2\mathbf{r}_\perp \Delta q(x, \mathbf{b}_\perp) e^{-i\Delta_\perp \cdot \mathbf{r}_\perp} \quad (9)$$

where for example (again we suppress spin indices for simplicity)

$$q(x, \mathbf{r}_\perp) = \sum_N \sum_j \int [dx] \int [d\mathbf{b}_\perp] \delta(x - x_j) \delta\left(\mathbf{r}_\perp - (\mathbf{b}_{\perp,j} - \mathbf{R}_\perp)\right) \left|\tilde{\Psi}_N(x, \mathbf{b}_\perp)\right|^2. \quad (10)$$

Thus, GPDs, in the limit of $\xi \to 0$, allow a simultaneous determination of the longitudinal momentum fraction and transverse impact parameter of partons in the target hadron in the IMF.

Eq. (9) is not only a re-derivation of the main result from Ref. [7] using LF Fock wave functions, but it also clearly illustrates that the impact parameter in the impact parameter dependent PDs entering Eq. (9) is measured w.r.t. $\mathbf{R}_\perp$. There is a striking similarity between this observation and the fact that the Fourier transform of form factors in nonrelativistic (NR) systems yields charge distributions measured w.r.t. $\vec{R}_{CM} = \sum_i m_i \vec{r}_i / M$. This should not come as a surprise, since there is a residual Galilei invariance under the purely kinematic $\perp$ boosts in the LF framework

$$x_i \longrightarrow x_i' = x_i \quad , \qquad \mathbf{k}_{i\perp} \longrightarrow \mathbf{k}_{i\perp}' \equiv \mathbf{k}_{i\perp} + x_i \Delta \mathbf{P}_\perp, \quad (11)$$

which resembles very much NR boosts

$$\vec{k}_i \longrightarrow \vec{k}_i' = \vec{k}_i + m_i \Delta \vec{v} = \vec{k}_i + \frac{m_i}{M} \Delta \vec{P}. \quad (12)$$

The above observation about the $\perp$ center of momentum has one immediate consequence for the $x \to 1$ behavior of $q(x, \mathbf{b}_\perp)$. Since the weight factors in the definition of $\vec{R}_\perp$ are the momentum fractions, any parton i that carries a large fraction x_i of the target's momentum will necessarily have a $\perp$ position $\vec{r}_{i\perp}$ that is close to $\vec{R}_\perp$. Therefore the transverse profile (i.e. the dependence on $\mathbf{b}_\perp$) of $q(x, \mathbf{b}_\perp)$ will necessarily become more narrow as $x \to 1$, i.e. we expect that partons become very localized in $\perp$ position as $x \to 1$. By Fourier transform, this also implies that the slope of $H(x, 0, t)$ w.r.t. t at $t = 0$, i.e.

$$\langle \vec{b}_\perp^2 \rangle \equiv 4 \frac{\frac{d}{dt} H(x, 0, t)|_{t=0}}{H(x, 0, 0)} \quad (13)$$

should in fact vanish for $x \to 1$!

EXTRAPOLATING TO $\xi \to 0$

From the experimental point of view, $\xi = 0$ is not directly accessible in DVCS since one needs some longitudinal momentum transfer in order to convert a virtual photon into a real photon. There are several ways around this difficulty. First of all, one can access $\xi = 0$ in real wide angle Compton scattering [8]. However, it should also be possible to perform DVCS experiments at finite ξ and to extrapolate to $\xi = 0$. This extrapolation is greatly facilitated by working with moments since the ξ dependence of the moments of GPDs is in the form of polynomials [3]. For example, for the even moments $H_n(\xi, t) \equiv \int_{-1}^{1} dx x^{n-1} H(x, \xi, t)$ one finds [4]

$$H_n(\xi, t) = A_{n,0}(t) + A_{n,2}(t)\xi^2 + \ldots + A_{n,n-2}(t)\xi^{n-2} + C_n(t)\xi^n, \tag{14}$$

i.e. for example

$$\int_{-1}^{1} dx x H(x, \xi, t) = A_{2,0}(t) + C_2(t)\xi^2. \tag{15}$$

Since the H_n have a known functional dependence on ξ, one can use measurements of the moments of GPDs at nonzero values of ξ to determine (fit) the "form factors" $A_{n,2i}(t)$ and $C(t)$. After determining these invariant form factors, one can evaluate Eq. (14) for $\xi = 0$, yielding $H_n(0, t) = A_{n,0}(t)$, and the impact parameter dependence of the $n-th$ moment of the parton distribution in the target reads

$$q_n(\mathbf{b}_\perp) \equiv \int_{-1}^{1} dx x^{n-1} q(x, \mathbf{b}_\perp) = \int d^2 q_\perp A_{n,0}(-\Delta_\perp^2) e^{i\Delta_\perp \mathbf{b}_\perp}. \tag{16}$$

A very similar procedure can be applied to the moments of spin dependent GPDs

$$\tilde{H}_n(\xi, t) \equiv \int_{-1}^{1} dx x^{n-1} \tilde{H}(x, \xi, t) = \tilde{A}_{n,0}(t) + \tilde{A}_{n,2}(t)\xi^2 + \ldots + \tilde{A}_{n,n-1}(t)\xi^{n-1}. \tag{17}$$

Similarly in the unpolarized case, one can extract the $\frac{n+1}{2}$ form factors of the nth moment from measurements of $\tilde{H}$ for $\frac{n+1}{2}$ different values of ξ (and the same values of t), yielding for the impact parameter dependence of the nth moment of the polarized PD

$$\Delta q_n(\mathbf{b}_\perp) \equiv \int_{-1}^{1} dx x^{n-1} \Delta q(x, \mathbf{b}_\perp) = \int d^2 \mathbf{\Delta}_\perp \tilde{A}_{n,0}(-\Delta_\perp^2) e^{i\Delta_\perp \mathbf{b}_\perp}. \tag{18}$$

Of course, this procedure becomes rather involved for high moments, but the steps outlined above seem to be a viable way of determining the impact parameter dependence of low moments of parton distributions from DVCS data.

SUMMARY AND OUTLOOK

GPDs for $\xi \to 0$, i.e. where the off-forwardness is only in the $\perp$ direction, can be identified with the Fourier transform of impact parameter dependent PDs. Here the impact parameter $\mathbf{b}_\perp$. is defined as the $\perp$ distance from the center of (longitudinal) momentum in the IMF. This identification of GPDs with Fourier transforms of impact parameter dependent PDs is very much analogous to the identification of the charge form factor with the Fourier transform of a charge distribution in position space.

Although the $\xi \to 0$ limit of GPDs cannot be probed directly in DVCS, one can use the know polynomial ξ-dependence of the x-moments to extrapolate to $\xi = 0$.

Knowing the impact parameter dependence allows one to gain information on the spatial distribution of partons inside hadrons and to obtain new insights about the nonperturbative intrinsic structure of hadrons. For example, the pion cloud of the nucleon is expected to contribute more for large values of $\mathbf{b}_\perp$. Shadowing of small x parton distributions, is probably stronger at small values of $\mathbf{b}_\perp$ since partons in the geometric center of the nucleon are more effectively shielded by the surrounding partons than partons far away from the center. Geometric models for the small x behavior of the PDs in the nucleon suggest that polarized PDs may be more spread out in $\mathbf{b}_\perp$ than unpolarized ones [9]. These and many other models and intuitive pictures for the parton structure of hadrons give rise to predictions for the impact parameter dependence of PDs that reflect the underlying microscopic dynamics of these models. Our results may also play an important role in the modeling of the t dependence in GPDs, which may in turn be relevant for fitting GPDs to experimental data for DVCS amplitudes. Finally, combining information about the impact parameter dependence with information about longitudinal correlations in position space [10] may lead to further insights into non-perturbative hadron structure.

Acknowledgments: This work was supported by the DOE (DE-FG03-95ER40965) and by TJNAF. I would like to thank R.L. Jaffe, S. Brodsky and R. Jakob for useful and encouraging comments.

REFERENCES

1. X. Ji, *Phys. Rev. Lett.* **78**, 610 (1997); *Phys. Rev.* **D55**, 7114 (1997).
2. A.V. Radyushkin, *Phys. Rev.* **D56**, 5524 (1997).
3. X. Ji, W. Melnitchouk, and X. Song, *Phys. Rev.* **D56**, 5511 (1997).
4. P. Hoodbhoy, X. Ji, and W. Lu, *Phys. Rev.* **D59**, 014013 (1998).
5. M. Burkardt, *Phys. Rev.* **D62**, 94003 (2000), hep-ph/0005209.
6. M. Diehl et al., hep-ph/0009255.
7. M. Burkardt, *Phys. Rev.* **D62**, 71503 (2000), hep-ph/0005108; hep-ph/0008051.
8. A.V. Radyushkin, *Phys. Rev.* D **58**, 114008 (1998).
9. S.M. Troshin and N.E. Tyurin, *Phys. Rev.* D **57**, 5473 (1998).
10. G. Piller et al., *Nucl. Phys.* **A663**, 328 (2000).

An Energy Recovery Electron Linac-On-Ring Collider

L. Merminga*, G. A. Krafft*, V. A. Lebedev*, and Ilan Ben-Zvi[†]

*Jefferson Laboratory, 12000 Jefferson Ave., Newport News, VA 23606, USA

[†]Brookhaven National Laboratory, P. O. Box 5000, Upton, NY, 11973-5000, USA[1]

ABSTRACT

We present the design of high-luminosity electron-proton/ion colliders in which the electrons are produced by an Energy Recovering Linac (ERL). Electron-proton/ion colliders with center of mass energies between 14 GeV and 100 GeV (protons) or 63 GeV/A (ions) and luminosities at the 10^{33} (per nucleon) level have been proposed recently as a means for studying hadronic structure. The linac-on-ring option presents significant advantages with respect to: 1) spin manipulations 2) reduction of the synchrotron radiation load in the detectors 3) a wide range of continuous energy variability. Rf power and beam dump considerations require that the electron linac recover the beam energy. Based on extrapolations from actual measurements and calculations, energy recovery is expected to be feasible at currents of a few hundred mA and multi-GeV energies. Luminosity projections for the linac-ring scenario based on fundamental limitations are presented. The feasibility of an energy recovery electron linac-on-proton ring collider is investigated and four conceptual point designs are shown corresponding to electron to proton energies of: 3 GeV on 15 GeV, 5 GeV on 50 GeV and 10 GeV on 250 GeV, and for gold ions with 100 GeV/A. The last two designs assume that the protons or ions are stored in the existing RHIC accelerator. Accelerator physics issues relevant to proton rings and energy recovery linacs are discussed and a list of required R&D for the realization of such a design is presented.

INTRODUCTION

Electron-proton/ion colliders with center of mass energies between 14 GeV and 100 GeV (protons) or 63 GeV/A (ions) and luminosities at the 10^{33} (per nucleon) level have been proposed recently as a means for studying hadronic structure. Electron beam polarization appears to be crucial for many of the experiments. Two

---

[1] Done under the auspices of the U.S. Department of Energy under contract number DE-AC02-98CH10886

accelerator design scenarios have been examined in detail: colliding rings and recirculating linac-on-ring. Although the linac-on-ring scenario is not as well developed as the ring-ring scenario, comparable luminosities appear feasible. The linac-on-ring option presents significant advantages with respect to: 1) spin manipulations 2) reduction of the synchrotron radiation load in the detectors 3) a wide range of continuous energy variability. Rf power and beam dump considerations require that the electron linac recover the beam energy. This technology has been demonstrated at Jefferson Lab's IR FEL with cw current up to 5 mA and beam energy up to 50 MeV. Based on extrapolations from actual measurements and calculations, energy recovery is expected to be feasible at higher currents (a few hundred mA) and higher energies (a few GeV) as well.

We begin with a brief overview of Jefferson Lab's experience with energy recovery and summarize its benefits. Luminosity projections for the linac-ring scenario based on fundamental limitations are presented next. The feasibility of an energy recovery electron linac-on-proton ring collider is investigated and four conceptual point designs are shown corresponding to electron to proton energies of: 3 GeV on 15 GeV, 5 GeV on 50 GeV and 10 GeV on 250 GeV, and for gold ions with 100 GeV/A. The last two designs assume that the protons or ions are stored in the existing RHIC accelerator. Accelerator physics issues relevant to proton rings and energy recovery linacs are discussed next and a list of required R&D for the realization of such a design is presented.

Energy Recovery Linacs

Energy recovery is the process by which the energy invested in accelerating a beam is returned to the rf cavities by decelerating the beam. To date, energy recovery has been realized in a number of different ways. The first energy recovery experiment was done at Stanford's superconducting rf (srf) linac where the recirculated electron beam was injected into the linac in such a phase that it lost energy to the cavity fields [1]. This is an example of the so-called same-cell energy recovery. All of the linac beam energy was recovered. Another experimental demonstration of energy recovery took place at Los Alamos where >70% of the beam energy was recovered [2]. Following acceleration, the beam was transported around a 180° bend and passed through a decelerating structure. The decelerators were coupled to the accelerators through the resonant bridge couplers. The rf power generated by the beam was shared with the accelerators through the couplers. An energy recovery experiment was also performed in the CEBAF Injector. Energy recovery was demonstrated in cw mode and at the same time multipass Beam Breakup (BBU) experiments were carried out [3].

Same-cell energy recovery with cw current up to 5 mA and energy up to 50 MeV has been demonstrated at Jefferson Lab's (JLab) IR FEL and it is used routinely for the operation of the FEL as a User Facility [4]. The machine layout is shown in Figure 1. Microbunches with an rms bunch length of ~20 psec are produced in a DC photocathode gun and accelerated to 320 kV. The bunches are compressed by a copper buncher cavity operating at 1497 MHz. They pass through a pair of srf cavities operating at an average gradient of 10 MV/m. The output beam at ~10 MeV is injected into an 8-cavity srf cryomodule where it is accelerated up to ~48 MeV. The beam then passes through the wiggler. Afterward it is recirculated through two isochronous, achromatic bends separated by a quadrupole transport line, back through the cryomodule in the decelerating rf phase and dumped at the injection energy of ~10 MeV.

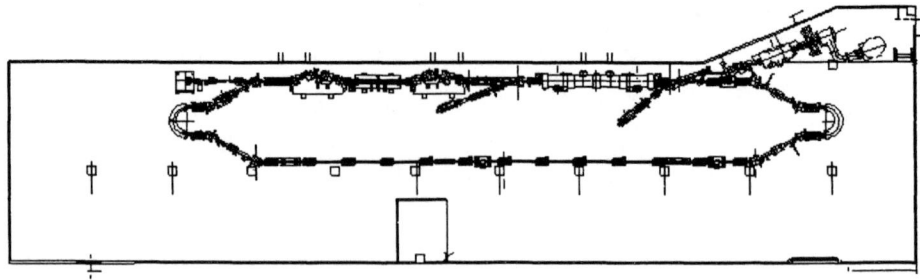

Figure 1. Machine layout of Jefferson Lab's IR FEL.

The first indicators of energy recovery, which subsequently were used as diagnostic tools, are the klystron drive signals for the gradient feedback loop, shown in Figure 2, for four of the linac rf cavities. When a 200 μsec beam pulse is injected in these cavities, in the absence of energy recovery, the gradient drive signals reach ~2 V to compensate for beam loading. With energy recovery, these signals are close to 0 V (where 0 V corresponds to the DC voltage required to drive the accelerating field in the cavity), as the decelerating and accelerating beam vectors cancel each other resulting in nearly zero net beam loading.

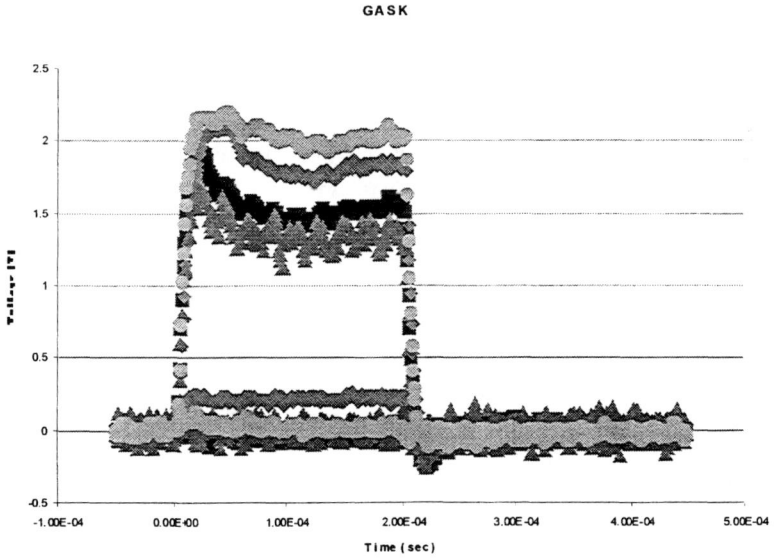

Figure 2. Response of the gradient loop drive signals, in four rf cavities, to a 200 μsec beam pulse, with and without energy recovery.

Another indication that energy recovery works is demonstrated in Figure 3 where the rf power required to accelerate up to 3.5 mA of cw beam current is compared to the power required for no beam, in each of the 8 cavities. Notice that the required rf power is nearly independent of beam current. An additional benefit is that the overall system efficiency is increased.

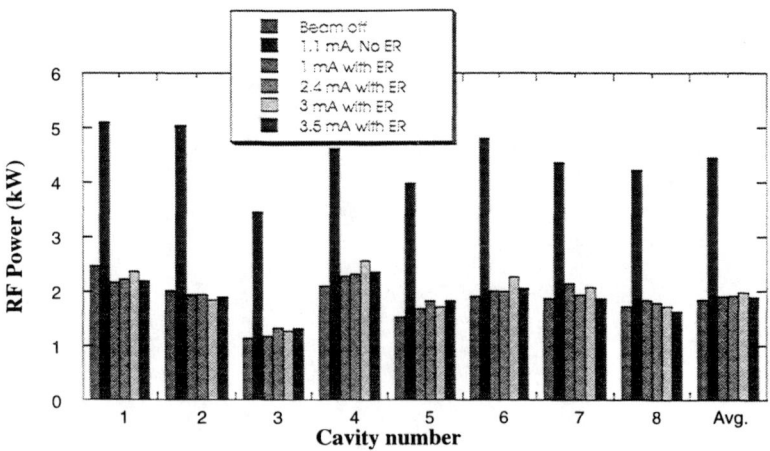

Figure 3. Rf power requirements in the linac cavities for a range of beam currents.

The HERA electron storage ring in DESY circulates a 27 GeV electron beam through a large number of superconducting cavities at an average current of 40 mA. While this is not strictly an energy recovery system (and not a linac), it provides a demonstration that a high energy, high current beam can be circulated through a very large number of passes in a long superconducting structure.

In summary, the benefits of energy recovery are:
1) The required rf power becomes nearly independent of beam current.
2) The overall system efficiency is increased.
3) The electron beam power to be disposed of at the beam dumps is reduced by the ratio of the final to injected energy.
4) The induced radioactivity (and therefore the shielding problem) is reduced, if the beam is dumped below the neutron production threshold.

Fundamental Limitations Of Colliders

The luminosity of an electron linac-on-proton ring collider, assuming both beams are round gaussians at the interaction point (IP), is given by

$$L = \frac{N_e N_p}{4\pi \sigma^{*2}} f_c \quad (1)$$

where N_e is the number of electrons per bunch, N_p is the number of protons per bunch, f_c is the bunch collision frequency, and σ^* is the rms beam size at the IP. The luminosity is proportional to the product of the intensity of one beam Nf_c/ε, and the number of particles per bunch in the other beam, where ε is the beam emittance. A number of effects impose fundamental limitations on the intensity, which together with limitations on the number of particles per bunch can ultimately limit the luminosity. We discuss intensity limits next.

a) Luminosity at the Laslett and Beam-Beam Tuneshifts Limit

The Laslett tuneshift Δv_L is given by:

$$\Delta v_L = \frac{N_p}{\varepsilon_p \sigma_z^p} \frac{r_p}{4\pi \gamma_p^3} \frac{C}{\sqrt{2\pi}} \tag{2}$$

where C is the ring circumference and r_p the classical radius of proton 1.534 x 10^{-18} m. Typically, in proton ring designs, the bunch length is chosen to be approximately equal to the beta function at the IP in order for the luminosity not to deteriorate too much within one collision. In this approximation, one concludes that the Laslett tuneshift imposes a fundamental limit on the ratio of N_p/σ_p^{*2}.

The beam-beam tuneshift of the proton beam given by:

$$\xi_p = \frac{N_e r_p \beta_p^*}{4\pi \gamma_p \sigma_e^{*2}} \tag{3}$$

where β_p^* is the beta function at the IP, also imposes a fundamental limit on N_e/σ_e^{*2}. One can write an expression for the luminosity in the limit of Laslett and beam-beam tuneshifts:

$$L = \left(\frac{4\pi\sqrt{2\pi}}{r_p^2}\right) \xi_p \Delta v_L \frac{\gamma_p^4}{C} \frac{\sigma^{*2}}{\beta^*} f_c \ . \tag{4}$$

As an example, we can assume a beta function at the IP $\beta^* = 10$ cm, an rms beam size at the IP $\sigma^* = 40$ μm and collision frequency $f_c = 150$ MHz. Figure 4 is the plot of luminosity vs. proton beam energy E_p, given by Eq. (4). The ring circumference C has been minimized subject to the engineering constraint of maximum magnetic field (in this case B=4 Tesla). The two curves correspond to: a) $\Delta v_L = 0.004$, which is a safe and generally accepted value for the Laslett tuneshift and b) $\Delta v_L = 0.04$, which is a more aggressive value, yet consistent with the value assumed in the ring-ring scenario presented in this document [5].

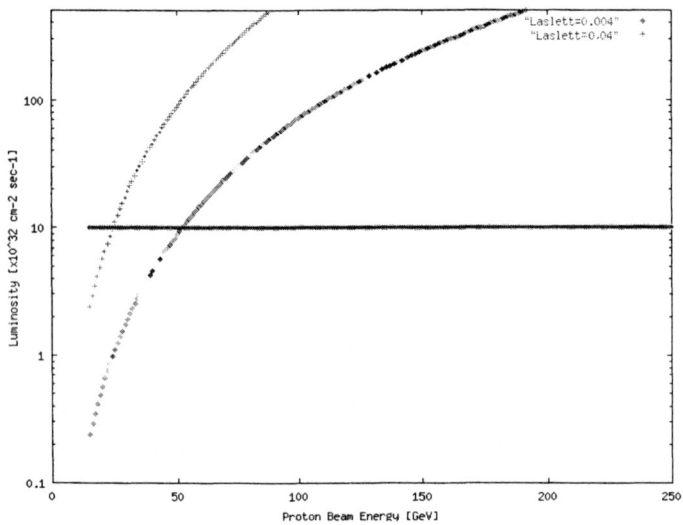

Figure 4. Luminosity vs. proton beam energy at the Laslett and beam-beam tuneshift limits, for two values of the Laslett tuneshift: 0.004 and 0.04. In both cases the beam-beam tuneshift is 0.004.

In both cases $\xi_p = 0.004$, consistent both with [5] and with the value assumed in the eRHIC ring-ring scenario [6]. Of course, in practice N_p and N_e are further limited by a number of other effects such as collective instabilities, and we will examine these later. The horizontal line in Fig. 4 corresponds to luminosity equal to 1.0×10^{33} cm^{-2} sec^{-1}.

b) Luminosity at the Beam-Beam Induced Head-Tail Instability Limit

The beam-beam induced head-tail instability is an additional effect, which could potentially impose a limit on the luminosity of linac-ring colliders. Presently this instability is the subject of focused investigation at Jefferson Lab. The effect is analyzed later on in this report, using a linearized model, according to which the stability condition can be expressed as

$$D_e \xi_p \leq 4 v_s \tag{5}$$

where D_e is the electron beam disruption parameter given by

$$D_e = \frac{N_p r_e \sigma_z^p}{\gamma_e \sigma_p^{*2}} \qquad (6)$$

and v_s is the synchrotron tune of the proton beam. One can re-write the luminosity in the limit of the head-tail instability, as

$$L = \left(\frac{4}{r_e r_p}\right) \gamma_e \gamma_p v_s \sigma^{*2} f_c \quad . \qquad (7)$$

Figure 5 displays the luminosity given by Eq. (7) as function of proton energy for four values of the electron beam energy: 3, 5, 7, 10 GeV. The synchrotron tune has been set equal to 1×10^{-3}, the rms angular divergence of the beam at the IP, $\sigma'' = \sigma^*/\beta^* = 40\mu m/10cm = 0.4$ mrad and the collision frequency $f_c = 150$ MHz. Superimposed are the luminosity curves from Figure 4 for comparison. Figure 5 demonstrates that above a certain proton beam energy, increasing the Laslett tuneshift beyond the generally accepted values does not benefit the luminosity. A second conclusion is that, assuming that the larger value of the Laslett tuneshift is attainable, then the luminosity is limited by the head-tail instability, over most of the energy range of protons and electrons.

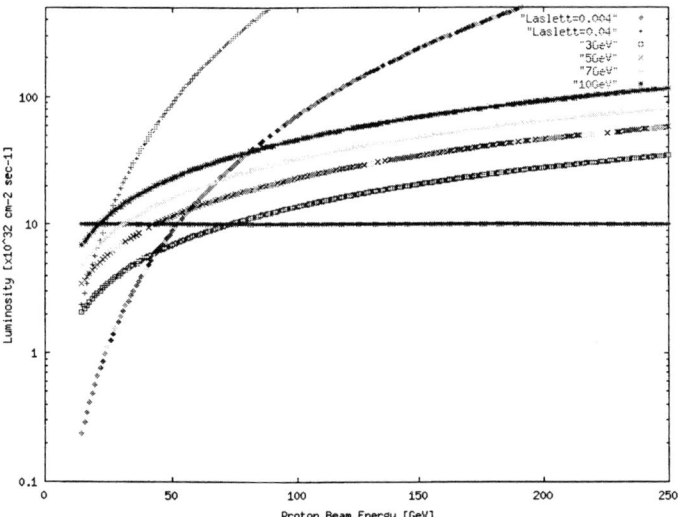

Figure 5. Luminosity vs. proton beam energy, at the stability limit of the beam-beam induced head-tail effect (linear approximation).

The above expressions ought to be compared to the luminosity expressions for a ring-ring collider based on fundamental limitations. In a ring-ring design these are the beam-beam tuneshifts of both electrons and protons. The luminosity in the beam-beam tuneshift limit is given by:

$$L = \left(\frac{4\pi}{r_e r_p}\right) \gamma_e \gamma_p \xi_e \xi_p \sigma'^{*2} f_c \quad . \tag{8}$$

Ring-ring limitations and comparisons with linac-ring will be the subject of a follow-up article.

Conceptual Point Designs

We now turn our attention to specific point designs that span a proton energy range from 15 to 250 GeV and an electron energy range from 3 to 10 GeV. As was discussed above, the head-tail instability is presently under investigation. The condition presented earlier was derived using a linear approximation, which clearly does not reflect the true complexity of the problem. The next steps include a more sophisticated analysis and simulation studies. In the absence of more rigorous results, we will develop a reasoning for the point designs without taking the head-tail instability into account and will defer the more complete study to a later document.

We present reasoning that allows us to develop a self-consistent sets of parameters. Four sets of input parameters will be considered: a) Proton beam energy of 15 GeV colliding with electron beam energy of 3 GeV b) Proton beam energy of 50 GeV and electron beam energy of 5 GeV c) proton beam energy of 250 GeV and electron beam energy of 10 GeV d) gold ion beam of 100 GeV/A and electron beam energy of 10 GeV. The high-energy designs are based on the existing RHIC storage ring.

Since the linac technology assumed is the same for all designs, we will briefly discuss it here. The technology of the electron linac is well established. Jefferson Laboratory has significant expertise in srf linacs and the srf accelerating structures are commercially available from a number of manufacturers.

For the sake of this report we assume that the linac structures will be identical to the well-known TESLA style cavities. These cavities and the ancillary equipment (cryostats, couplers, tuners, HOM loads etc.) have been optimized cost-wise and performance-wise. The TESLA cavities have a shunt impedance R/Q=1036 Ohms and structure length 1.038m. The residual resistance is 3 nΩ, equivalent to a Q of 10^{11}. Considering demonstrated performance from a number of manufacturers, we will assume, conservatively, a Q_0 of 1.5×10^{10} at 2K and accelerating gradient of 20MV/m.

At these values the refrigeration power is 26 W/structure. Thus a 10 GeV linac, for example, will require 500 cavities with a dissipation (excluding standing losses) of 13 kW. TESLA optimization was driven towards high gradient, not low Q. We can expect improvement in Q_{BCS}. See, for example, Kneisel's report on doubling Q_{BCS} by furnace baking in Ref. [7]. Figure 6 demonstrates the performance of TESLA style cavities.

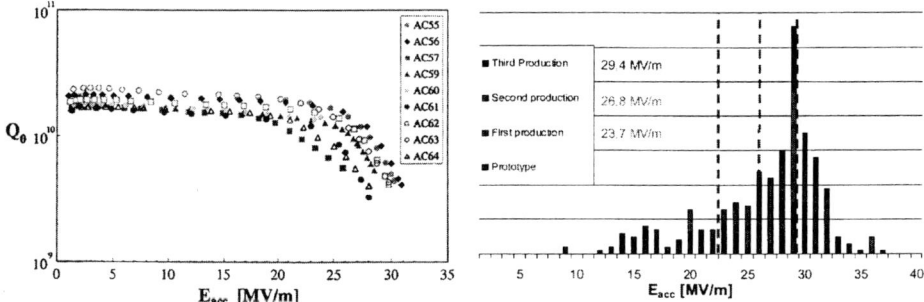

Figure 6. Plots of Q_0 vs. accelerating gradient and distribution of gradients in the Tesla 9-cell cavities.

1. 3 GeV electrons on 15 GeV protons

  1A: $\Delta v_L \leq 0.004$

In this case we assume that both the Laslett and beam-beam tuneshifts can not exceed 0.004. To arrive at a self-consistent set of parameters and a luminosity estimate, we first set the electron beam size at the IP based on projected electron source performance. Then the proton beam parameters are set at the Laslett tuneshift limit. The maximum number of electrons per bunch is determined at the beam-beam tuneshift limit of the protons. Finally effects that influence the choice of the bunch collision frequency are discussed.

An rms normalized emittance of 60 μm for electrons at a bunch charge of approximately 1 nC is assumed, yielding a geometric emittance of 10 nm at the IP (3 GeV). For a beta function of 12 cm, which is discussed later, the rms electron beam size at the IP is 35 μm. (Round beams are assumed for electrons and protons.)

In order for the luminosity not to degrade, typically the beta function for the proton beam at the IP is set approximately equal to the rms proton bunch length. In this approximation, the Laslett tuneshift given by Eq. (2) can be written as

$$\Delta v_L = \frac{N_p}{\sigma_p^{*2}} \frac{r_p}{4\pi\gamma_p^3} \frac{C}{\sqrt{2\pi}} \ . \qquad (9)$$

Clearly the tuneshift sets a limit on the ratio of N_p/σ_p^{*2}. Assuming a proton beam rms normalized emittance of 3 μm (consistent with LHC and RHIC experience) and $\Delta v_L = 0.004$, the rms beam size for protons at the IP is 107 μm, for a beta function of 6 cm. Then the number of protons per bunch at the Laslett tuneshift limit is 3×10^{10}.

The number of electrons per bunch can be limited either by the beam-beam tuneshift of the proton beam or by single-bunch transverse Beam Breakup in the linac [8]. Beam-beam tuneshift of the protons is given by Eq. (3), and we assume that it can not exceed 0.004. This value of the beam-beam tuneshift sets the number of electrons per bunch equal to 1.1×10^{10}. One can obtain a simple estimate for the emittance growth due to single bunch BBU in the linac by using the following expression [8],

$$\eta = \frac{L r_e N_e W_0}{k_0 (\gamma_f - \gamma_0)} \ln \frac{\gamma_f}{\gamma_0} \tag{10}$$

where k_0 is the betatron wavenumber, W_0 is the transverse wake function, and r_e the classical radius of the electron. For an rms bunch length of 1 mm and betatron wavelength in the linac of 50 m, the amplification parameter η remains less than ~1 if the number of electrons per bunch does not exceed 1.5×10^{11}. Should this effect become a serious limit, BNS damping can be used. Therefore, in this case the limit on N_e is set by the beam-beam tuneshift, and not the single-bunch BBU.

The bunch collision frequency should be maximized subject to the constraints of parasitic collisions, user requirements and possibly the electron cloud effect in the proton ring. We have assumed a bunch separation of 6.66 nsec or 150 MHz repetition rate. Note that the luminosity scales linearly with the frequency.

For the case of unequal electron-proton bunch sizes, the luminosity is given by

$$L = \frac{N_e N_p f_c}{2\pi [\sigma_e^{*2} + \sigma_p^{*2}]} \tag{11}$$

For $N_e = 1.1 \times 10^{10}$, $N_p = 3.0 \times 10^{10}$, $f_c = 150$ MHz, $\sigma_e^* = 35$ μm and $\sigma_p^* = 107$ μm, the luminosity is equal to 6.2×10^{31} cm^{-2} sec^{-1}.

1B: $\Delta v_L \leq 0.05$

We now consider a point design assuming that the Laslett tuneshift can be as high as 0.05. In this case, the electron beam parameters remain the same and again the Laslett tuneshift sets the ratio of N_p/σ_p^{*2}. However the optimization now proceeds as follows: We first determine the limit on N_p and then set the minimum spot size at the IP, at the Laslett tuneshift limit.

The number of protons per bunch can be limited by collective instabilities or by the emittance growth of the electron beam due to a single round-beam collision with the protons. We set $N_p = 1 \times 10^{11}$, similar to LHC and RHIC, and we will examine these limiting effects later. Then at the Laslett tuneshift, the rms beam size of the protons is 58 μm, and for $\beta^* = 10$ cm, the normalized rms emittance is 0.54 μm. Note that at $N_e = 1.1 \times 10^{10}$, the beam-beam tuneshift is 0.0068. These parameters yield luminosity equal to 5.7×10^{32} cm^{-2} sec^{-1}.

2. *5 GeV electrons on 50 GeV protons*

Following similar arguments for the case of 5 GeV electrons on 50 GeV protons, we arrive at the two sets of parameters outlined in Table 1. Note that luminosity at the 10^{33} level is attainable at these energies, for average current in the linac of 0.264 A and average current in the ring of 2.4 A.

Table 1. Parameters for point designs 1 and 2.

Parameter	Units	Point Design 1A	Point Design 1B	Point Design 2A	Point Design 2B
E_e	GeV	3	3	5	5
E_p	GeV	15	15	50	50
N_e	ppb	1.1×10^{10}	1.1×10^{10}	1.1×10^{10}	1.1×10^{10}
N_p	ppb	3.0×10^{10}	1.0×10^{11}	1.0×10^{11}	1.0×10^{11}
f_c	MHz	150	150	150	150
σ_e^*	μm	35	35	25	25
σ_p^*	μm	107	58	60	25
ε_e^*	nm	10	10	6	6
ε_p^*	nm	200	33.6	36	6.25
β_e^*	cm	12	12	10	10
β_p^*	cm	6	10	10	10
σ_z^p	cm	6	10	10	10
σ_z^e	mm	1	1	1	1
ξ_p	–	.004	.0068	.004	.004
$\Delta\nu_L$	–	.004	.05	.004	.024
I_e	A	.264	.264	.264	.264
I_p	A	.720	2.4	2.4	2.4
L	cm^{-2} sec^{-1}	6.2×10^{31}	5.7×10^{32}	6.2×10^{32}	2.1×10^{33}

3. 10 GeV electrons on 250 GeV protons based on the RHIC storage ring [6]

The third and fourth point designs presented here are for 10 GeV electrons colliding with 250 GeV protons or 100 GeV/A ions using the existing RHIC storage ring. The lower energy point designs may also be implemented in RHIC, but we skip this detail.

Since the design of the Interaction Point (IP) (in particular the size of the detector) depends on the energy, it is reasonable to assume that at least two detectors will be required. The RHIC machine has two independent ion rings and thus could support one (or more) IP per ring, with collisions taking place with two different energies. The electron linac could be designed also to provide simultaneously two energies, but the detailed description of this mode will not be addressed in this report.

The presented parameters are consistent with the RHIC layout. A schematic layout of the linac-ring collider is shown in Figure 7. A schematic layout for the ring-ring scenario, also based on the RHIC accelerator, is shown in Figure 8. Table 2 summarizes the linac parameters, common to both protons and gold.

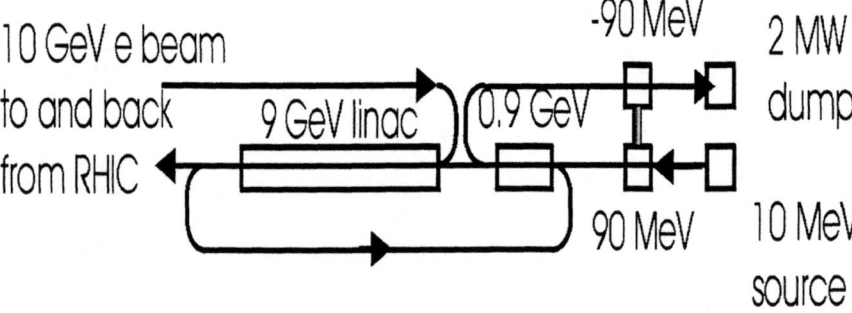

Figure 7. Schematic layout of the RHIC-based linac-ring collision scenario. (See Ref. [6]).

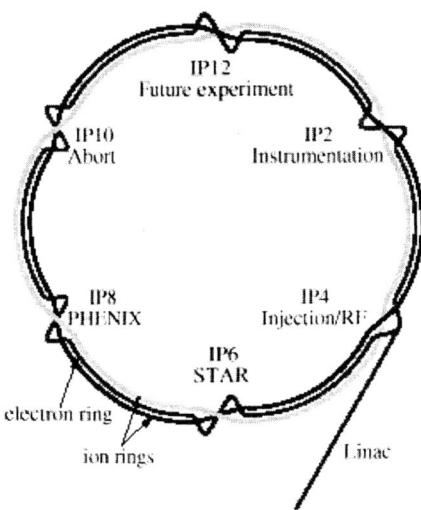

Figure 8. Schematic layout of the electron ring and the two ion rings in the RHIC-based ring-ring collision scenario (See Ref. [6]).

Table 2. Electron linac parameters, common for gold and protons.

Parameter	Units	Value
Electron energy	GeV	10
Electron average current	Amperes	0.27
Collision frequency	MHz	56
Electron bunch population N_e		3×10^{10}
Electron rms emittance, ε_e	μm	0.003

As seen in Figure 7, all the acceleration is done virtually in a straight line, to avoid emittance growth and synchrotron radiation loss in the accelerated beam. The recuperating beam is bent to return to the appropriate sections. The electron source has an injector linac that accelerates the beam to 10 MeV. The power invested (at 0.27 amperes) for this purpose is 2.7 MW. This section has no recuperation. Next is a low gradient 90 MeV (energy gain) pre-accelerator. Here energy recuperation may be done in a dedicated linac section, and the recovered energy would then be fed through waveguides to the accelerating section, shown schematically in Fig. 7 as a connection between the accelerating and decelerating linacs. The 100 MeV beam from the pre-accelerator is fed into a 0.9 GeV (energy gain) intermediate linac with energy recovery done in the same section. Last is the main linac, with an energy gain of 9 GeV. The 10 GeV beam is taken to the collision point. We do not show the details of this in Fig. 7, since the beam may be introduced into a ring-like transport for multiple interaction points, a single IP or anything in-between. The beam is returned at 10 GeV to the entrance of the main linac for deceleration and energy recovery. The beam is decelerated to 1 GeV, and then sent to the intermediate energy linac for deceleration to 100 MeV. In recuperating the energy of the beam in the same linac structure we conform to the conservative limit of the Douglas principle [9] of keeping the energy ratio of the two beam under 10. Deceleration to 10 MeV follows in a dedicated 90 MeV pre-dump linac.

The 10 MeV beam is sent to the beam dump, rated at 2.0 MW (a power level of 2 MW was demonstrated in SLAC beam dumps), assuming a synchrotron radiation loss of 0.7 MW. It is possible to decelerate the beam to a lower energy should the beam dump rating be below 2 MW, but this RF power will not be used for acceleration. Note that any synchrotron radiation power loss (anywhere in the high energy transport) will subtract from the power deposited in the beam dump. The beam dumping is done at under 10 MeV (lower energy corresponds to synchrotron radiation losses) has very minimal activation of the beam dump. An even lower beam energy at the dump and dump power can be obtained by 'braking' the beam in a linac structure, by generating RF power that can be disposed off in dummy loads.

Tables 3 and 4 summarize the electron-gold and electron-proton collision parameters. We take the 720 bunch case, which corresponds to a bunch collision frequency of 56 MHz, and assume a minimal electron cooling of the ion beam. That results in smaller ion beam emittances and allows for larger tune shifts in the electron beam. Other than that, the RHIC parameters are mostly the same as for the ring-ring case. In the case of gold the limit is set by the beam-beam tune shift for the ions. We do not take advantage of the large possible increase in the beam-beam tune shift due to the cooling. That may account for a further increase in the luminosity.

The high-energy IP of RHIC is extremely generous in size, with free space (no accelerator component) for the detectors extending ±10 m from the IP center. A low energy IP (using the second storage ring) can be designed with accelerator components extending much closer to the IP center, thus boosting the luminosity at low energies.

The calculation is done for an electron energy of 10 GeV, however the performance would be unaffected by a much lower (or higher) electron energy. Thermal loading in magnet chambers is not a limitation for the relatively low electron beam current. In addition, larger radii of curvature in the IP optics are possible due to the removal of spin rotation optics from that area, further reducing the thermal loads relative to a ring-ring case.

We conclude that good luminosities can be obtained using a linac-ring collider with modest electron cooling of the ions in the ring. Further increases in the luminosity are possible. In the gold case, the increase would come from pushing the Laslett tune shift to higher values, taking advantage of the cooling. In the proton case the luminosity can be improved by going to a higher current in the electron linac by pushing on the beam-beam tune shift in the ring, once again taking advantage of the cooling.

Table 3. Electron-gold collision parameters, assuming electron cooling of RHIC

Parameter	Units	Value
Gold bunch population		1.9×10^9
Number of bunches		720
Gold 95% normalized emittance	μm	11
RMS beam size at the IP, σ^*	mm	0.05
Electron IP beta function	m	0.72
Electron beam-beam tune shift ξ_e		0.287
Au beam-beam tune shift ξ_i		0.0046
Ion Laslett tune shift, ξ_L		0.008
Luminosity, L	[$cm^{-2}s^{-1}$]	1.2×10^{31}

Table 4. Electron-proton collision parameters, assuming electron cooling of RHIC

Parameter	Units	Value
Proton bunch population		0.93×10^{11}
Number of bunches		720
Proton 95% normalized emittance	μm	9.5
RMS beam size at the IP, σ^*	mm	0.03
Electron IP beta function	m	0.36
Electron beam-beam tune shift ξ_e		0.5
P beam-beam tune shift ξ_i		0.0046
P Laslett tune shift, ξ_L		0.0046
Luminosity, L	[$cm^{-2}s^{-1}$]	2.6×10^{33}

In summary, a linac has a number of distinct advantages over a ring, some of which are specific to collisions with the RHIC beam, most of which are general. Some of the advantages are:

- A linac, in principle, avoids the limitation of the electron beam-beam tune shift inherent in a ring-ring scenario. That allows one to reduce the beam size of the ion storage ring and increase its charge per bunch considerably. However, further study of the beam-beam head-tail instability is required to determine how large an advantage truly exists in the linac-ring collider scenarios.

- A linac has a very low emittance. This leads to a small collision point beam size with a relatively large beta function, increasing luminosity and simplifying the optics of the interaction point.

- The fact that the electrons are used only once means that complicated spin rotation conditions are relaxed. Thus a linac-based collider can provide a polarized electron beam at any energy, while a storage ring is limited to a very narrow energy range.

- The interaction point optics of the linac is much simpler (since the polarization may be prepared well in advance) thus the bending radii are larger. This removes a significant synchrotron-radiation power restriction from the maximum beam current.
- A linac can operate over a wide energy range without sacrificing performance.
- The polarization of a linac is high and can be alternated rapidly at will.
- A linac produces a naturally round beam, to match well with the RHIC beam.
- A linac can be extended to higher energies with a cost that is linear in length whereas an storage ring faces an increase in RF power that goes with the fourth power of its energy.

4. Interaction Region (V. Lebedev)

The minimum achievable beta-function at the IP is determined by the following limitations. First, making β_i^* below rms bunch length does not increase the luminosity and we put $\beta_i^* \geq \sigma_z^i$. Second, decrease of β_i^* causes an increase of beta-functions in the final focus quadrupole. That brings an increase of final focus chromaticity and, consequently, an increase in the sensitivity of the machine tunes to the quadrupole settings and a loss of stability for betatron motion of particles with large momentum deviations. The final focus chromaticity per interaction point can be estimated using the following formula

$$v' \approx \frac{1}{\pi} \frac{F}{\beta^*} \qquad (12)$$

where F is the final focus focusing distance. Usually, quad currents are stabilized better than 0.01% and the increase of machine tunes sensitivity to the quadrupole settings does not cause a problem. Although the total machine chromaticity can be compensated in arcs, the chromaticity of beta-functions due to final focus normally cannot. That limits the total maximum tune shift to about 0.2 per IP. Choosing the energy spread at 5σ to be $5 \cdot 10^{-3}$ we obtain that $v' = \Delta v/(\Delta p/p) \approx 40$. For the focal distance of 6 m we obtain $\beta_i^* \approx 5$ cm. Third, the aperture is limited at the final focus quadrupoles. If superconducting quads are used at the IP, the aperture limitation is not a problem for the case with the electron cooling because of sufficiently small emittance, but it limits the beta-functions at the IP to about 10 cm for the case without cooling. Figure 9 shows an example of the interaction region with $\beta_i^* = 6$ cm. Only one parasitic collision occurs per bunch crossing.

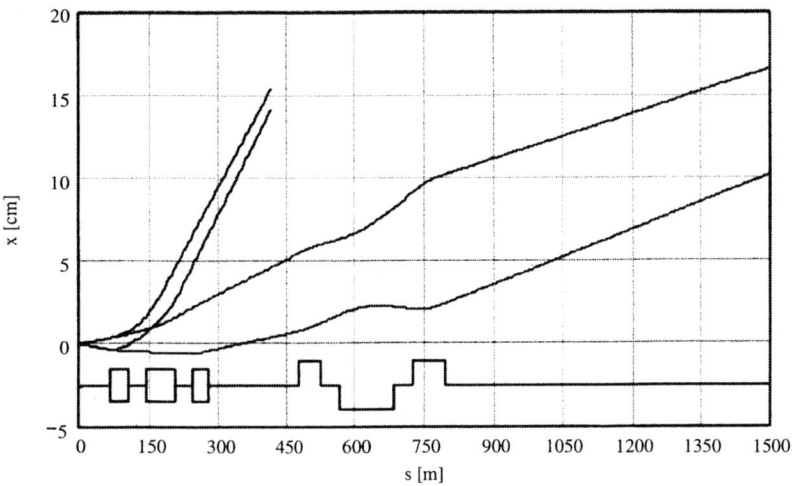

Figure 9. Example of the interaction region design for $\beta^* = 6$ cm.

Accelerator Physics Issues Of The Proton Ring

We will examine transverse and longitudinal intrabeam scattering and collective effects. Estimates of the emittance growth of the electron beam due to collisions with the protons, although not a proton issue, will also be given here, as it may impose a limit on the proton bunch population.

a) Intrabeam Scattering (IBS)
The diffusion time for *transverse* IBS is given by

$$\tau_{tr} = \frac{8}{\Lambda} \gamma_p^3 \frac{R^2}{r_p^2} \left(\frac{\varepsilon_x}{\beta_{av}} \right)^{5/2} \frac{\sigma_z}{c} \frac{1}{N_p} \quad (13)$$

where Λ is the Coulomb logarithm defined by:

$$\Lambda = \ln \left[\sqrt{\left(\frac{\pi}{2}\right)^{1/2} \gamma_p^7 \beta_p^6 \frac{\varepsilon_x^4 v_x^2 \sigma_z}{C^2 r_p^3} \frac{1}{N_p}} \right] . \quad (14)$$

At 15 GeV, for the parameters of design 1A, the diffusion time is approximately 4 hours and for the parameters of design 1B the diffusion time is 1.7 minutes, which implies that electron cooling with cooling rate of similar magnitude would be required. It is expected that cooling rates of order 1 min are feasible, therefore this number may be acceptable. At 50 GeV the diffusion time is 36 minutes for design 2A and 32 seconds for design 2B.

The *longitudinal* IBS also imposes stringent requirements on the cooling rate. The expression for the longitudinal IBS diffusion rate is

$$\tau_{long} = \frac{8\gamma_p^3 \sigma_z \varepsilon_x^{3/2} (\sigma_E/E)^2}{c r_p^2 N_p \Lambda} \sqrt{\frac{C}{2\pi v_x}} \quad . \tag{15}$$

For 15 GeV protons and an rms energy spread equal to 3×10^{-3} the rate for design 1A is ~1 hour. For design 1B, the rate is ~2.5 minutes. At 50 GeV and an rms energy spread of 3×10^{-3}, the rate for design 2A is ~3 hours and for design 2B is 14 minutes.

b) Collective Instabilities

We first examine the longitudinal mode coupling or microwave instability. The threshold is given by

$$\left|\frac{Z_\parallel}{n}\right| = \sqrt{\frac{\pi}{2}} \frac{Z_0 \alpha \gamma \sigma_\delta^2 \sigma_z}{N_p r_p} \quad . \tag{16}$$

For $\left|\frac{Z_\parallel}{n}\right|$ ~0.25 Ω (from LHC [10] and the Tevatron [11]), the number of protons per bunch is limited to ~6×10^{12}. The threshold for the transverse mode coupling instability, in the regime where the bunch length is greater than the beam pipe aperture, can be expressed as [12],

$$|Z_\perp| = Z_0 \frac{\pi \gamma \omega_s \sigma_z}{3 N_p r_p \beta_{av} \omega_0} \quad . \tag{17}$$

For $|Z_\perp|$ ~ 5×10^4 Ω (scaled from LHC), the proton bunch population is limited to ~1.8×10^{12}. In conclusion, both types of instabilities occur at proton bunch populations above the design parameters we chose.

c) Electron Beam Emittance growth due to a Single Collision (G. Krafft)

A single collision disrupts the electron beam and causes emittance growth. In an energy recovering linac, the electron beam with degraded phase space has to be recirculated for energy recovery. Deceleration in the linac cavities can result in scraping and beam loss due to adiabatic antidamping. Therefore, the amount of tolerable beam loss at the linac exit (where the beam size is largest) imposes a limit on the tolerable emittance growth due to a single collision. This, in turn, imposes a limit on the number of protons per bunch.

Let us assume that the maximum tolerable beam loss is 4×10^{-6} which corresponds to 1 µA out of 250 mA, based on Jefferson Lab experience. Assuming a gaussian distribution of the electrons, aperture size of 7 cm and an average beta function in the linac of ~50 m, then the maximum rms normalized emittance, consistent with this amount of beam loss is calculated to be 800 µm. In the small disruption limit, the emittance growth of the electron beam due to a single collision with the proton beam of intensity N_p is given by:

$$\varepsilon_n^2 = \varepsilon_{0,n}^2 + (0.194 r_e N_p)^2 . \tag{18}$$

For $\varepsilon_{0,n}$=60 µm, ε_n = 800 µm, N_p has to be less than 1.5×10^{12} particles per bunch, well below our design specification.

Accelerator Physics Issues Of Energy Recovering Linacs

The list of topics that may need to be examined in an energy recovered linac (ERL) include accelerator transport, coherent synchrotron radiation, Higher Order Mode (HOM) power dissipation and Beam Breakup phenomena. The analysis of the head-tail instability, in a linear approximation, is also outlined here. We will now examine each topic in turn.

a) Accelerator Transport

Topics of longitudinal and transverse matching and beam loss may all be relevant. Depending on the bunch length requirements at the IP, longitudinal gymnastics may be necessary and it is routinely and successfully done at JLab IRFEL [13]. Transverse phase space matching into the IP is almost invariably required for good overlap to maximize luminosity, and maybe required in order to transport and energy recover the disrupted electron beam after the collisions. Adiabatic antidamping imposes additional constraints in the linac optics. Another issue specific to ERLs has to do with the dynamic range of the linac and the ability to confine two beams of different energies in the same focusing structure and it may impose a constraint on the ratio of injected to final beam energies.

Understanding the origin of and being able to control beam loss are crucial in an ERL with the parameters outlined above. In the JLab IRFEL several indicators place an upper limit on the amount of beam loss in the recirculator to 2 µA out of 5 mA. This amount of loss, although extremely small is unacceptable for the ERL designs discussed here, as it can potentially give rise to hundreds of kW of uncontrolled, lost beam power. More work is required to understand both the origin of the loss and possible cures.

b) Coherent Synchrotron Radiation

When the beam travels around a bend it radiates, and when the radiation wavelength is longer than the bunch length, it radiates coherently and interacts to deteriorate the beam quality. Both transverse and longitudinal self-forces can cause emittance growth, which is potentially serious for high brightness beam quality preservation. A self-consistent, two-dimensional code has been developed [14] and is being verified against experimental data obtained from the JLab IRFEL and the CTF II Facility at CERN.

c) HOM Power Dissipation

Power in HOMs, primarily longitudinal, depends on the product of bunch charge and average current and could present a serious enough constraint that engineering choices are imposed for improved cryogenic efficiency. The power dissipated by the beam in HOMs is given by

$$P_{diss} = k_{\|} Q \bar{I} \tag{19}$$

For CEBAF cavities, the calculated loss factor is equal to 5.4 V/pC for 1mm rms bunch length [15], therefore the power dissipated by the beam for 0.264 A average current is approximately 5 kW per cavity. It is important to address the question of where these losses go. A simple model was developed [16] that shows that a) the fraction of the power dissipated on the *cavity walls* is a strong function of bunch length and b) most of the power is actually extracted into loads, or propagates out of the cavity through the beam pipe openings. With reasonable assumptions for the extraction efficiency of the HOM power (Q~2000), a small fraction of the power, of the order of a few Watts, is expected to be deposited on the walls, for the parameters quoted here. Engineering studies on HOM cooled absorbers between cavities or cryomodules are recommended.

d) Beam Breakup

BBU refers to a variety of collective phenomena that can limit the performance of srf energy recovering linacs. These coherent effects include single-bunch, single-pass phenomena which limit the charge per bunch, and multi-bunch phenomena which limit the average current. Single bunch effects include energy spread induced by variation of the longitudinal wakefield across the bunch, and emittance growth induced by transverse wakefields across the bunch. The induced relative energy spread at 3 GeV and 0.264 A average current is 5×10^{-4}. The single bunch BBU emittance growth was discussed earlier and we concluded that it does not become an issue until the electron bunch population is increased by approximately an order of magnitude.

Multipass, multi-bunch BBU occurs when a recirculating beam through a linac cavity leads to a transverse instability. Transverse beam displacement on successive recirculations can excite HOMs that further deflect the initial beam. The recirculated beam and cavities form a feedback loop, which, for beam current greater than the threshold current of the instability, can be driven unstable. The effect is worse in srf cavities because of the higher Q's of the HOMs. The threshold current depends on various cavity and lattice parameters, including the Q's, frequencies and R/Q's of the HOMs, the beam energy, the beta functions and phase advance in both planes and the recirculation path length.

A two-dimensional simulation code, TDBBU [17] has been developed for the calculation of the threshold current in an actual machine configuration. Recent TDBBU simulations of an ERL with 10 MeV injection energy, 5 GeV final energy, average beta functions in the linac of approximately 15 m in both planes (60 m maximum) and HOM data from the 9-cell TESLA cavities [18], resulted in a threshold current of about 225 mA [19]. Furthermore, the typical growth rate of the instability just above threshold is in the msec range, allowing for the possibility of controlling the instability with feedback.

e) Beam-beam induced Head-Tail Instability (V. Lebedev)

For the case when the electron bunches are much shorter than the proton bunches, one can make a simple estimate of the effect in the linear approximation. To find the corresponding transverse impedance one can consider that a small fraction of the proton beam of length Δs is displaced by x. This will deflect the electron beam by the angle equal to

$$\Delta\theta_e = \frac{Zr_e}{\gamma_e \sigma_i^{*2}} \frac{dN_i}{ds} x\Delta s \qquad (20)$$

where dN/ds is the charge per unit length in the proton beam. Then, the electron beam starts performing a betatron oscillation in the field of proton beam. For sufficiently small ξ_e one can neglect this reverse attraction for the electron bunch and consider the electron bunch moving along a straight line, but this increasing distance between centers of electron and proton bunches causes a deflection of protons in the bunch tail by an angle

$$\Delta\theta_i(s) = \frac{Zr_p N_e}{\gamma_p \sigma_e^{*2}}(s-s')\Delta\theta_e = \frac{Z^2 r_e r_p N_e}{\gamma_e \sigma_i^{*2} \gamma_i \sigma_e^{*2}} \frac{dN_i}{ds}(s-s')x\Delta s \quad . \qquad (21)$$

This yields the following expressions the transverse wake function,

$$W(s) = \frac{r_e N_e}{\gamma_e \sigma_i^{*2} \sigma_e^{*2}} s \, , \quad (22)$$

and the transverse impedance,

$$Z_\perp(s) = i \frac{Z_0}{4\pi} \frac{r_e N_e}{\gamma_e \sigma_i^{*2} \sigma_e^{*2}} \frac{c^2}{\omega^2} \, , \quad (23)$$

where $Z_0 = 377 \, \Omega$. The threshold for the strong head-tail instability can be written as

$$N_{th} \approx \frac{16\pi \gamma_i \sigma_{si} v_s}{r_p Z^2 \, \mathrm{Im}\langle \beta Z_\perp(c/\sigma_{si}) \rangle} \, . \quad (24)$$

Combining the above two formulas one finally obtains the limitation:

$$D_e \xi_i \leq 4 v_s \, , \quad (25)$$

where

$$\xi_i = \frac{Z r_p N_e \beta_i^*}{4\pi \gamma_i \sigma_e^{*2}} \quad (26)$$

is the betatron tune shift of the ion beam and

$$D_e = \frac{Z r_e N_i \sigma_{si}}{\gamma_e \sigma_i^{*2}} \quad (27)$$

is the electron beam disruption parameter.

Eq. (25) can impose a severe limit on possible collider parameters and luminosity estimates in this limit were given at the beginning of this report. Nevertheless, one needs to keep in mind that the estimate was obtained in the linear approximation while the actual transverse field of the bunch is strongly non-linear across the bunch. Although it cannot eliminate the instability, it should move the threshold to higher beam currents. Note that the estimate itself is not good enough to claim accuracy better than factor of two. An independent study using a two-particle model arrived at the same eq. (25) within a factor of two [20]. Performing more detailed study and computer simulations of the instability is going to be one of the high priority tasks at Jefferson Lab in the near future. We also note that the proton bunch is quite long and a transverse feedback system for suppression of the instability's lowest modes may be feasible, which can increase the instability threshold by a factor of 2 or 3. Thus, we currently believe that additional increase in the instability threshold by a factor of 3-10 is possible.

Technological Issues

a) *High current source of polarized electron (C. Sinclair)*

The generation of high average current, high polarization electron beams is a significant technological issue. The state-of-the art in polarized electron sources was reviewed in the PAC99 article by Sinclair [21]. The prospects for sources of high average current polarized electrons was presented in the Proceedings of the 2nd eRHIC Workshop [22]. Presently, polarized sources at JLab have cathode operational lifetimes one order of magnitude greater than those reported by Sinclair at PAC99. Cathode operational lifetime in these sources is limited only by ion backbombardment, and now exceeds 100,000 Coulombs/cm^2. While construction of a high average current polarized source with modest polarization (~ 37%) is probably within reach, a source with a high polarization (~ 75%) faces a number of serious technological challenges. Significant R&D would be required before one could plan on a source delivering a high average current at high polarization.

b) *Electron Cooling*

This is an important topic, which is mentioned here only for completeness. It is thoroughly addressed in a separate section of this report.

Concluding Remarks And Outlook

Preliminary results of a feasibility study of an energy recovering electron linac on a proton ring collider are presented. Luminosities at the 10^{33} level appear attainable and the linac-on-ring scenario presents a significant advantage with respect to spin manipulations, energy variability and synchrotron radiation power loading of the detectors.

R&D topics that would be required before such a facility is designed and built have been identified and include:
1. High current polarized electron source
2. High current (~100 mA) demonstration of energy recovery
3. Theoretical and, if possible, experimental investigation of the beam-beam induced head-tail instability and feedback
4. Electron cooling and its ramifications on Laslett and beam-beam tuneshifts
5. Development of multipass BBU feedback

Recently, recirculating, energy-recovering linacs have attracted much attention and are being considered for a number of applications, such as drivers for synchrotron radiation sources, and high average power FELs. A number of the listed

R&D topics, especially those related to the energy recovery of high average currents, are being pursued by these communities, so it is safe to assume that progress will be rapid in these directions.

REFERENCES

1. T. I. Smith, et al., "Development of the SCA/FEL for use in Biomedical and Materials Science Experiments," **NIM** A259 (1987) 1-7
2. D. W. Feldman et al., "Energy Recovery in the Los Alamos Free Electron Laser," **NIM** A259 (1987) 26-30
3. N. R. Sereno, University of Illinois, Ph.D. Thesis (1994)
4. G. R. Neil, et al., "Sustained Kilowatt Lasing in a Free Electron Laser with Same-Cell Energy Recovery," **Physical Review Letters**, Volume 84, Number 4 (2000)
5. I. A. Koop, et al., "Conceptual Design Study of the Electron-Proton Storage Ring Collider with Polarized Beams," this document and Proc. of the 2nd EPIC Workshop, MIT, September 2000
6. I. Ben-Zvi, J. Kewisch, J. Murphy and S. Peggs, "Accelerator Physics Issues in eRHIC," Proc. of the 2nd eRHIC Workshop, Yale University, April 2000
7. P. Kneisel, "Preliminary Experience with "In-Situ" Baking of Niobium Cavities," Proc. of RF Superconductivity Workshop, Santa Fe, November 1999.
8. A. W. Chao, B. Richter and C.-Y. Yao, "Beam Emittance Growth Caused by Transverse Deflecting Fields in a Linear Accelerator," **NIM** 178 (1980) 1-8
9. D. Douglas, Jefferson Lab Technical Note, TN-98-040, 1998
10. L. Vos, Private communication
11. K.-Y. Ng, Private communication
12. A. W. Chao, "Physics of Collective Beam Instabilities in High Energy Accelerators," Wiley Series in Beam Physics and Accelerator Technology, 1993
13. G. A. Krafft and D. Douglas, Private communication
14. R. Li, "The Impact of Coherent Synchrotron Radiation on the Beam Transport of Short Bunches," Proc. of the 1999 Particle Accelerator Conference, New York, 1999
15. B. Yunn, Private communication
16. L. Merminga, et al., "Specifying HOM-power Extraction Efficiency in a High Average Current, Short Bunch Length SRF Environment," Proc. of Linac 2000 Conference, Monterey, August 2000
17. G. A. Krafft and J. J. Bisognano, "Two dimensional simulations of multipass beam breakup," Proc. of 1987 Particle Accelerator Conference, pg. 1356
18. J. Sekutowicz, "Higher Order Mode Coupler for TESLA," DESY Report, TESLA 94-07, February 1994
19. I. Bazarov, Private communication
20. R. Li and J. Bisognano, Private communication
21. C. K. Sinclair, "Recent Advances in Polarized Electron Sources," Proc. of the 1999 Particle Accelerator Conference, New York, 1999
22. P. Hartmann et al., "Polarized Electron Linac Sources," Proc. of the 2nd eRHIC Workshop, Yale University, April 2000

Wide Angle Compton Scattering

Rainer Jakob

Fachbereich Physik, Universität Wuppertal, D-42097 Wuppertal, Germany

Abstract. We present the handbag contribution to Wide Angle Compton Scattering (WACS) at moderately large momentum transfer obtained with a proton distribution amplitude close to the asymptotic form. In comparison it is found to be significantly larger than results from the hard scattering (pQCD) approach.

INTRODUCTION

Compton scattering off nucleons provides us with valuable information on the nucleon structure. In general, the Compton process involves the relativistic propagation of an excited, composite system, a bound state, which is extremely difficult to describe. Fortunately, in special kinematic situations the description of the process becomes much simpler. In this contribution we focus on Compton scattering off protons with large momentum transfer and small (or zero) photon virtuality, i.e. Compton scattering at wide angles (WACS). Here the process receives its main contribution from the lowest Fock state of the nucleon, i.e. a three quark configuration. The propagation of hard gluons and quarks is described perturbatively. We argue that in the region of large, but not yet asymptotically large, momentum transfer the process is dominated by the handbag contribution involving new Compton form factors defined from skewed parton distributions. We compare the handbag contribution calculated from the overlap of soft wave fuctions [1,2] with the results obtained in the hard scattering approach [3].

HARD VS. SOFT SCATTERING MECHANISM

It is generally accepted that the hard scattering mechanism in the context of perturbative QCD [4] provides the correct description for processes at asymptotically large momentum transfer. In this picture a hard scattering amplitude describes the redistribution of the transfered momentum among partons via hard gluon exchange, and the non-perturbative part is given as distribution amplitudes (DAs), i.e. light-cone wave functions (LCWFs) integrated over transverse momenta. There is a minimal number of hard gluons required to connect all parton lines, the lowest Fock state gives the dominant contribution. In Fig. 1 the diagrams **c** and **d** (and

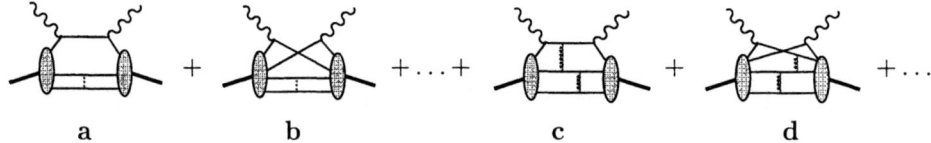

FIGURE 1. Some of the diagrams contributing to WACS. The dots between **b** and **c** indicate an intermediate class of diagrams with one exchanged hard gluon, whereas the dots behind **d** stand for diagrams with a higher number of gluons (α_s corrections), and diagrams from higher Fock states (power corrections).

diagrams obtainable from them by permutations) stand for the hard scattering picture.

For exclusive processes at moderately large momentum transfer, on the other hand, there is a longstanding debate on the question which is the dominant reaction mechanism. The diagrams **a** and **b** in Fig. 1 stand for the Feynman mechanism, to be calculated from the direct overlap of soft LCWFs. No redistribution of the momentum transfer is necessary, if the momentum fraction of the active parton line is large. The wave functions suppress such asymmetric configurations; asymptotically these diagrams will be power corrections to the hard scattering ones. At intermediate momentum transfer, however, the relative importance cannot be judged *a priori*. Explicit numerical calculations are necessary.

WACS IN THE HARD SCATTERING APPROACH

For a comparison we will quote the results from a recent leading order calculation of real WACS off protons in the hard scattering approach [3] superseding (and partly correcting) earlier calculations [5,6].

The unpolarised cross section, scaled by s^6, obtained with different DAs is shown in Fig. 2 in comparison to the data. To minimize the influence of choices for the $\alpha_s(\mu)$ argument, the same quantity normalized to the factor $(Q^4 F_1^p(Q^2))^2$, where F_1^p is the Dirac form facor of the proton, is also shown. Uncertainties are expected to cancel from this ratio to a large extent.

Clearly, the results of the hard scattering (pQCD) contributions to the unpolarised cross sections fall short to describe the available data, which are admittedly at rather lowish momentum transfers. From the ratio shown on the RHS the authors of [3] conclude that '...*it seems unlikely that the elastic proton form factor and the Compton scattering amplitudes are both described by pQCD at presently accessible energies*'.

These results suggest, that hard scattering is not the dominant reaction mechanism at the intermediate large momentum transfer, a situation very similar to the elastic nucleon (and pion) form factors.

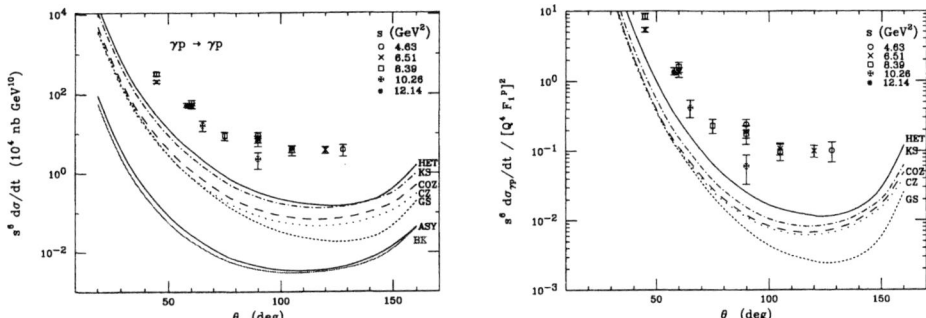

FIGURE 2. Left: The hard scattering contribution to the unpolarised differential cross section for WACS obtained with different DAs compared to the data [7]. Right: The same quantity scaled by the pQCD result for $(Q^4 F_1^p(Q^2))^2$. Figures reproduced from [3] with one additional curve (BK) provided by the same authors.

WACS IN THE SOFT PHYSICS APPROACH

The contribution from the Feynman mechanism to WACS is described by a handbag diagram [8]. The diagram factorises in a hard photon-parton amplitude and a non-perturbative part [1], the latter described by a skewed parton distribution at vanishing 'skewedness', which can be calculated from the overlap of LCWFs, as indicated by diagram **a** in Fig. 1. The cat-ears diagram **b** was shown to be power suppresssed relativ to the handbag diagram [1]. The (unpolarised) differential cross section can be written

$$\frac{d\sigma}{dt} = \frac{2\pi\alpha_{em}^2}{s^2}\left[-\frac{u}{s}-\frac{s}{u}\right]\left\{\frac{1}{2}\left(R_V^2(t)+R_A^2(t)\right) - \frac{us}{s^2+u^2}\left(R_V^2(t)-R_A^2(t)\right)\right\} \quad (1)$$

with new form factors [8] specific to Compton scattering depending on $-t$ only

$$\sum_a e_a^2 \int_0^1 \frac{dx}{x} p^+ \int \frac{dz^-}{2\pi} e^{i\,xp^+z^-}\,\langle p'|\,\overline{\psi}_a(0)\,\gamma^+\,\psi_a(z^-) - \overline{\psi}_a(z^-)\,\gamma^+\,\psi_a(0)\,|p\rangle$$

$$= R_V(t)\,\bar{u}(p')\,\gamma^+ u(p) + R_T(t)\,\frac{i}{2m}\,\bar{u}(p')\,\sigma^{+\nu}\Delta_\nu\,u(p)\,. \quad (2)$$

R_T being related to nucleon helicity flips is neglected in Eq. (1). An analogous definition holds for R_A involving the axialvector nucleon matrix element.

Assuming a Gaussian model for transverse parton momenta the form factors factorise in ordinary parton distribution functions (PDFs) and a (t,x) dependent exponential for each N parton Fock state separately

$$R_V^{(N)}(t) = \int_0^1 \frac{dx}{x}\exp\left[\frac{a_N^2 t}{2}\frac{1-x}{x}\right]$$
$$\times \left\{e_u^2\,[u_v^{(N)}(x)+2\bar{u}^{(N)}(x)] + e_d^2\,[d_v^{(N)}(x)+2\bar{d}^{(N)}(x)] + e_s^2\,2\bar{s}^{(N)}(x)\right\}, \quad (3)$$

and analogously for $R_V \to R_A$, $q(x) \to \Delta q(x)$. With a phenomenologically based model for the x-dependence of the LCWF, the BK distribution amplitude [9] for the lowest Fock state, the form factors R_V and R_A can be calculated. The results are shown in Fig. 3 together with estimates for the additional contributions from the next higher Fock states (N=4,5), and an estimate for the the effect of all Fock states based on parametrizations for PDFs [10]. For details see Ref. [1]. Inserting

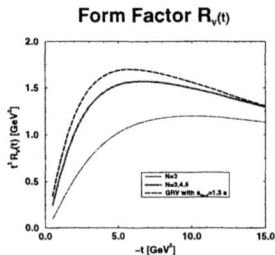
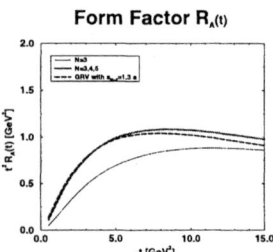

FIGURE 3. CS Form factors calculated from the valence Fock state only (thin solid line), N=3,4,5 Fock states (thick solid line), and with an additional estimate for higher Fock states (dashed line). Right: $R_V(t)$. Left: $R_A(t)$.

the form factors into Eq. (1) the cross section of WACS are obtained as displayed in Fig. 4. A comparison shows that the predictions for cross sections from the

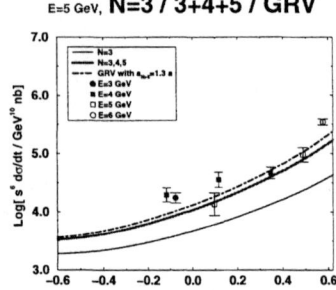
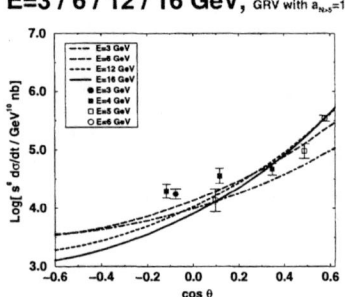

FIGURE 4. Cross sections for WACS obtained from the handbag diagram. Data are from [7]. Left: for $E = 5$ GeV the contributions from the valence Fock state, the N=3,4,5 Fock states, and for all Fock states are compared. Right: the cross sections for different photon energies obtained by the 'all Fock state estimate'. Data from [7].

handbag diagram are much higher then the corresponding ones obtained in the hard scattering picture and can fairly well describe the present data. Note that for a direct comparison a curve was added in Fig. 2 (left) obtained within the hard scattering approach in exactly the same way (same value for the parameter f_N) as the other curves in Fig. 2, but with the BK distribution amplitude as input.

Of particular interest is the initial state helicity correlation

$$A_{\mathrm{LL}} \frac{d\sigma}{dt} = \frac{1}{2} \left(\frac{d\sigma(\mu=+1, \nu=+1/2)}{dt} - \frac{d\sigma(\mu=+1, \nu=-1/2)}{dt} \right) \quad (4)$$

where μ, ν are the helicities of the incoming photon and proton, respectively. The prediction for this quantity from the handbag diagram is distincively different from predictions obtained in the hard scattering approach, or the diquark model [11].

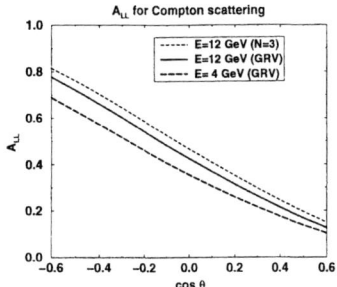 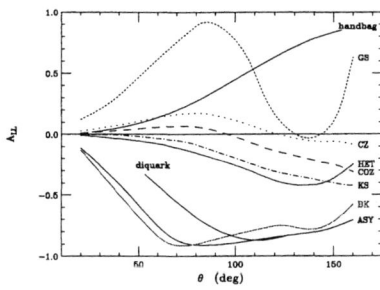

FIGURE 5. Initial state helicity correlation. Left: Predictions of the handbag contribution for different energies. Right: Comparison of different predictions at $E = 4\,\mathrm{GeV}$ (taken from [3]).

ACKNOWLEDGMENTS

This contribution is based on work done in collaboration with M. Diehl, Th. Feldmann, and P. Kroll. I am grateful to T. Brooks and L. Dixon for the permission to reproduce their figures for comparison and for providing the curve BK in Fig 2.

REFERENCES

1. M. Diehl, T. Feldmann, R. Jakob and P. Kroll, Eur. Phys. J. **C8** (1999) 409.
2. M. Diehl, T. Feldmann, R. Jakob and P. Kroll, Phys. Lett. **B460** (1999) 204.
3. T. C. Brooks and L. Dixon, hep-ph/0004143.
4. G. P. Lepage and S. J. Brodsky, Phys. Rev. **D22** (1980) 2157;
   A. V. Efremov and A. V. Radyushkin, Theor. Math. Phys. **42** (1980) 97.
5. A. S. Kronfeld and B. Nizic, Phys. Rev. **D44** (1991) 3445.
6. M. Vanderhaeghen, P. A. Guichon and J. Van de Wiele, Nucl. Phys. **A622** (1997) 144c.
7. M. A. Shupe et al., Phys. Rev. **D19** (1979) 1921.
8. A. V. Radyushkin, Phys. Rev. **D58** (1998) 114008.
9. J. Bolz and P. Kroll, Z. Phys. **A356** (1996) 327.
10. M. Glück, E. Reya and A. Vogt, Eur. Phys. J. **C5** (1998) 461;
    M. Glück, E. Reya, M. Stratmann and W. Vogelsang, Phys. Rev. **D53** (1996) 4775.
11. P. Kroll, M. Schürmann and W. Schweiger, Int. J. Mod. Phys. **A6** (1991) 4107.

Structure of the Goldstone Bosons

Roy J. Holt and Paul E. Reimer

Argonne National Laboratory
Argonne, Illinois 60439

Abstract. The feasibility of measuring the pion and kaon structure functions has been investigated. A high luminosity electron-proton collider would make these measurements feasible. Also, it appears feasible to measure these structure functions in a nuclear medium. Simulations using the RAPGAP Monte Carlo of a possible pion structure function measurement are presented.

Understanding hadron structure from the underlying quark and gluon degrees of freedom and understanding modifications of hadrons in nuclear matter are two of the most important goals of nuclear physics. The light mesons have a central role in nucleon and nuclear structure. The masses of the lightest hadrons, the mesons, are believed to arise from explicit chiral symmetry breaking. In particular, the light mesons are the Goldstone bosons of quantum chromodynamics [1]. The pion, being the lightest meson, is particularly interesting not only because of its importance in chiral perturbation theory, but also because of its importance in explaining the quark sea in the nucleon and the nuclear force in nuclei.

THE PION STRUCTURE FUNCTION

At present, the pion is believed to contain a valence quark and antiquark as well as a partonic sea. Several theoretical calculations are aimed at explaining the pion structure function in the valence region. These include Dyson-Schwinger [2] and Nambu Jona-Lasinio models [3,4]. Lower order moments of the structure function were determined in lattice gauge calculations [5]. Typical agreement with the pion structure function is shown in Fig. 1. Here, a curve from the Dyson-Schwinger model is compared with the data from a pionic Drell-Yan experiment [6] in the valence region. The general features of the valence structure of the pion are qualitatively understood. However, there is no good understanding of the pion sea.

Recently, measurements of the pion structure function [7] at very low x, in the region of the pion sea have been performed. The results of this work show two interesting findings: (1) the sea in the pion has the same shape in x as the sea in the proton, and (2) the pion sea has approximately one-third of the magnitude as

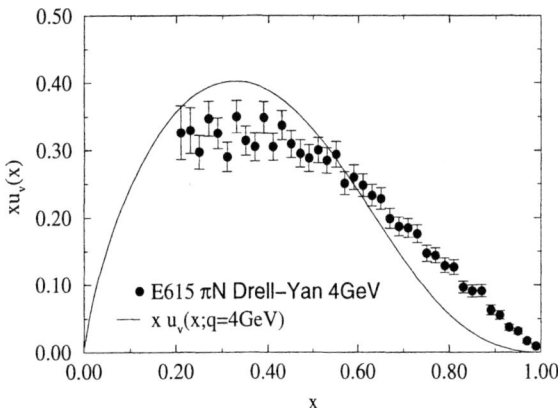

FIGURE 1. Existing data for the pion structure function from Drell-Yan scattering [6]. The solid curve represents a calculation of Hecht et al. [2].

the sea in the proton. This latter result is especially surprising since one expects that the pion sea to be two-thirds the value of the proton sea. These findings are even more surprising from the viewpoint of a chiral quark model [8]. This model predicts that the pion sea carries a larger momentum fraction than the proton sea. It appears that a comparison of the sea in the pion and the proton is a clue to understanding the nonperturbative structure of constituent quarks. Thus, it is essential to measure the pion structure function, especially the sea component, throughout the x region from 0.02, the highest value of the HERA data, up to 0.3, the lower value of the Drell-Yan data. This will map out the sea region where models should have a high degree of validity. Also, since there appears to be a discrepancy between the data and the theoretical calculation at very high x, another measurement using a different technique at high x would be important.

KAON STRUCTURE FUNCTION

The valence structure of the kaon is comprised of a light u or d quark/antiquark and a strange quark/antiquark. If our understanding of the meson structure is correct, then the large difference between the strange quark and u or d quark masses gives rise to a very interesting effect for the kaon structure function. In this case, the strange quark, because of its large mass, carries more of the kaon's momentum than the u quark, say. Then, the u_v quark distribution in the kaon should be shifted lower in x than that in the pion.

A Nambu Jona-Lasinio calculation [3,4] exhibits this behavior as shown in Fig. 2. Here, the ratio of the valence u quark in the kaon to that in the pion is shown as the curve in the figure. Drell-Yan measurements [9] of the ratio of K^- to π^- shows a consistency with unity over most of the x region, with a suggestion that the ratio

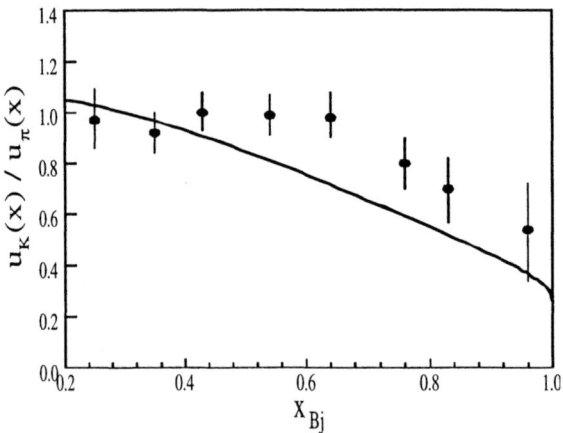

FIGURE 2. Existing data for K^-/π^- ratio from Drell-Yan scattering [9]. The solid curve represents a calculation of Suzuki [12,3].

is dropping at high x. However, the data are not of sufficient quality to verify our understanding of this process. Thus, it is essential to measure the structure function at high x as well as in the sea region.

MESON STRUCTURE IN THE NUCLEAR MEDIUM

The role of pions, and particularly, pion excess in the nuclear medium have been long-standing issues in nuclear physics. The pion excess has not been observed in either Drell-Yan experiments [10] at FNAL or in $A(e,e'\pi)$ reactions [11] at Jefferson Lab. The main question is whether the pion or kaon structure function in a nuclear medium is modified from the free structure function. Calculations within the framework of a Nambu Jona-Lasinio model [12] indicate that the medium modification for the pion structure function should be small. However, if one invokes Brown-Rho scaling [13], then the effect is large. These calculations are shown in Fig. 3. Here the NJL calculation is the solid curve, while the dashed curve represents the Brown-Rho scaling.

MEASUREMENT OF THE MESON STRUCTURE FUNCTIONS

The scattering process illustrated in Fig. 4 was simulated using the RAPGAP Monte Carlo program [14]. RAPGAP models processes in which there is a large rapidity gap between a fast outgoing nucleon and the remainder of the inelastic scattering fragments. These processes include DIS from an exchanged pion [15–17]

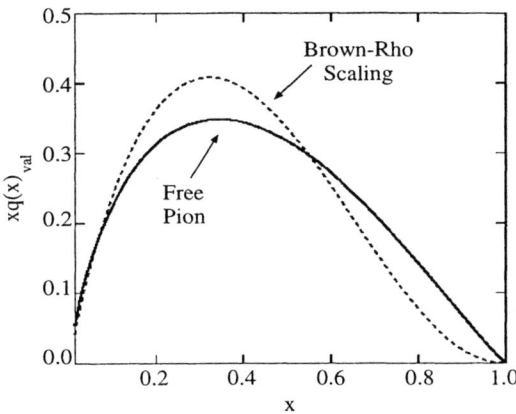

FIGURE 3. The solid curve represents the NJL calculation in a nuclear medium, while the dashed curve gives the effect of Brown-Rho scaling in nuclear matter [12].

or pomeron.. A comparison of results from RAPGAP with HERA data show reasonable agreement for fast outgoing neutrons [7].

For a 5 GeV electron beam on a 25 GeV proton beam, RAPGAP calculates 22 nb cross section for the $eq \to e'q'$ process. The expected accuracy of the experiments was calculated using events in which a "spectator" neutron was identified. In general, the neutron is scattered less than 50 mrad from the nominal proton beam axis. Events were cut on $-q^2 > 1$ GeV^2. The expected errors are shown in Fig. 5. A luminosity of 10^{32} cm^{-2}s^{-1} was assumed for a run lasting 10^6 s.

The K^+ structure function can be measured by considering deep inelastic scattering from the kaon cloud surrounding a proton. The basic Feynman diagram would be the same as in Fig. 4 with the pion replaced by a kaon and the neutron replaced by a Λ. The probability for scattering from the K^+ cloud surrounding the proton should be comparable to that for the π^+ since the KNΛ coupling constant is comparable to that of the πNN vertex. In fact, one would only expect about a factor of two reduction in the vertex function for the kaon compared to the pion.

The difficulty with this process is in the detection of the Λ. The Λ decays predominantly (64%) to a proton and a π^-. Thus, a special forward proton spectrometer as well as a forward pion spectrometer would be necessary. This should

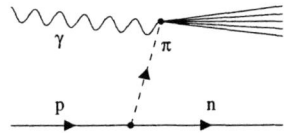

FIGURE 4. Deep inelastic scattering from the pion cloud surrounding a proton.

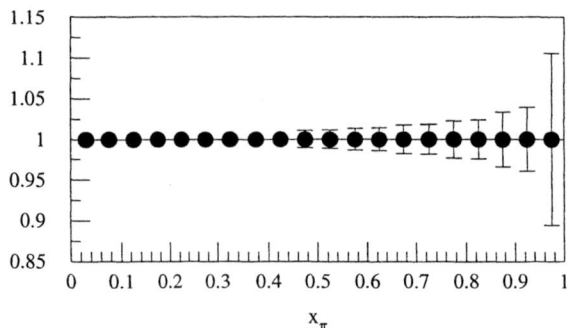

FIGURE 5. Simulated errors for DIS events using a 5 GeV electron beam on a 25 GeV proton beam with a luminosity of 10^{32}cm^{-2}s^{-1} and 10^6s of running.

be feasible since the ZEUS and H1 experiments at HERA have already successfully employed forward proton spectrometers. When designing the ring magnets, one should take into account the possibility of detecting both positively and negatively charged forward going hadrons. Simulations for this part of the experiment would be necessary to optimize the detection efficiency.

A collider should also render these studies feasible for nuclei up to ^{4}He. For the pion case, the idea would be to detect all of the forward going nucleons from the deep inelastic scattering from the pion. In the case of a deuterium target, for example, one would detect both forward going neutrons or forward going protons, depending on whether the DIS occurred from the π^+ or the π^-, respectively.

SUMMARY

In summary, a measurement of the pion and kaon structure functions over a large $x_{\pi/K}$ region was shown to be feasible with an electron proton collider, where the electron energy is 5 GeV, the proton energy is 25 GeV, and the luminosity is 10^{32} cm^{-2}s^{-1}. A collider will also open up other very interesting possibilities such as a measurement of the meson structure function in the nuclear medium.

ACKNOWLEDGMENTS

We wish to thank C. Roberts, G. Levman, M. Derrick and T.-S. H. Lee for very useful discussions. In addition, we thank H. Jung and D. H. Potterveld for valuable assistance with RAPGAP. This work was supported by the U. S. Department of Energy, Nuclear Physics Division, under contract No. W-31-109-ENG-38.

REFERENCES

1. Maris, P., Roberts, C.D., Tandy, P.C., *Phys. Lett.* **B420**, 287 (1998).
2. Hecht, M. B., Roberts, C. D., Schmidt, S. M., preprint (2000), nucl-th/0008049.
3. Shigetani, T., Suzuki, K., Toki, H. *Phys. Lett.* **B308**, 383 (1993).
4. Davidson, R. M., Arriola, E. Ruiz *Phys. Lett* **B348**, 163 (1995).
5. Best, C. *et al. Phys. Rev.* **D56**, 2743 (1997).
6. Conway, J. S. *et al. Phys. Rev.* **D39**, 39 (1989).
7. Adloff, C. *et al.* (H1 Collaboration) *Eur. Phys. J. C* **6**, 587 (1999).
8. Suzuki, K., Weise, W. *Nucl. Phys.* **A634**, 141 (1998).
9. Badier, J., *et al. Phys. Lett.* **B93**, 354 (1980).
10. Alde, D. M. *et al. Phys. Rev. Lett.* **64**, 2479 (1990).
11. Jackson, H. E. (NucPi Collaboration) *Sixteenth Int'l Conf. on Few Body Systems* Taipei, preprint (2000).
12. Suzuki, K. *Phys. Lett.* **B368**, 1 (1996).
13. Brown, G. E., Rho, M. *Phys. Rev. Lett.* **66**, 2720 (1991).
14. Jung, H., *Comp. Phys. Commun.* **86**, 147 (1995).
15. D'Alesio, U. and Pirner, H. J. *Eur. Phys. J. A* **7**, 109 (2000).
16. Holtmann, H. *et al. Nucl. Phys.* **A569**, 631 (1996).
17. Kopeliovich, H. *et al. Z. Phys.* **6**, 587 (1999).

Calculation of Fragmentation Functions in Two-hadron Semi-inclusive Processes

A. Bianconi[a], S. Boffi[b], D. Boer[c], R. Jakob[d]
and M. Radici[b]

[a] *Dipartimento di Chimica e Fisica per i Materiali e per l'Ingegneria, Università di Brescia, I-25133 Brescia, Italy*
[b] *Dipartimento di Fisica Nucleare e Teorica, Università di Pavia, and Istituto Nazionale di Fisica Nucleare, Sezione di Pavia, I-27100 Pavia, Italy*
[c] *RIKEN/BNL Research Center, Physics Department, NY 11973 Upton, USA*
[d] *University of Wuppertal, Theoretical Physical Department, D-42097 Wuppertal, Germany*

Abstract. We investigate the properties of interference fragmentation functions arising from the emission of two leading hadrons inside the same jet for inclusive lepton-nucleon deep-inelastic scattering. Using an extended spectator model for the mechanism of the hadronization, we give a complete calculation and numerical estimates for the examples of a proton-pion pair produced with invariant mass on the Roper resonance, and of two pions produced with invariant mass close to the ρ mass. We discuss azimuthal angular dependence of the leading order cross section to point up favourable conditions for extracting transversity from experimental data.

INTRODUCTION

Because of the still lacking rigorous explanation of confinement, the nonperturbative nature of quarks and gluons inside hadrons can be explored by extracting information from distribution (DF) and fragmentation functions (FF) in hard scattering processes. There are three fundamental DF that completely characterize the quark inside hadrons at leading twist with respect to its longitudinal momentum and spin: the momentum distribution f_1, the helicity distribution g_1 and the transversity distribution h_1. At variance with the first two ones, h_1 is difficult to address because of its chiral-odd nature. A complementary information can come from the analysis of the hadrons produced by the fragmentation process of the final quark, namely from FF. So far, only the leading unpolarized FF, D_1, is partly known, which is the counterpart of f_1. The basic reason for such a poor knowledge is related to the difficulty of measuring more exclusive channels in hard processes. The new generation of machines (HERMES, COMPASS, RHIC) and planned projects (ELFE, EPIC) allow for a much more powerful final-state identification and, therefore, for a wider and deeper analysis of FF, particularly when

Final State Interactions (FSI) are considered. In this context, naive "T-odd" FF naturally arise because the existence of FSI prevents constraints from time-reversal invariance to be applied to the fragmentation process [1]. This new set of FF includes also chiral-odd objects that become the natural partner needed to isolate h_1.

The presence of FSI allows that in the fragmentation process there are at least two competing channels interfering through a nonvanishing phase. However, this is not enough to generate naive "T-odd" FF. Excluding *ab initio* any mechanism breaking factorization, there are basically two ways to describe the residual interactions of the leading hadron inside the jet: assume the hadron moving in an external effective potential, or model microscopically independent interaction vertices that lead to interfering competing channels. In the former case, introduction of an external potential in principle breaks the translational and rotational invariance of the problem. Further assumptions can be made about the symmetries of the potential, but at the price of loosing interesting contributions to the amplitude such as those coming from naive "T-odd" FF [2]. In the latter case, the difficulty consists in modelling a genuine interaction vertex that cannot be effectively reabsorbed in the soft part describing the hadronization. This poses a serious difficulty in modelling the quark fragmentation into one observed hadron because it requires the ability of modelling the FSI between the hadron itself and the rest of the jet [2]. Therefore, here we will consider a hard process, semi-inclusive Deep-Inelastic Scattering (DIS), where the hadronization leads to two observed hadrons inside the same jet. A new set of interference FF arise at leading twist [2] and their symmetry properties are briefly reviewed. For the hadron pair being a proton and a pion with invariant mass equal to the Roper resonance, we have already estimated these FF using an extended version of the diquark spectator model [3]. In this case, FSI come from the interference between the direct production of the two hadrons and the decay of the Roper resonance. Here, we present results for the hadron pair being two pions with invariant mass around the ρ mass.

QUARK-QUARK CORRELATION FUNCTION

In analogy with semi-inclusive hard processes involving one detected hadron in the final state [4], the simplest matrix element for the hadronisation into two hadrons is the quark-quark correlation function describing the decay of a quark with momentum k into two hadrons P_1, P_2, namely

$$\Delta_{ij}(k; P_1, P_2) = \sum_X \int \frac{d^4\zeta}{(2\pi)^4} e^{ik\cdot\zeta} \langle 0|\psi_i(\zeta) a^\dagger_{P_2} a^\dagger_{P_1}|X\rangle \langle X|a_{P_1} a_{P_2} \overline{\psi}_j(0)|0\rangle, \qquad (1)$$

where the sum runs over all the possible intermediate states involving the two final hadrons P_1, P_2. Since the three external momenta k, P_1, P_2 cannot all be collinear at the same time, we choose for convenience the frame where the total pair momentum

$P_h = P_1 + P_2$ has no transverse component. By generalizing the Collins-Soper light-cone formalism [5] for fragmentation into multiple hadrons, the cross section for two-hadron semi-inclusive emission is a linear combination of projections $\Delta^{[\Gamma]}$ of Δ by specific Dirac structures Γ, after integrating over the (hard-scale suppressed) light-cone component k^+ and, consequently, taking ζ as light-like [2]. At leading order, we get

$$\Delta^{[\gamma^-]}(z_h, \xi, \mathbf{k}_T^2, \mathbf{R}_T^2, \mathbf{k}_T \cdot \mathbf{R}_T) \equiv D_1$$

$$\Delta^{[\gamma^-\gamma_5]}(z_h, \xi, \mathbf{k}_T^2, \mathbf{R}_T^2, \mathbf{k}_T \cdot \mathbf{R}_T) \equiv \frac{\epsilon_T^{ij} R_{Ti} k_{Tj}}{M_1 M_2} G_1^\perp$$

$$\Delta^{[i\sigma^{i-}\gamma_5]}(z_h, \xi, \mathbf{k}_T^2, \mathbf{R}_T^2, \mathbf{k}_T \cdot \mathbf{R}_T) \equiv \frac{\epsilon_T^{ij} R_{Tj}}{M_1 + M_2} H_1^{\sphericalangle} + \frac{\epsilon_T^{ij} k_{Tj}}{M_1 + M_2} H_1^\perp , \qquad (2)$$

where $\epsilon_T^{\mu\nu} = \epsilon^{-+\mu\nu}$. The functions $D_1, G_1^\perp, H_1^{\sphericalangle}, H_1^\perp$ are the interference FF that depend on how much of the fragmenting quark momentum k is carried by the hadron pair ($z_h = z_1 + z_2$), on the way this momentum is shared inside the pair ($\xi = z_1/z_h$), and on the "geometry" of the pair, namely on the transverse relative momentum of the two hadrons ($\mathbf{R}_T^2$) and on the relative orientation between the pair plane and the quark jet axis ($\mathbf{k}_T^2, \mathbf{k}_T \cdot \mathbf{R}_T$, see also Fig. 1). The different

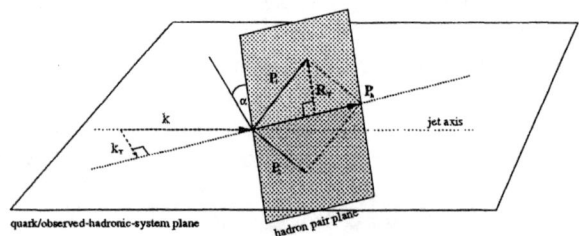

FIGURE 1. Kinematics for a fragmenting quark jet containing a pair of leading hadrons.

Dirac structures Γ are related to different spin states of the fragmenting quark and lead to the nice probabilistic interpretation at leading order [2]: D_1 is the probability for an unpolarized quark to produce a pair of unpolarized hadrons; $G_1^\perp$ is the difference of probabilities for a longitudinally polarized quark with opposite chiralities to produce a pair of unpolarized hadrons; $H_1^{\sphericalangle}$ and $H_1^\perp$ both are differences of probabilities for a transversely polarized quark with opposite spins to produce a pair of unpolarized hadrons. $G_1^\perp, H_1^{\sphericalangle}$ and $H_1^\perp$ are (naive) "T-odd" and do not vanish only if there are residual interactions in the final state. In this case, the constraints from time-reversal invariance cannot be applied. $G_1^\perp$ is chiral even; $H_1^{\sphericalangle}$ and $H_1^\perp$ are chiral odd and can, therefore, be identified as the chiral partners needed to access the transversity h_1. Given their probabilistic interpretation, they can be considered as a sort of "double" Collins effect [6].

NUMERICAL RESULTS

In order to make quantitative predictions, we adopt the formalism of the spectator model, specializing it to the emission of a hadron pair. The basic idea is to replace the sum over the complete set of intermediate states in Eq. (1) with an effective spectator state with a definite mass M_D, momentum P_D. Consequently, the correlator simplifies to

$$\Delta_{ij}(k; P_1, P_2) \sim \frac{\theta(P_D^+)}{(2\pi)^3} \delta\left((k - P_h)^2 - M_D^2\right) \langle 0|\psi_i(0)|P_1, P_2, D\rangle\langle D, P_2, P_1|\overline{\psi}_j(0)|0\rangle , \tag{3}$$

where the additional δ function allows for a completely analytical calculation of the Dirac projections (2). For the hadron pair being a proton and a pion with invariant mass the mass of the Roper resonance, results have been published in Ref. [3]. In this case, the spectator state has the quantum numbers of a scalar or axial diquark. FSI arise from the interference between the direct production and the decay of the Roper resonance. Here, we show results for the hadron pair being two pions with invariant mass in the range $[m_\rho - \Gamma_\rho, m_\rho + \Gamma_\rho]$, $m_\rho = 768$ MeV and $\Gamma_\rho \sim 250$ MeV. The spectator states becomes an on-shell quark with mass $m_q = 340$ MeV. The quark decay is specialized to the set of diagrams shown in Fig. 2, and their hermitean conjugates, where the naive "T-odd" FF now arise from the interference between the direct production of the two π and the decay of the ρ. Diagram 2a

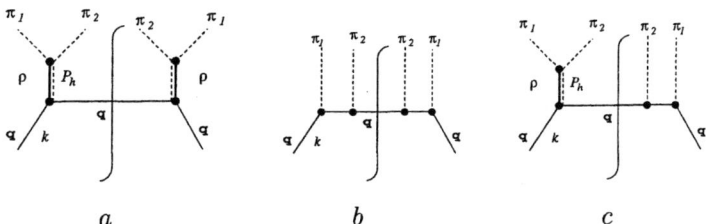

FIGURE 2. Diagrams for quark q decay into two pions through a direct channel or a ρ resonance.

accounts for almost all the strength of $\pi - \pi$ production in the relative P-channel. We have explicitly checked that diagram 2b reproduces the experimental transition probability for $\pi - \pi$ production in the relative S-channel. Hence, we believe this choice represents most of the $\pi - \pi$ strength for invariant mass in the considered interval. We define Feynman rules for the $\rho\pi\pi$, $q\pi q$ and $q\rho q$ vertices introducing cut-offs to exclude large virtualities of the quark while keeping the asymptotic behaviour of FF at large z_h consistent with the quark counting rule. We infer the vertex form factors from previous works on the spectator model [7]. However, there numbers should be taken as indicative, since the ultimate goal is to verify that nonvanishing "T-odd" FF occur, particularly when integrating on some of the kinematical variables and possibly washing all interference effects out. Results of analytical calculation of Eq.(3) show that $H_1^\perp = 0$ and $H_1^{\sphericalangle} = -2m_q G_1^\perp/m_\pi$. After integrating over $\mathbf{k}_T, \mathbf{R}_T^2$ while keeping $\mathbf{R}_T$ in the horizontal plane of the lab (usually identified with the scattering plane), we still get nonvanishing FF. As an example, $H_1^{\sphericalangle}(z_h, M_h)$ is shown in Fig. 3 for the fragmentation $u \to \pi^+\pi^-$, where M_h is the invariant mass of the two pions. The cross section for the deep-inelastic

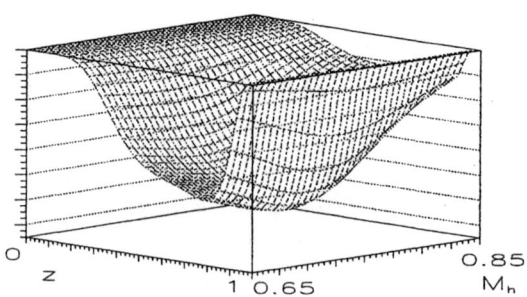

FIGURE 3. $H_1^{\sphericalangle}(z_h, M_h)$ for the fragmentation of a quark u into π^+ and π^-.

scattering of an unpolarized electron on a polarized proton target where two pions are detected in the final state, contains, after integrating all the transverse dynamics ($\mathbf{P}_{hT}, \mathbf{k}_T, \mathbf{R}_T^2$), an unpolarized contribution proportional to $D_1(z_h, M_h)$ and a term proportional to $H_1^{\sphericalangle}(z_h, M_h)$ which depends on the transverse target polarization S_T. Therefore, by flipping the polarization of the target, it is possible to build the following azimuthal asymmetry

$$\mathcal{A} \propto \frac{S_T}{2m_\pi} \sin(\phi_{R_T} + \phi_{S_T}) \frac{h_1(x)}{f_1(x)} \frac{H_1^{\sphericalangle}(z_h, M_h)}{D_1(z_h, M_h)}, \qquad (4)$$

where ϕ_{R_T}, ϕ_{S_T} are the azimuthal angles of $\mathbf{R}_T, \mathbf{S}_T$ with respect to the scattering plane, respectively. The asymmetry shows indeed the familiar sinusoidal azymuthal dependence. Noteworthy is the factorization of the chiral-odd, naive "T-odd" $H_1^{\sphericalangle}$ from the chiral-odd transversity h_1. Therefore, such asymmetry measurement allows for the extraction of h_1 using a model input for the FF.

REFERENCES

1. De Rújula A., Kaplan J.M., and de Rafael E., *Nucl. Phys.* **B35**, 365 (1971).
2. Bianconi A., Boffi S., Jakob R., and Radici M., *Phys. Rev. D* **62**, 034008 (2000).
3. Bianconi A., Boffi S., Jakob R., and Radici M., *Phys. Rev. D* **62**, 034009 (2000).
4. Mulders P.J., and Tangerman R.D., *Nucl. Phys. B* **461**, 197 (1996).
5. Collins J.C., and Soper D.E., *Nucl. Phys. B* **194**, 445 (1982).
6. Collins J.C., *Nucl. Phys. B* **396**, 161 (1993).
7. Jakob R., Mulders P.J., and Rodrigues J., *Nucl. Phys. A* **626**, 937 (1997).

TESLA-N
Electron Scattering with Polarized Targets at TESLA

V. Korotkov
on behalf of the TESLA-N Study Group

DESY Zeuthen, D-15738 Zeuthen, Germany
IHEP, RU-142284 Protvino, Russia

Abstract. Measurements of polarized eN scattering can be realized at the TESLA linear collider facility at DESY with luminosities that are about two orders of magnitude higher than those expected for other experiments at comparable energies. A large variety of polarized parton distribution and fragmentation functions can be determined with unprecedented accuracy, many of them for the first time.

The Experiment

The basic TESLA-N [1] idea is to use one of the arms of the e^+e^- collider TESLA to organize collisions of longitudinally polarized electrons with a solid-state fixed target that may be either longitudinally or transversely polarized. The electrons for TESLA-N will be accelerated together with the positrons in the north arm of the TESLA main accelerator. This 'opposite charge option' was chosen to be able to realize a separation between the e^--beam for eN physics and the main e^+-beam by a static magnet system.

The given TESLA duty cycle of 0.5 % in conjunction with the basic machine frequency of 1.3 GHz has most severe consequences for the proposed experiment. The time structure as foreseen for the e^+e^- experiment, i.e. one bunch of 20 ps length crossing the target every 337 ns, would result in an unacceptable event rate within these 20 ps. To minimize the number of multiple events while maximizing the luminosity, it is foreseen to fill every bucket of the bunch train (one every 0.77 ns), while limiting the beam current to 20 nA. This corresponds to 20k electrons per bunch and to 6.2 million bunches per second crossing the TESLA-N target.

The basic layout for the proposed eN-experiment within the TESLA infrastructure is shown in Figure 1. A separate electron gun system is required for TESLA-N at the north end of the TESLA machine. It is envisaged to use a laser driven strained GaAs SLAC-type gun that must be made capable to deliver 20k highly polarized ($P \sim 90\%$) electrons per 0.77 ns. It must be followed by a separate preac-

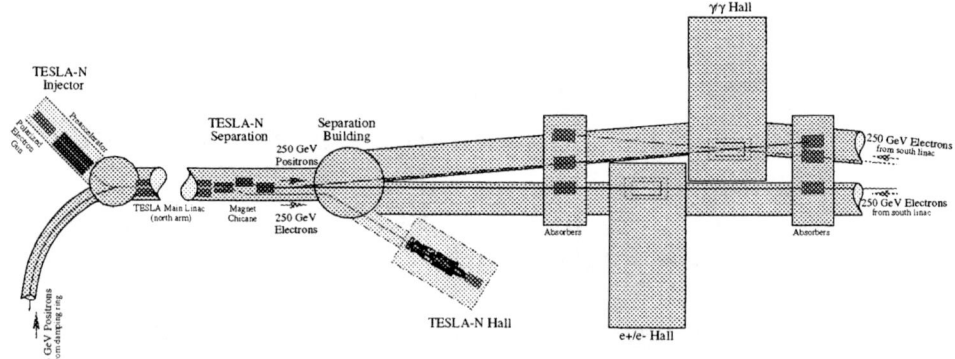

FIGURE 1. Schematic top view of the machine-related elements.

celerator whose end energy and type are under discussion. A short extra tunnel is required from the collider separation building (North) to the TESLA-N hall.

One of the main ingredients of the TESLA-N apparatus is the polarized target. To reach the required high luminosity a polarized solid state target of about 1 g/cm^2 areal density will be chosen. The polarized target will consist of a ^{4}He evaporator cryostat, a 5 T Helmholtz-type magnet and a 140 GHz microwave system for permanent dynamic nuclear polarization. A symmetric Helmholtz design of the magnet combines excellent homogeneity with large opening angles for both longitudinal and transverse polarization. Presently, the target materials NH$_3$ (P_T = 0.8, f = 0.176) and ^{6}LiD (P_T = 0.3, f = 0.44) appear as the best choices to study electron scattering off polarized protons and deuterons, respectively.

The overall dimensions of the TESLA-N apparatus are expected to be comparable to those of COMPASS [2] because the kinematics of both experiments are similar.

Assuming a combined up-time of accelerator and experiment of 0.33 in conjunction with an efficiency of the experiment of 0.75, results in a maximum achievable integrated luminosity for TESLA-N of 600 fb^{-1} per effective year.

All physics projections presented below are based on an integrated luminosity of 100 fb^{-1}. This represents a conservative estimate for *one* year of data taking.

Physics Prospects

A quark of a given flavor in the nucleon is characterized by three twist-two quark distributions: the number density distribution $q(x)$, the helicity distribution $\Delta q(x)$, and the transversity distribution $\delta q(x)$, which have never been measured yet. For non-relativistic quarks $\delta q(x) = \Delta q(x)$ can be expected, but generally they are independent distribution functions. The first moments of $\delta q(x, Q^2) - \delta \bar{q}(x, Q^2)$ are known as quark tensor charges $\delta q(Q^2)$.

The transversity distributions $\delta q(x)$ are not accessible in inclusive DIS, because they are chiral-odd and only occur in combinations with other chiral-odd objects.

In semi-inclusive DIS of unpolarized leptons off transversely polarized nucleons several methods have been proposed to access $\delta q(x)$ via specific single target-spin asymmetries.

An observation of the Collins effect [3] offers the experimentally most direct access to $\delta q(x)$. Recent HERMES measurements [6] indicate that the polarized fragmentation function $H_1^{\perp(1)}(z)$ may be quite sizeable. An appropriately weighted cross-section asymmetry can be expressed as a flavor-sum where each distribution function δq enters in combination with a fragmentation function $H_1^{\perp(1)q}(z)$ of the same flavor [4,5]:

$$A_T(x, Q^2, z) = P_T \cdot D_{nn} \cdot \frac{\sum_q e_q^2 \, \delta q(x, Q^2) \, H_1^{\perp(1)q}(z)}{\sum_q e_q^2 \, q(x, Q^2) \, D_1^q(z)} \quad (1)$$

Here D_{nn} is the polarization transfer coefficient, P_T is the size of the nucleon's transverse polarization, and $D_1^q(z)$ is the unpolarized quark fragmentation function.

Measurements of asymmetries in the production of positive and negative pions on proton and deuteron targets ($A_{p,d}^{\pi^+,\pi^-}$) allow, under reasonable assumptions, the simultaneous reconstruction of the shapes of the unknown functions $\delta q(x, Q^2)$ and $H_1^{\perp(1)}(z)/D_1(z)$. The relative normalization cannot be fixed without a further assumption. This ambiguity can be resolved by relating $\delta q(x)$ to $\Delta q(x)$ at small values of Q^2. Following to ref. [7], the normalization ambiguity is resolved by assuming $\delta u(x_0, Q_0^2) = \Delta u(x_0, Q_0^2)$ at $x_0 = 0.25$.

The projected statistical accuracy for the measurement of $\delta u_v(x, Q^2)$ at TESLA-N is shown in Figure 2. A broad range of $0.003 < x < 0.7$ can be accessed in conjunction with $1 < Q^2 < 100$ GeV^2. A simultaneous reconstruction of the quark transversity distributions δd_v, $\delta \bar{u}$, and $\delta \bar{d}$ is attained with a somewhat lower accuracy. Projections for the accuracies of a measurement of the u- and d-quark tensor charges are ± 0.01 and ± 0.02 at the scale of 1 GeV^2, respectively. At the same time, precise values would be measured for the ratio of polarized and unpolarized favored quark fragmentation functions $H_1^{\perp(1)}(z)/D_1(z)$.

The polarized gluon distribution function $\Delta G(x)$ in the nucleon is essentially unknown as of today. There is a variety of approaches to determine $\Delta G(x)$ from polarized eN collisions. The direct methods rely on the photon-gluon fusion subprocess measurement (see e.g. ref. [8]). Indirectly, the NLO Q^2-evolution of the structure function $g_1(x, Q^2) = \frac{1}{2}\sum_q e_q^2 \Delta q(x, Q^2)$ is sensitive to $\Delta G(x, Q^2)$. No NLO QCD fit to the existing data has been able to deliver a statistically convincing determination of the first moment ΔG yet. A precision measurement of $g_1(x, Q^2)$ at TESLA-N will dramatically enlarge the accuracy and the kinematic range, as can be concluded from Figure 3. To obtain a projection for the measurement accuracy of $\int_0^1 dx \Delta G(x)$, first a QCD NLO fit was performed in the $\overline{\text{MS}}$ scheme using all published DIS data, giving a result of 0.43 ± 0.21 (stat.), at the scale of 1 GeV^2. The resulting structure function $g_1(x, Q^2)$ was then evolved into the kinematical region

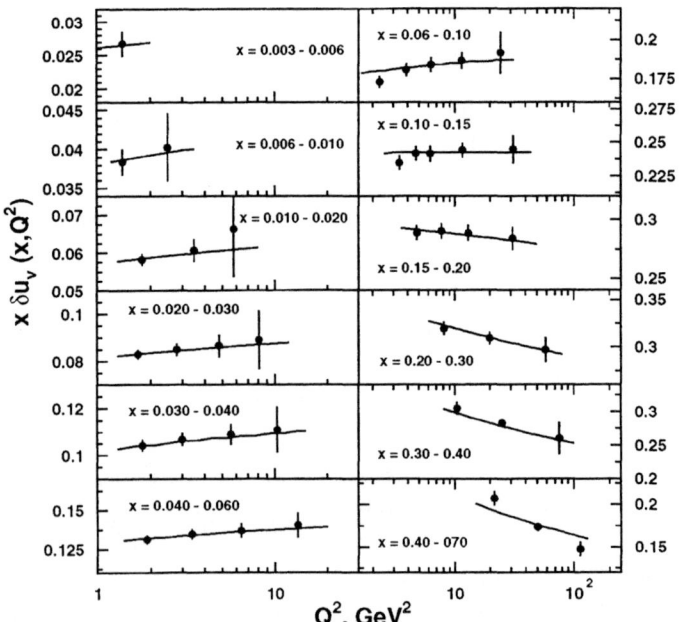

FIGURE 2. The valence u-quark transversity distribution as a function of x and Q^2 as it would be measured at TESLA-N, based on an integrated luminosity of 100 fb^{-1}. The curves show the LO Q^2-evolution of δu_v obtained with a fit to the simulated asymmetries.

of TESLA-N and then used as additional input data for two new fits. Adding data that correspond to 100 fb^{-1} using a proton target improves the statistical accuracy down to ±0.06. An additional data set obtained with 100 fb^{-1} on a deuteron target yields a further improvement down to ±0.04. This additional deuteron data set considerably improves the statistical accuracy in the determination of the non-singlet quark distribution in the neutron, when comparing to existing data.

Summary

A possible layout for a fixed-target eN scattering experiment at the TESLA facility is presented. The physics potential of the TESLA-N project is demonstrated with two examples showing that an accurate measurement of the x- and Q^2-dependence of both the helicity and the transversity quark distributions would be possible. Complemented by results on the polarized gluon distribution most of the components of the angular momentum structure of the nucleon will be determined with high precision. Other important possible measurements with TESLA-N include:

- Deeply Virtual Compton Scattering and other hard exclusive reactions;

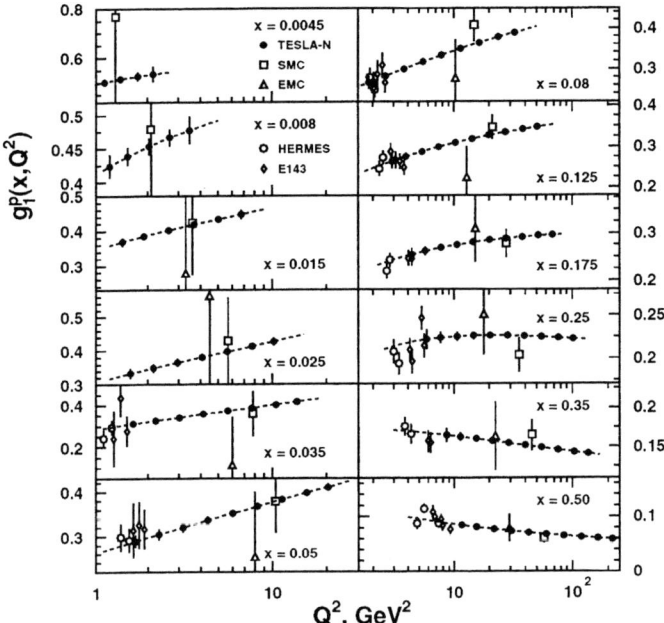

FIGURE 3. Projected statistical accuracy for a measurement of $g_1^p(x, Q^2)$ at TESLA-N, based on a luminosity of 100 fb^{-1} and a minimum detector acceptance of 5 mrad.

- the polarized structure function $g_2(x)$;
- specific deuteron structure functions $b_{1,2}(x)$ and $\Delta(x)$.

Hence, the measurements envisaged for TESLA-N will constitute one of the most comprehensive and precise investigations of the predictive power of QCD in the polarized sector. The possibilities of using unpolarized targets and of experiments with a real photon beam turn TESLA-N into a versatile next-generation facility at the intersection of particle and nuclear physics.

REFERENCES

1. http://www.ifh.de/hermes/future/
2. COMPASS Coll., CERN/SPSLC 96-14 (1996).
3. J.C. Collins, *Nucl. Phys.* **B396**, 161 (1993);
   J.C. Collins, S.F. Heppelmann, G.A. Ladinsky, *Nucl. Phys.* **B420**, 565 (1994)
4. A.M. Kotzinian, P.J. Mulders, *Phys. Lett.* **B406**, 373 (1997).
5. P.J. Mulders, R.D. Tangermann, *Nucl.Phys.* **B461**, 197 (1996);
   D. Boer and P.J. Mulders, *Phys. Rev.* **D57**, 5780 (1998)
6. HERMES Coll., A. Airapetian et al., *Phys. Rev. Lett.* **80**, 4047 (2000).
7. V.A. Korotkov, W.-D. Nowak, K.A. Oganessyan, DESY 99-176, hep-ph/0002268.
8. HERMES Coll., A. Airapetian et al., *Phys. Rev. Lett.* **84**, 2584 (2000)

Hadronic wave functions and exclusive processes

Eckart Stein

Institut für Theoretische Physik, Universität Regensburg, D-93040 Regensburg, Germany

Abstract. The role of hadronic wave functions and distribution amplitudes in hard exclusive processes are discussed. Special attention is paid to the nucleon and its higher twist distribution amplitudes.

INTRODUCTION

Hard exclusive processes are coming to the forefront of high energy nuclear and particle physics. This is already visible at TJNAF, COMPASS, HERMES that go for more and more exclusive channels, and this will be even more the case also for EPIC. One main reason for that is the growing understanding that exclusive reactions provide unmatched opportunities to study the hadron structure, as demonstrated by recent interest to deeply virtual Compton scattering (DVCS) and hard diffractive meson production. All future plans also call for very high luminosity and would therefore be perfectly suited for the investigation of exclusive and semi-exclusive reactions with and without polarization.

The classical theoretical framework for the calculation of hard exclusive processes in QCD was developed in [1], see [2] for a review. This approach introduces a concept of hadron distribution amplitudes as fundamental nonperturbative functions describing the hadron structure in rare parton configurations with a minimum number of Fock constituents (and at small transverse separations). The distribution amplitude is related to the hadrons Bethe Salpeter wave function by

$$\Phi(x_i, \mu) \sim \int^{|k_\perp|<\mu} dk_\perp \; \Phi_{\rm BS}(x_i, k_\perp). \tag{1}$$

Distribution amplitudes are equally important and to a large extent complementary to conventional parton distributions which correspond to one-particle probability distributions for the parton momentum fraction in an *average* configuration.

For a theorist, the main challenge is to make the QCD description of hard exclusive reactions fully quantitative. Although the leading-twist QCD factorization approach correctly reproduces the power dependence of exclusive amplitudes on the large momentum, there exist many indications that soft end-point and higher-twist corrections might dominate over the naive QCD prediction over the range of accessible values of Q^2. The prospect to perform exclusive measurements of nucleon properties, such as form factors, at moderate values of Q^2 therefore calls for the knowledge of nucleon's higher twist distribution amplitudes. While in the meson case higher twist amplitudes are known for some time, the problem to identify the higher twist distribution amplitudes of the nucleon was solved only recently [3].

NUCLEON DISTRIBUTION AMPLITUDES OF HIGHER TWIST

The notion of hadron distribution amplitudes in general refers to hadron-to-vacuum matrix elements of nonlocal operators built of quark and gluon fields at light-like separations. In the nucleon case the easiest (and probably dominating) contribution comes from the the three-quark matrix element

$$\langle 0|\varepsilon^{ijk} u_\alpha^{i'}(a_1 z) [a_1 z, a_0 z]_{i',i}\, u_\beta^{j'}(a_2 z) [a_2 z, a_0 z]_{j',j}\, d_\gamma^{k'}(a_3 z) [a_3 z, a_0 z]_{k',k} |P(P,\lambda)\rangle \quad (2)$$

where $|P(P,\lambda)\rangle$ denotes the proton state with momentum P, $P^2 = M^2$ and helicity λ. u, d are the quark-field operators. The Greek letters α, β, γ stand for Dirac indices, the Latin letters i, j, k refer to color. z is an arbitrary light-like vector, $z^2 = 0$, the a_i are real numbers. The gauge-factors $[x, y]$ render the matrix element gauge-invariant. To simplify the notation we will not write the gauge-factors explicitly in what follows but imply that they are always present.

To perform a twist decomposition of the above matrix element, one decomposes the quark fields in 'plus' and 'minus' components $q = q^+ + q^-$. The leading twist amplitude is identified as the one containing three 'plus' quark fields while each 'minus' component introduces one additional unit of twist. Up to possible complications due to isospin, one expects, therefore, to find eight independent three quark nucleon distribution amplitudes: One corresponding to the twist-3 operator $(u^+ u^+ d^+)$, three related to the possible twist-4 operators $(u^+ u^+ d^-), (u^+ u^- d^+), (u^- u^+ d^+)$, three more amplitudes of twist-5 of the type $(u^- u^- d^+), (u^- u^+ d^-), (u^+ u^- d^-)$ and one amplitude of twist-6 having the structure $(u^- u^- d^-)$.

The physics interpretation of the distribution amplitudes $\Phi(x_i)$ is most transparent in terms of quark fields of definite chirality $q^{\uparrow(\downarrow)} = \frac{1}{2}(1 \pm \gamma_5) q$. In terms of the quarks light-cone components and spin projections all 8 distribution amplitudes can be given. We give here twist-3 and twist-4. The leading twist-3 distribution, the $(u^+ u^+ d^+)$ is well known and can be represented as (cf. [4]):

$$\langle 0|\varepsilon^{ijk}\left(u_i^\uparrow(a_1 z) C \not{z} u_j^\downarrow(a_2 z)\right) \not{z} d_k^\uparrow(a_3 z)|P\rangle = -\frac{1}{2} pz \not{z} N^\uparrow \int \mathcal{D}x\, e^{-ipz \sum x_i a_i}\, \Phi_3(x_i). \quad (3)$$

The twist-4 distributions allow for the following representation correspond to the light cone projections $(u^+ u^+ d^-), (u^+ u^- d^+), (u^- u^+ d^+)$:

$$\langle 0|\varepsilon^{ijk}\left(u_i^\uparrow(a_1z)C\not{z}u_j^\downarrow(a_2z)\right)\not{p}d_k^\uparrow(a_3z)|P\rangle = -\frac{1}{2}pz\,\not{p}N^\uparrow\!\int\mathcal{D}x\,e^{-ipz\sum x_ia_i}\,\Phi_4(x_i)\,,$$

$$\langle 0|\varepsilon^{ijk}\left(u_i^\uparrow(a_1z)C\not{z}\gamma_\perp\not{p}u_j^\downarrow(a_2z)\right)\gamma^\perp\!\!\not{z}d_k^\downarrow(a_3z)|P\rangle = -pzM\,\not{z}N^\uparrow\!\int\mathcal{D}x\,e^{-ipz\sum x_ia_i}\,\Psi_4(x_i)\,,$$

$$\langle 0|\varepsilon^{ijk}\left(u_i^\uparrow(a_1z)C\not{p}\not{z}u_j^\uparrow(a_2z)\right)\not{z}d_k^\uparrow(a_3z)|P\rangle = \frac{1}{2}pzM\,\not{z}N^\uparrow\!\int\mathcal{D}x\,e^{-ipz\sum x_ia_i}\,\Xi_4(x_i) \quad (4)$$

To model the distribution amplitudes we make use of conformal symmetry of QCD: The conformal expansion of light-cone distribution amplitudes is the field-theoretic analogue to the partial wave expansion in quantum mechanics. The idea, in both cases, is to use the symmetry of the problem to introduce a set of separated coordinates. In quantum mechanics, spherical symmetry of the potential allows to separate dependence on radial coordinates from angular ones. All angular dependence is included in spherical harmonics which form an irreducible representation of the symmetry group O(3), while the dependence on radial coordinates is governed by a one-dimensional Schrödinger equation.

In the same spirit conformal symmetry of the QCD Lagrangian can be used to study the distribution amplitudes as it allows to separate longitudinal degrees of freedom from transverse ones [5]. The dependence on longitudinal momentum fractions is taken into account by a set of orthogonal polynomials that form an irreducible representation of the collinear subgroup $SL(2,R)$ of the conformal group which describes Möbius transformations on the light-cone. Transverse coordinates are replaced by the renormalization scale, the dependence on which is governed by the renormalization group. Since the renormalization group equations to leading logarithmic accuracy are driven by tree-level counterterms, they have the conformal symmetry. As a consequence, components in the distribution amplitudes with different conformal spin, dubbed conformal partial waves, do not mix under renormalization to this accuracy. To next-to-leading conformal spin accuracy the distribution amplitudes above can be expanded for twist-3 as:

$$\Phi_3(x_i,\mu) = 120x_1x_2x_3\left[\phi_3^0(\mu) + \phi_3^-(\mu)(x_1-x_2) + \phi_3^+(\mu)(1-3x_3)\right]. \quad (5)$$

For twist-4 one gets correspondingly:

$$\Phi_4(x_i,\mu) = 24x_1x_2\left[\phi_4^0(\mu) + \phi_4^-(\mu)(x_1-x_2) + \phi_4^+(1-5x_3)\right],$$

$$\Psi_4(x_i,\mu) = 24x_1x_3\left[\psi_4^0(\mu) + \psi_4^-(\mu)(x_1-x_3) + \psi_4^+(1-5x_2)\right],$$

$$\Xi_4(x_i,\mu) = 24x_2x_3\left[\xi_4^0(\mu) + \xi_4^-(\mu)(x_2-x_3) + \xi_4^+(1-5x_1)\right]. \quad (6)$$

All the non-perturbative parameters $\phi_3^0, \phi_3^-, \ldots$ are not independent and can be related to a set of parameters that can be obtained from QCD sum rules calculation. Besides those known from the leading twist analysis one has

$$\langle 0|\varepsilon^{ijk}\left[u^i(0)C\gamma_\mu u^j(0)\right]\gamma_5\gamma^\mu d^k(0)|P\rangle = \lambda_1 MN(P)\,,$$

$$\langle 0|\varepsilon^{ijk}\left[u^i(0)C\sigma_{\mu\nu}u^j(0)\right]\gamma_5\sigma^{\mu\nu}d^k(0)|P\rangle = \lambda_2 MN(P)\,. \quad (7)$$

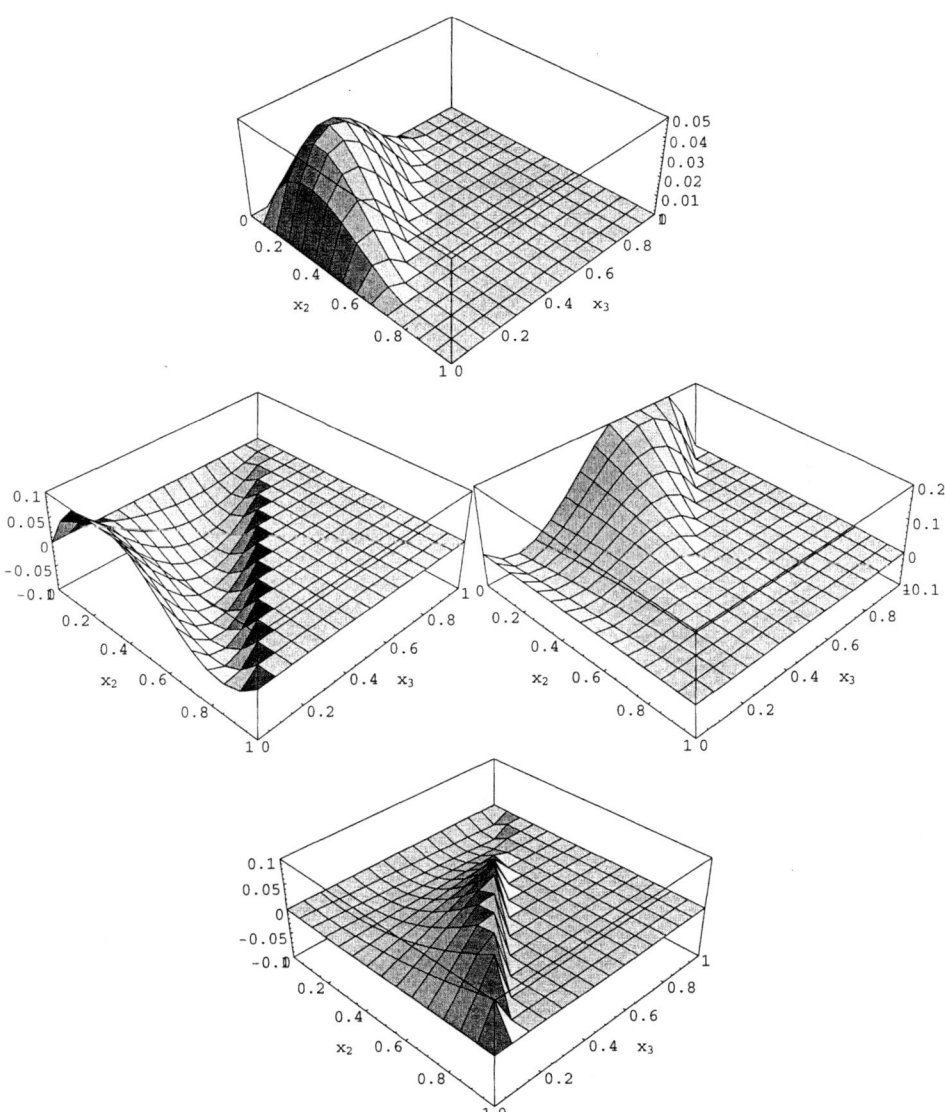

FIGURE 1. Twist-3 distribution amplitude $\Phi_3(x_i)$ in the first line. Twist-4 distribution amplitudes $\Phi_4(x_i)$ in the second line to the left, $\Psi_4(x_i)$ in the second line to the right, and $\Xi(x_i)$ in the third line

for the leading conformal spin (asymptotic expansion) and with one derivative

$$\langle 0| \, (u(0)C\gamma_\mu u(0)) \, \gamma_5 \gamma^\mu \not{z} (iz \, \vec{D} \, d)(0)|P\rangle = \lambda_1 f_1^d(pz) M \not{z} N(P),$$
$$\langle 0| \, (u^a(0)C\sigma_{\mu\nu} u(0)) \, \gamma_5 \sigma^{\mu\nu} \not{z} (iz \, \vec{D} \, d)(0)|P\rangle = -\lambda_2 f_2^d(pz) M \not{z} N(P),$$
$$\langle 0| \left(u(0) C\gamma_\mu \gamma_5 iz \, \overleftrightarrow{D} \, u(0) \right) \gamma^\mu \not{z} d(0)|P\rangle = -\lambda_1 f_1^u(pz) M \not{z} N(P). \tag{8}$$

necessary to describe higher conformal conformal spin. Relating these non-perturbative parameters to the conformal coefficients in the expansion allows to model the wave functions [3]. Results are plotted in the figures.

CONCLUSIONS

We have carried out a systematic study of the higher-twist light-cone distribution amplitudes of the nucleon in QCD and found that a generic three-quark matrix element on the light-cone can be parametrized in terms of eight independent nucleon distribution amplitudes. The distribution amplitudes presented in this paper can immediately be applied to calculations of exclusive reactions at moderate momentum transfers in the framework of QCD light-cone sum rules. First encouraging results from the calculation of the nucleon form factors show that soft contributions are large, and higher twist distribution amplitudes describe known data at low momentum transfer [6].

Acknowledgements

I would like the organizers for the invitation and providing the nice and stimulating atmosphere. V. Braun, R.J. Fries, A. Lenz, and N. Mahnke I would like to thank for fruitful collaboration.

REFERENCES

1. V.L. Chernyak and A.R. Zhitnitsky, JETP Lett. **25** (1977) 510;
   A.V. Efremov and A.V. Radyushkin, Phys. Lett. B **94** (1980) 245;
   G.P. Lepage and S.J. Brodsky, Phys. Lett. B **87** (1979) 359;
2. S.J. Brodsky and G.P. Lepage, in: *Perturbative Quantum Chromodynamics*, ed. by A.H. Mueller, p. 93, World Scientific (Singapore) 1989.
3. V. Braun, R. J. Fries, N. Mahnke and E. Stein, Nucl. Phys. **B589**, 381 (2000)
4. V.L. Chernyak and I.R. Zhitnitsky, Nucl. Phys. **B246** (1984) 52;
5. S.J. Brodsky et al., Phys. Lett. **B91** (1980) 239;
   I.I. Balitsky and V.M. Braun, Nucl. Phys. **B311** (1989) 541.
6. V. Braun, A. Lenz, N. Mahnke and E. Stein, in preparation.

Λ and $\bar{\Lambda}$ polarization as a measurement of distribution and fragmentation functions

M. Boglione[1], M. Anselmino[2], F. Murgia[3]

[1] *Division of Physics, Faculty of Sciences, Vrije Universiteit van Amsterdam*
De Boelelaan 1081, 1081 HV Amsterdam
[2] *Dipartimento di Fisica Teorica, Università di Torino and*
INFN, Sezione di Torino, Via P. Giuria 1, I-10125 Torino, Italy
[3] *Istituto Nazionale di Fisica Nucleare, Sezione di Cagliari*
and Dipartimento di Fisica, Università di Cagliari
C.P. 170, I-09042 Monserrato (CA), Italy

Abstract. A combined analysis of the polarization vector of the Λ baryons produced in DIS processes may give a relevant insight of the hadronization process which governs the transition from partons to physical hadrons and precise indications on the mechanisms of *spin* transfer from partons to hadrons. We give here a short review of some interesting results.

INTRODUCTION

We consider Λ baryon production in polarized Deep Inelastic Scattering (DIS) processes, where lepton and proton may or may not be polarized. The idea is in principle quite simple, in fact the Λ polarization can easily be measured by looking at the angular distribution of the $\Lambda \to p\pi$ decay (in the Λ helicity rest frame) and the fragmenting parton polarization is determined by the elementary Standard Model interactions, provided one knows the initial parton spin state.

Our calculation is performed in the $\ell - p$ center of mass frame: the lepton moves along the z-axis and the proton moves in the opposite direction, with four-momenta ℓ and p respectively; xz is the lepton-hadron production plane. We denote by S_L the (longitudinal) nucleon spin oriented along the z-axis, and by S_N the (normal) spin oriented along the y-axis. We will only consider longitudinally polarized leptons with spins $\pm s_L$ which correspond respectively to $\pm$ helicities.

The three components of the baryon polarization vector, $P_i(B)$, as measured in its helicity rest frame, are related to the components of the helicity density matrix of a spin 1/2 baryon: $P_i(B) = \text{Tr}[(\sigma^i \rho(B)]$ ($i = x, y, z$). See Ref. [1] for details and definitions.

Different Notations

Before we give the specific expressions for the Λ and $\bar{\Lambda}$ polarization vector components, we would like to spend a few words about different notations used by different groups to denote distribution and fragmentation functions, and how these are related. For the distribution functions the following identities hold:

$$q(x) = f_{q/p}(x) = f_1^q(x) \tag{1}$$

$$\Delta q(x) = g_1^q(x) \tag{2}$$

$$\Delta_T q(x) = \delta q = h_1^q(x) \tag{3}$$

$$\Delta^N f_{q/p^\uparrow}(x, \boldsymbol{k}_T) = 2\,\frac{|k_T|\sin\phi}{M}\, f_{1T}^{\perp q}(x, k_T) \tag{4}$$

where ϕ is the angle of the quark transverse momentum with respect to the proton spin. See Ref. [2] for more details and a pictorial representation of the physical content of these functions. For the fragmentation functions, the corresponding relations are

$$D_{h/q}(z) = D_1^q(z) \tag{5}$$

$$\Delta D_{h/q}(z) = G_1^q(z) \tag{6}$$

$$\Delta^N D_{h^\uparrow/q}(z, \boldsymbol{k}_T)\, d^2\boldsymbol{k}_T = -2\,\frac{|k_T|\sin\phi}{M_h}\, H_1^\perp(z, k_T')\, d^2\boldsymbol{k}_T' \tag{7}$$

where, in this case, ϕ is the relative azimuthal angle of the outgoing hadron momentum (see Ref. [2] for details). In what follows we will make use of the first set of notations.

Λ AND $\bar{\Lambda}$ POLARIZATION

Let's consider a DIS process in which the lepton is not polarized, whereas the proton target has positive helicity. In general, for a spin 1/2 baryon, we find:

$$P_z^{(0,+)} = P_z^{(0,-S_L)} = \frac{\sum_q e_q^2 \Delta q(x) \Delta D_{B/q}(z)}{\sum_q e_q^2 q(x) D_{B/q}(z)}. \tag{8}$$

In the case in which the produced hadrons are $\Lambda + \bar{\Lambda}$, and using the assumptions proposed in Ref. [3], this expression simply becomes:

$$P_z^{(0,+)} = \frac{\Delta Q'(x)}{Q(x)}\,\frac{\Delta D_{\Lambda/s}(z)}{D_{\Lambda/q}(z)} \tag{9}$$

where

$$D_{\Lambda/u} = D_{\Lambda/d} = D_{\Lambda/s} = D_{\Lambda/\bar{u}} = D_{\Lambda/\bar{d}} = D_{\Lambda/\bar{s}} \equiv D_{\Lambda/q}$$
$$\Delta D_{\Lambda/u}(z, Q_0^2) = \Delta D_{\Lambda/d}(z, Q_0^2) = N_u\, \Delta D_{\Lambda/s}(z, Q_0^2) \tag{10}$$

are the fragmentation functions of a quark into a Λ baryon, unpolarized and longitudinally polarized respectively [1,3]. The combinations Q, ΔQ, Q' and $\Delta Q'$ are defined as follows

$$Q \equiv 4(u + \bar{u}) + (d + \bar{d}) + (s + \bar{s})$$
$$\Delta Q \equiv 4(\Delta u + \Delta \bar{u}) + (\Delta d + \Delta \bar{d}) + (\Delta s + \Delta \bar{s})$$
$$Q' \equiv [4(u + \bar{u}) + (d + \bar{d})] N_u + (s + \bar{s})$$
$$\Delta Q' \equiv [4(\Delta u + \Delta \bar{u}) + (\Delta d + \Delta \bar{d})] N_u + (\Delta s + \Delta \bar{s}) \tag{11}$$

For DIS processes where the lepton is longitudinally polarized and the target is unpolarized, we have:

$$P_z^{(+,0)} = P_z^{(S_L,0)} = \frac{\sum_q e_q^2 q(x) \Delta D_{B/q}(z)}{\sum_q e_q^2 q(x) D_{B/q}(z)} \hat{A}_{LL}(y) \tag{12}$$

for any spin 1/2 baryon, whereas for $\Lambda + \bar{\Lambda}$ production we obtain

$$P_z^{(+,0)} = \frac{Q'(x)}{Q(x)} \frac{\Delta D_{\Lambda/s}(z)}{D_{\Lambda/q}(z)} \hat{A}_{LL}(y), \tag{13}$$

where $\hat{A}_{LL}(y)$ is the dynamical factor

$$\hat{A}_{LL}(y) = \frac{d\hat{\sigma}_q^{++} - d\hat{\sigma}_q^{+-}}{d\hat{\sigma}_q^{++} + d\hat{\sigma}_q^{+-}} = \frac{y(2-y)}{1+(1-y)^2}. \tag{14}$$

In the events in which both lepton and target proton are polarized, we have

$$P_z^{(+,\pm)} = P_z^{(s_L,\mp S_L)} = \frac{\sum_q e_q^2 [q(x)\hat{A}_{LL}(y) \pm \Delta q(x)]\Delta D_{B/q}(z)}{\sum_q e_q^2 [q(x) \pm \hat{A}_{LL}(y)\Delta q(x)]D_{B/q}(z)}, \tag{15}$$

and for $\Lambda + \bar{\Lambda}$ production

$$P_z^{(+,\pm)} = \frac{Q'(x)\hat{A}_{LL}(y) \pm \Delta Q'(x)}{Q(x) \pm \Delta Q(x)\hat{A}_{LL}(y)} \frac{\Delta D_{\Lambda/s}(z)}{D_{\Lambda/q}(z)}. \tag{16}$$

When the final hadrons are produced from *transversely* polarized protons we find

$$P_y^{(0,S_N)} = \frac{\sum_q e_q^2 \Delta_T q(x) \Delta_T D_{B/q}(z)}{\sum_q e_q^2 q(x) D_{B/q}(z)} \hat{D}_{NN}(y) \tag{17}$$

which, for $\Lambda + \bar{\Lambda}$ production and under the same assumptions (10) for the fragmentation of transversely polarized quarks, reduces to

$$P_y^{(0,S_N)} = \frac{\Delta_T Q'(x)}{Q(x)} \frac{\Delta_T D_{\Lambda/s}(z)}{D_{\Lambda/q}(z)} \hat{D}_{NN}(y) \tag{18}$$

where

$$\hat{D}_{NN}(y) = \frac{d\hat{\sigma}^{\ell q^\uparrow \to \ell q^\uparrow} - d\hat{\sigma}^{\ell q^\uparrow \to \ell q^\downarrow}}{d\hat{\sigma}^{\ell q^\uparrow \to \ell q^\uparrow} + d\hat{\sigma}^{\ell q^\uparrow \to \ell q^\downarrow}} = \frac{2(1-y)}{1+(1-y)^2} \qquad (19)$$

and

$$\Delta_T Q' \equiv [4(\Delta_T u + \Delta_T \bar{u}) + (\Delta_T d + \Delta_T \bar{d})] N_u + (\Delta_T s + \Delta_T \bar{s}). \qquad (20)$$

Notice that these equations hold within QCD factorization theorem at leading twist and leading order in the coupling constants; the intrinsic $\boldsymbol{k}_T$ of the partons have been integrated over and collinear configurations dominate both the distribution and the fragmentation functions. Furthermore, for simplicity of notations we have not indicated the Q^2 scale dependences in f and D.

The measurable components of the Λ polarization vector depend on different combinations of distribution functions, elementary dynamics and fragmentation functions: each of these terms predominantly depends on a single variable, respectively x, y and z, and a careful analysis of different situations can yield precious information. For example, the longitudinal polarization induced to the baryon by a longitudinally polarized nucleon, $P_z^{(0,+)}$, does not depend on the elementary dynamics, but only on the quark and spin distribution and fragmentation properties (see Eq.(8)). So, neglecting Q^2 evolution, it does not depend on the variable y but only on x and z. On the other side, longitudinal polarization induced to the baryon by a longitudinally polarized lepton, or by polarized nucleon and lepton at the same time, $P_z^{(+,0)}$ and $P_z^{(+,+)}$, depend also on the y variable through $\hat{A}_{LL}(y)$, but in differently weighted combinations in the various possible cases.

More specific conclusions can be drawn if one, for example, starts by exploiting $P_z^{(+,0)}$ and $P_z^{(0,+)}$ to determine N_u and $\Delta D_{\Lambda/s}(z)$. It is then possible to apply this knowledge to $P_z^{(+,\pm)}$, which can be used to test the validity of the model or suggest its weaknesses and merits.

In the case of *transverse* polarization of the proton, if $P_y^{(0,S_N)}$ is measured, very important information not only on the fragmentation function $\Delta_T D$, but also on the transversity distribution function $\Delta_T q(x)$ could be achieved.

Very interesting numerical estimates can be performed by using existing models and parameterizations for the distribution and fragmentation functions which appear in Eqs (9),(13),(15) and (18). For details, plots and results see Ref. [1].

One last comment before we finish. In the case in which one could use data on single cross sections instead of ratios, much more information could be gained from this kind of experiments. A very extensive work which exploits these new ideas is currently in preparation [4].

Conclusions

The study of the angular distribution of the $\Lambda \to p\pi$ decay allows a simple and direct measurement of the components of the Λ polarization vector. For Λ's produced in the current fragmentation region in DIS processes, the component of the polarization vector are related to spin properties of the quark inside the nucleon, to spin properties of the quark hadronization, and to spin dynamics of the elementary interactions.

We have discussed all different polarization states of baryons, obtainable in the fragmentation of a quark in DIS with polarized initial leptons and nucleons, showing how they can reveal different quark features, weighted and shaped by elementary dynamics.

Acknowledgements

M. Boglione would like to thank the organizers of the workshop for creating interesting and stimulating discussion opportunities.

REFERENCES

1. M. Anselmino, M. Boglione, F. Murgia, *Phys. Lett.* **B481**, 253 (2000).
2. M. Boglione, P.J. Mulders, *Phys. Rev.* **D60**, 054007 (1999).
3. D. De Florian, M. Stratmann, W. Vogelsang, *Phys. Rev.* **D57**, 5811 (1998).
4. M. Anselmino, M. Boglione, U. D'Alesio, E. Leader, F. Murgia, *in preparation*.

Single Spin Asymmetries in Semi-Inclusive Electroproduction: Access to Transversity

K.A. Oganessyan*[†1], N. Bianchi*, E. De Sanctis*, W.-D. Nowak[‡]

*INFN-Laboratori Nazionali di Frascati, via Enrico Fermi 40, I-00044 Frascati, Italy
[†]DESY, Notkestrasse 85, 22603 Hamburg
[‡]DESY Zeuthen, Platanenallee 6, D-15738 Zeuthen, Germany

Abstract. We discuss the quark transversity distribution function and a possible way to access it through the measurement of single spin azimuthal asymmetry in semi-inclusive single pion electroproduction on a transversely polarized target.

At leading order in $1/Q$, the cross section for a hard scattering process is given by the convolution of a hard part and a soft part. The former describes the scattering among elementary constituents and can be calculated perturbatively in the framework of QCD. The latter accounts for the processes in which either partons are produced from the initial hadrons (parton distribution functions) or final hadrons are produced from partons (parton fragmentation functions) which result from the hard elementary scattering.

For every quark flavor, besides the well-known parton distribution $f_1(x)$ and the longitudinal spin distribution $g_1(x)$, there is a third twist-two distribution function, the transversity distribution function $h_1(x)$ which was first discussed by Ralston and Soper [1] in double transverse polarized Drell-Yan scattering. The transversity distribution $h_1(x)$ measures the probability to find a transversely polarized quark in a transversely polarized nucleon. It is equally important for the description of the spin structure of nucleons as the more familiar function $g_1(x)$; their information being complementary. In the non-relativistic limit, where boosts and rotations commute, $h_1(x) = g_1(x)$; then difference between these two functions may turn out to be a measure for the relativistic effects within nucleons. On the other hand, there is no gluon analog on $h_1(x)$. This may have interesting consequences for ratios of transverse to longitudinal asymmetries in polarized hard scattering processes (see e.g. Ref. [2]).

The transversity distribution $h_1(x)$ remains still unmeasured. The reason is that it is a chiral odd function, and consequently it is suppressed in inclusive deep

---

[1] E-mail: kogan@hermes.desy.de

inelastic scattering (DIS) [3]. Since electroweak and strong interactions conserve chirality, $h_1(x)$ cannot occur alone, but has to be accompanied by a second chiral odd quantity.

In principle, transversity distributions can be extracted from cross section asymmetries in polarized processes involving a transversely polarized nucleon. In the case of hadron-hadron scattering these asymmetries can be expressed through a flavor sum involving a product of two chiral-odd transversity distributions. This is one of the main goals of the spin program at RHIC [4]. An evaluation of the corresponding asymmetry was carried out [5] by assuming the saturation of Soffer's inequality [6] for the transversity distribution: the maximum possible asymmetry at RHIC energies was estimated to be about 2%. At smaller energies ($\sqrt{s} \simeq 40$ GeV), e.g. for a possible fixed-target hadron-hadron spin experiment at the proposed HERA-$\vec{N}$ facility [7] the asymmetry is expected to be higher (about 4%).

In the case of *semi-inclusive* deep inelastic lepton scattering[2] (SIDIS) off transversely polarized nucleons there exist several methods to access transversity distributions. One of them, the twist-3 pion production [8], uses longitudinally polarized leptons and measures a double spin asymmetry. The other methods do not require a polarized beam, and rely on the *polarimetry* of the scattered transversely polarized quark. They consist on:

- the measurement of the transverse polarization of Λ's in the current fragmentation region [9,10],

- the observation of a correlation between the transverse spin vector of the target nucleon and the normal to the two-meson plane [11,12],

- the observation of the "Collins effect" in quark fragmentation through the measurement of pion single target-spin asymmetries [13–15].

In the following we will mainly focus on the last method.

To access the transversity in SIDIS off transversely polarized nucleons, one can measure the azimuthal angular dependences in the production of spin-0 or (on average) unpolarized hadrons. This production is described by the intrinsic transverse momentum dependent fragmentation function $H_1^\perp(z)$ which is also chiral odd and, moreover, T-odd, i.e., non-vanishing only due to final state interactions. Collins [13] was the first to propose such a spin dependent fragmentation function. It can be obtained, for example, in two-hadron production in e^+e^- annihilation [16]. In the cross section of SIDIS off transversely polarized nucleons it shows up as a $\sin(\phi_h + \phi_S)$ dependence, where ϕ_h is the azimuthal angle of the outgoing hadron (with non-zero transverse momentum P_{hT}) around the virtual-photon direction, and ϕ_S is the azimuthal angle of the target spin vector, both in relation to the lepton scattering plane.

---

[2] The relevant kinematics is: $Q^2 = -q^2$ where $q = k_1 - k_2$, k_1 (k_2) being the 4-momentum of the incoming (outgoing) charged lepton is the 4-momentum of the virtual photon. P (P_h) is the momentum of the target (final hadron), $x = q^2/2(Pq)$, $y = (Pq)/(Pk_1)$, $z = (PP_h)/(Pq)$

The $\sin(\phi_h + \phi_S)$ moment in the SIDIS cross-section can be related to the parton distribution and fragmentation functions involved in the parton level description of the underlying process [14,15]. This moment is defined as the appropriately weighted integral over P_{hT} (the transverse momentum of the observed hadron) of the cross section asymmetry:

$$\langle \frac{|P_{hT}|}{M_h} \sin(\phi_h + \phi_S) \rangle_{UT} \equiv \frac{\int d^2 P_{hT} \frac{|P_{hT}|}{M_h} \sin(\phi_h + \phi_S) \left(d\sigma^\uparrow - d\sigma^\downarrow \right)}{\frac{1}{2} \int d^2 P_{hT} \left(d\sigma^\uparrow + d\sigma^\downarrow \right)}, \quad (1)$$

where $\uparrow$ ($\downarrow$) denotes the up (down) transverse polarization of the target in the virtual-photon frame, M (M_h) is the mass of the target (final hadron), and the subscripts U and T indicate unpolarized beam and transversely polarized target, respectively.

This asymmetry is given by [13–15]:

$$\langle \frac{|P_{hT}|}{M_h} \sin(\phi_h + \phi_S) \rangle_{UT} = 4 S_T \frac{(1-y) h_1(x) z H_1^{\perp(1)}(z)}{(1 + (1+y)^2) f_1(x) D_1(z)}. \quad (2)$$

This weighted single target-spin asymmetry is related to the unweighted one through the following relation:

$$A_{UT}^{\sin(\phi_h + \phi_S)} \approx \frac{M_h}{\langle P_{hT} \rangle} \langle \frac{P_{hT}}{M_h} \sin(\phi_h + \phi_S) \rangle_{UT}. \quad (3)$$

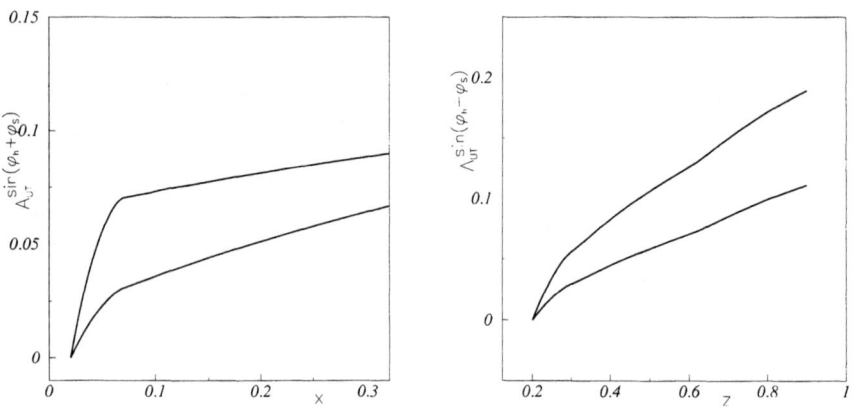

FIGURE 1. The single target-spin asymmetry $A_{UT}^{\sin(\phi_h + \phi_S)}$ for π^+ production as a function of x and z, evaluated using $M_C = 2 m_\pi$ and $\eta = 0.8$ in Eq.(3). The two curves correspond to $h_1 = g_1$ (lower curve) and $h_1 = (f_1 + g_1)/2$ (upper curve).

Note that in the case of a longitudinally polarized target this asymmetry gives contribution to the $\sin\phi_h$ asymmetry [17], which was recently observed in semi-inclusive deep inelastic lepton scattering off a longitudinally polarized proton target at HERMES [18]. Describing this data by an approach where the transverse quark spin distribution in the longitudinally polarized nucleon is vanishing [20], only about 25% are contributed to $\sin\phi_h$ by the 'kinematic' term that is proportional to the transverse component of the nucleon spin vector with respect to the virtual-photon momentum.

For the numerical evaluation of the unweighted asymmetry $A_{UT}^{\sin(\phi_h+\phi_S)}$ the non-relativistic approximation $h_1(x) = g_1(x)$ is used as a lower limit and $h_1(x) = (f_1(x) + g_1(x))/2$ as an upper limit [6]. For the sake of simplicity, Q^2-independent parameterizations were chosen for the distribution functions $f_1(x)$ and $g_1(x)$ [22].

To obtain the T-odd fragmentation function $H_1^{\perp(1)}(z)$, the Collins ansatz [13] for the analyzing power of transversely polarized quark fragmentation was adopted:

$$A_C(z,k_T) \equiv \frac{|k_T|}{M_h} \frac{H_1^\perp(z,k_T^2)}{D_1(z,k_T^2)} = \eta \frac{M_C |k_T|}{M_C^2 + k_T^2}, \tag{4}$$

where η is taken as a constant, although, in principle it could be z-dependent, and M_C is a typical hadronic mass whose value ranges from $2m_\pi$ to M_p.

In our calculations we use $M_C = 2m_\pi$ and $\eta = 0.8$ as in Ref. [21], where a good agreement was found with the single spin asymmetries of the distribution in the azimuthal angle ϕ_h for semi-inclusive π^+ production on a *longitudinally* polarized hydrogen target observed at HERMES [18,19].

For the distribution of the final parton intrinsic transverse momentum, k_T, in the unpolarized fragmentation function $D_1(z, k_T^2)$, a Gaussian parameterization was used [23] with $\langle z^2 k_T^2 \rangle = b^2$ (in the numerical calculations $b = 0.36$ GeV was taken [24]). For $D_1^{\pi^+}(z)$, the parameterization from Ref. [25] was adopted.

In Fig. 1, the asymmetry $A_{UT}^{\sin(\phi_h+\phi_S)}$ of Eq.(3) for π^+ production on a transversely polarized proton target is presented as a function of x and z. The curves have been calculated by integrating over the HERMES kinematic range taking $\langle P_{hT} \rangle = 0.365$ GeV as input. The latter value is obtained in this kinematic region assuming a Gaussian parameterization of the distribution and fragmentation functions with $\langle p_T^2 \rangle = (0.44)^2$ GeV^2 [24].

From Fig. 1 one sees that the single transverse-target-spin asymmetry is quite large. In the HERMES kinematics ($\langle x \rangle \approx 0.1, \langle z \rangle \approx 0.4$) it amounts to $(4 \div 7)\%$. The HERMES experiment using a transversely polarized proton target will be able to extract $h_1(x)$ in a simple way, e.g. as suggested in Ref. [26]. First results on transverse quark spin distribution can be expected from combined results of HERMES and COMPASS within 3-5 years from now, while a complete high precision mapping of their Q^2- and x-dependence requires next-generation facilities, such us TESLA-N [27], ELFE [28], eRHIC, EPIC, with high statistics measurements.

REFERENCES

1. J. Ralston, D.E. Soper, Nucl. Phys. B**152**, 109 (1979).
2. R.L. Jaffe, hep-ph/9602236; hep-ph/9603422.
3. R.L. Jaffe, X. Ji, Phys. Rev. Lett. **67**, 552 (1991) and Nucl. Phys. B **375**, 527 (1992).
4. D. Hill et al., RHIC Spin Collaboration: Letter of Intent, BNL, May 1991.
5. O. Martin et al., *Phys. Rev.* **D60**, 117502 (1999)
6. J. Soffer, *Phys. Rev. Lett.* **74**, 1292 (1995).
7. V.A. Korotkov and W.-D. Nowak, DESY 99-122; hep-ph/9908490.
8. R.L. Jaffe, X.J. Ji, *Phys. Rev. Lett.* **71**, 2547 (1993).
9. X. Artru, M. Mekhfi, *Zeit. Phys.* **C45**, 669 (1990).
10. R.L. Jaffe, *Phys. Rev.*, **D54**, 6581 (1996).
11. R.L. Jaffe, hep-ph/9710465.
12. R.L. Jaffe, X. Jin, J. Tang, *Phys. Rev. Lett*, **80**, 1166 (1998).
13. J.C. Collins, *Nucl. Phys.* **B396**, 161 (1993).
14. A. Kotzinian, *Nucl. Phys.* **B441**, 234 (1995).
15. P.J. Mulders, R.D. Tangerman, *Nucl. Phys.* **B461**, 197 (1996); Erratum - ibid. **B484**, 538 (1997).
16. D. Boer, R. Jakob, P.J. Mulders, Nucl. Phys. B **504** (1997) 345; Phys. Lett. B **424** (1998) 143.
17. K.A. Oganessyan, A.R. Avakian, N. Bianchi, and A.M. Kotzinian, hep-ph/9808368;
18. HERMES Collaboration, A. Airapetian, et.al, Phys. Rev. Lett. **84** (2000) 4047.
19. D. Hasch, for the HERMES collaboration, 35th Rencontres de Moriond: QCD and Hadronic Interactions, Les Arcs, France, March 18 - 25, 2000.
20. E. De Sanctis, W.-D. Nowak, and K.A. Oganessyan, Phys. Lett. B**483** (2000) 69.
21. K.A. Oganessyan, N. Bianchi, E. De Sanctis, W.-D. Nowak, hep-ph/0010261, and hep-ph/0010063.
22. S. Brodsky, M. Burkardt, and I. Schmidt, Nucl. Phys. B**441** (1995) 197.
23. A.M. Kotzinian, and P.J. Mulders, Phys. Lett. B**406** (1997) 373.
24. T. Sjostrand, Comp. Phys. Commun. **82** (1994) 74; CERN-TH.7112/93; hep-ph/9508391.
25. E. Reya, Phys. Rep. **69** (1981) 195.
26. V.A. Korotkov, W.-D. Nowak, K.A. Oganessyan, hep-ph/0002268; DESY 99-176.
27. M. Anselmino, et.al., "Electron scattering with polarized targets at TESLA", in preparation; www.ifh.de/hermes/future
28. M. Anselmino, et.al., ELFE (Electron Laboratory For Europe), Physics Motivation, NuPECC report in print, http://www-dapnia.cea.fr/Sphn/Elfe/Report

HEAVY QUARKS/
TARGET FRAGMENTATION

Quark-Hadron Duality in Electron Scattering

W. Melnitchouk

Jefferson Lab, 12000 Jefferson Avenue, Newport News, VA 23606, and Special Research Centre for the Subatomic Structure of Matter, Adelaide University, Adelaide 5005, Australia

Abstract. Quark-hadron duality addresses some of the most fundamental issues in strong interaction physics, in particular the nature of the transition from the perturbative to non-perturbative regions of QCD. I summarize recent developments in quark-hadron duality in lepton–hadron scattering, and outline how duality can be studied at future high-luminosity facilities such as the electron–hadron collider, EPIC.

I INTRODUCTION

Understanding the structure and interaction of hadrons in terms of the quark and gluon degrees of freedom of QCD is the greatest unsolved problem of the Standard Model of nuclear and particle physics. If one accepts QCD as the correct theory of the strong interactions, then the transition from quark-gluon to hadron degrees of freedom should in principle amount to a change of basis, with all physical quantities independent of which basis is used. However, although the duality between quark and hadron descriptions is formally exact, in practice the necessity of truncating any Fock state expansion means that the extent to which duality holds reflects the validity of the truncations under different kinematical conditions and in different physical processes. Quark-hadron duality is therefore an expression of the relationship between confinement and asymptotic freedom, and is intimately related to the nature of the transition from non-perturbative to perturbative QCD.

In nature, the phenomenon of duality is in fact quite general and can be studied in a variety of processes, such as $e^+e^- \to$ hadrons, or heavy quark decays [1]. One of the more intriguing examples, initially observed some 30 years ago, is in inclusive inelastic electron–nucleon scattering.

II BLOOM-GILMAN DUALITY

In studying inelastic electron scattering in the resonance region and the onset of scaling behavior, Bloom and Gilman [2] found that the inclusive F_2 structure function at low W generally follows a global scaling curve which describes high W data, to which the resonance structure function averages. More recently, high precision data on the F_2 structure function from Jefferson Lab [3], shown in Fig. 1, have confirmed the earlier observations, demonstrating that duality works remarkably well for each of the low-lying resonances, including the elastic, to rather low values of Q^2 (~ 0.5 GeV^2).

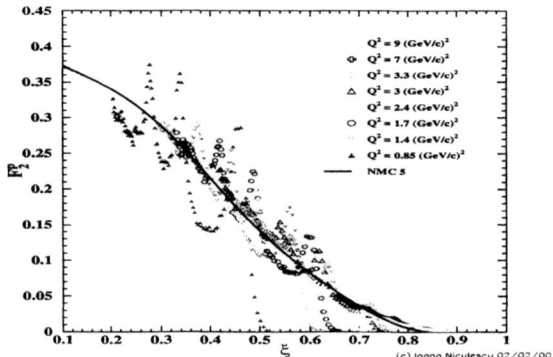

FIGURE 1. Proton F_2 structure function in the resonance region (data from a compilation [3] of JLab and SLAC data); solid line is a fit to large-W deep-inelastic data at $Q^2 = 5$ GeV^2.

Before the advent of QCD, Bloom-Gilman duality was initially interpreted in the context of finite-energy sum rules [4]. Formulated originally for hadron-hadron scattering, they relate the high-energy behavior of amplitudes, described within Regge theory in terms of t-channel Regge pole exchanges, to the behavior at low energy, which can be well described by a sum over a few s-channel resonances [5]. A generalization of the duality picture to include both resonant and non-resonant background contributions to cross sections was suggested by Harari [6], in which resonances were dual to non-diffractive Regge pole exchanges, while the non-resonant background was dual to Pomeron exchange. For electron scattering, this implies that resonances are dual to valence quarks (whose small $x \sim 1/s$ behavior is given in Regge theory by non-diffractive Reggeon exchanges), while the background is dual to sea quarks (for which the small-x behavior is determined by diffractive Pomeron exchange).

In QCD, Bloom-Gilman duality can be formulated in the language of the operator product expansion, in which QCD moments of structure functions are organized according to powers of $1/Q^2$ [7]. The leading terms are associated with free quark scattering, and are responsible for the scaling of the structure function, while the $1/Q^2$ terms involve interactions between quarks and gluons and hence reflect elements of confinement dynamics. The weak Q^2 dependence of the low moments of F_2 is then interpreted to indicate that the non-leading, $1/Q^2$-suppressed, interaction terms do not play a major role even at low Q^2 (≈ 1 GeV^2).

An important consequence of duality is that the strict distinction between the resonance and deep-inelastic regions is quite artificial. As observed by Ji and Unrau [8], at $Q^2 = 1$ GeV^2 around 70% of the total cross section comes from the resonance region, $W < W_{\rm res} = 2$ GeV, however, the resonances and the deep-inelastic continuum conspire to produce only about a 10% correction to the lowest moment of the scaling F_2 structure function at the same Q^2. The deep-inelastic and resonance regions are therefore intimately related, and properly averaged resonance data can help us understand the deep-inelastic region [9,10]. This has immediate implications for global analyses of parton distribution functions, in which the standard procedure is to omit from the data base the entire resonance region below $W = 2$ GeV. This is of practical relevance especially for the large-x region, where deep-inelastic data are scarce [11].

III TESTING THE BOUNDS OF DUALITY

Since the details of quark–hadron duality are process dependent, there is no reason to expect the accuracy to which it holds and the kinematic regime where it applies to be similar for different observables. In fact, there could be qualitative differences between the workings of duality in spin-dependent structure functions and spin-averaged ones [12,13], or for different hadrons — protons compared with neutrons, for instance.

At present there are data on the F_2 structure function of the proton and deuteron [3], but essentially no information at all exists on the spin-dependent g_1 and g_2 structure functions (which correspond to differences of cross sections), nor on the longitudinal to transverse structure function ration, R. It is vital for our understanding of duality and its practical exploitation that the spin and flavor dependence of duality, as well as its nuclear dependence, be established empirically.

Another largely unexplored domain with potentially broad applications is the production of mesons (M) in semi-inclusive electron scattering, $eN \to e'MX$. At high energy the scattering and production mechanisms factorize, with the cross section at leading order in QCD given by a simple product of the structure function and a quark $\to$ meson fragmentation function, as in Fig. 2. In terms of hadronic variables the same process can be described through the excitation of nucleon resonances, N^*, and their subsequent decays into mesons and lower lying resonances, $\widetilde{N}^*$. The hadronic description is rather elaborate, as the production of a fast outgoing meson in the current fragmentation region at high energy requires non-trivial cancellations of the angular distributions from various decay channels [9,10]. Heuristically, the duality between the quark and hadron descriptions of semi-inclusive meson production can be written as (see Fig. 2):

$$\sum_{N^*,\widetilde{N}^*} F_{\gamma^* N \to N^*}(Q^2, W^2)\, \mathcal{D}_{N^* \to \widetilde{N}^* M}(W^2, \widetilde{W}^2) \sim \sum_q e_q^2\, q(x)\, D_{q \to M}(z) \,,$$

where $D_{q \to M}$ is the quark $\to$ meson fragmentation function for a given $z = E_M/\nu$, $F_{\gamma^* N \to N^*}$ is the $\gamma^* N \to N^*$ transition form factor, which depends on the mass of

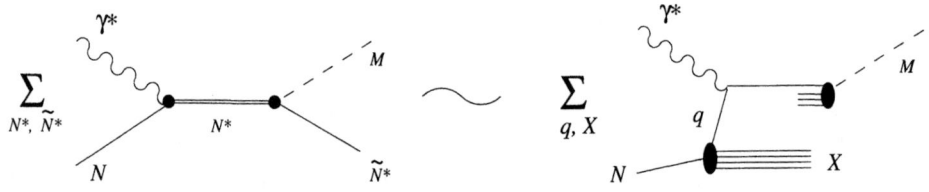

FIGURE 2. Duality between descriptions of semi-inclusive meson production in terms of quark (right) and nucleon resonance (left) degrees of freedom.

the virtual photon and the excited nucleon, $W = M_{N^*}$, and $\mathcal{D}_{N^* \to \widetilde{N}^* M}$ is a function representing the decay of the resonance $N^* \to \widetilde{N}^* M$, where $\widetilde{W}$ is the invariant mass of the final state $\widetilde{N}^*$.

The virtue of semi-inclusive production lies in its ability to identify individual quark species in the nucleon by tagging specific mesons in the final state, enabling both the flavor and spin of quarks and antiquarks to be systematically determined. To what extent factorization applies at lower energy is an open question, and the signatures of duality in the resonance region of semi-inclusive scattering are still under investigation. Confirmation of duality in inclusive hadron production would clearly open the way to an enormously rich semi-inclusive program in the pre-asymptotic regime, allowing unprecedented spin and flavor decomposition of quark distributions.

IV CONCLUSION

Quark-hadron duality offers the prospect of addressing the physics of the transition from the strong to weak coupling limits of QCD, where neither perturbative QCD, nor effective descriptions such as chiral perturbation theory, are applicable. While considerable insight into quark-hadron has already been gained from recent theoretical studies, it will be important in future to understand more quantitatively the features of the electron scattering data in the resonance region and the phenomenological N^* spectrum in terms of realistic models of QCD.

On the experimental side, the spin and flavor dependence of duality can be most readily accessed through semi-inclusive scattering, which requires a facility with both high luminosity and a high duty factor. Jefferson Lab at 12 GeV would be an ideal facility to study meson production in the current fragmentation region at moderate Q^2, allowing the onset of scaling to be tracked in the pre-asymptotic regime. On the other hand, the higher center of mass energy available at an electron-hadron collider, such as EPIC, would, despite a lower luminosity, enable measurement of semi-inclusive cross sections to larger values of Q^2 where perturbative QCD is more readily applicable, and factorization of the current and target fragmentation regions less problematic. Furthermore, unlike fixed-target facilities, a collider mode would allow systematic measurement of hadrons produced in the target fragmentation region. An understanding of duality for target fragments would be the next challenge for electron scattering experiments.

ACKNOWLEDGEMENTS

I would like to thank F.E. Close, R. Ent, N. Isgur, S. Jeschonnek, C. Keppel and J.W. Van Orden for many informative and stimulating discussions about duality. This work was supported by the Australian Research Council, and U.S. Department of Energy contract DE-AC05-84ER40150.

REFERENCES

1. M.B. Voloshin and M.A. Shifman, *Sov. J. Nucl. Phys.* **47**, 511 (1988); N. Isgur, *Phys. Rev.* **D 40**, 101 (1989); *Phys. Lett.* **B 448**, 111 (1999); M. Shifman, hep-ph/0009131.
2. E.D. Bloom and F.J. Gilman, *Phys. Rev. Lett.* **16**, 1140 (1970); *Phys. Rev.* **D 4**, 2901 (1971).
3. I. Niculescu, et al., *Phys. Rev. Lett.* **85**, 1182, 1186 (2000).
4. R. Dolen, D. Horn and C. Schmid, *Phys. Rev. Lett.* **19**, 402 (1967); *Phys. Rev.* **166**, 1768 (1968).
5. G. Veneziano, *Nuov. Cim.* **57 A**, 190 (1968).
6. H. Harari, *Phys. Rev. Lett.* **20**, 1395 (1969).
7. A. de Rújula, H. Georgi and H.D. Politzer, *Ann. Phys.* **103**, 315 (1975).
8. X. Ji and P. Unrau, *Phys. Rev.* **D 52**, 72 (1995).
9. N. Isgur, talk presented at *Workshop on Physics Opportunities with 12 GeV Electrons* (Jefferson Lab, January 2000).
10. N. Isgur, S. Jeschonnek, W. Melnitchouk and J.W. Van Orden, in preparation.
11. N. Isgur, *Phys. Rev.* **D 59**, 034013 (1999); W. Melnitchouk and A.W. Thomas, *Phys. Lett.* **B 377**, 11 (1996); I.R. Afnan, et al., *Phys. Lett.* **B** (in press), nucl-th/0006003.
12. C.E. Carlson and N.C. Mukhopadhyay, *Phys. Rev.* **D 58**, 094029 (1998); *Phys. Rev.* **D 41**, 2343 (1989); C.E. Carlson, hep-ph/0005169.
13. X. Ji and W. Melnitchouk, *Phys. Rev.* **D 56**, 1 (1997).

Λ Polarization in the Target Fragmentation Region of SIDIS

Aram Kotzinian

CERN, CH-1211, Geneva 23, Switzerland [1]

Abstract. I discuss theoretical approaches to Λ polarization in the target fragmentation region (TFR) of (anti)neutrino semi inclusive deep inelastic scattering (SIDIS) and compare them with existing experimental data. It is shown that the best description of data can be achieved in the polarized intrinsic strangeness model.

INTRODUCTION

Polarization measurements provide sensitive tests of strong–interaction dynamics, and have produced a number of surprising results. The largest amount of discussion has been stimulated by measurements of polarized structure functions in DIS experiments, which indicate that the angular momentum of the proton is not distributed among its parton constituents in the way expected in the naïve quark model (NQM). More information on nonperturbative QCD can be obtained by investigating polarization phenomena in SIDIS.

The self–analyzing properties of the Λ baryon make this particle particularly well suited for spin physics. Here I will consider the non-trivial longitudinal (with respect to the virtual boson direction) Λ polarization observed in the TFR ($x_F < 0$) of (anti)neutrino SIDIS [1]– [4].

MODELS FOR Λ POLARIZATION IN THE TFR

$SU(6)$ Quark-Diquark Model

Longitudinal polarization of Λ's produced in SIDIS was first considered in [5]. In the NQM after kicking out a lefthanded u or d-quark from an unpolarized nucleon we have the following relative probabilities for the polarization states of the remnant diquark:

$$\nu p : \ p \ominus d^\uparrow \Rightarrow \frac{1}{36}[2(uu)_{10} + 4(uu)_{1-1}]$$

---

[1] On leave of absence from Yerevan Physics Institute, Yerevan, Armenia and JINR, Dubna, Russia.

$$\nu n : n \ominus d^\uparrow \Rightarrow \frac{1}{36}[9(ud)_{00} + (ud)_{10} + 2(ud)_{1-1}]$$
$$\bar{\nu} p : p \ominus u^\uparrow \Rightarrow \frac{1}{36}[9(ud)_{00} + (ud)_{10} + 2(ud)_{1-1}] \quad (1)$$
$$\bar{\nu} n : n \ominus u^\uparrow \Rightarrow \frac{1}{36}[2(dd)_{10} + 4(dd)_{1-1}]$$

One assumes that during recombination with the unpolarized s-quarks the diquark does not changes its polarization. Since in the NQM the polarization of Λ's is equal to the s-quark polarization the directly produced Λ's will be unpolarized. However, the final state Λ's may also be produced indirectly via electromagnetic decay of Σ^0 or strong decay of Σ^*'s. In both cases, the nonstrange diquark changes its spin from one to zero, while the strange quark retains its polarization [7]. Using the $SU(6)$ wave functions of octet and decuplet baryons one gets for Λ polarization

$$P_\Lambda^{\nu p} = P_\Lambda^{\bar{\nu} n} \simeq -0.47 \quad (2)$$
$$P_\Lambda^{\nu n} = P_\Lambda^{\bar{\nu} p} \simeq 0.03, \quad (3)$$

yielding for an iso-scalar target

$$P_\Lambda^{\nu(p+n)} = P_\Lambda^{\bar{\nu}(p+n)} \simeq -0.22. \quad (4)$$

Meson Cloud Model

Some non-perturbative features of the nucleon structure, such as the deviation from the QCD-parton model inspired Gottfried sum rule, can be explained in the framework of meson cloud model. The pion cloud model provides a natural explanation of the isospin symmetry breaking in the unpolarized proton sea. In the case of polarized DIS, the scattering on lowest ΛK and ΣK component of the nucleon wave function provides a possible mechanism leading to a violation of the Ellis-Jaffe sum rule. The polarization of Λ's produced in the TFR in this model has been considered in [6].

It appears that Λ polarization is almost 100% anticorrelated with the target polarization. Thus, it is expected to be zero for an unpolarized target.

Polarized Intrinsic Strangeness Model

The polarized intrinsic strangeness model (for a review, see [8]) qualitatively reproduce experimental features of the ϕ production in $\bar{p}N$ annihilation. The model is based on the following major observations. First, the fact that the masses of pseudoscalar mesons are small as compared with the typical hadronic scale can be attributed to the existence of effective strong attraction in the $J^{PC} = 0^{-+}$ channel. Secondly, from phenomenological analyses of the quark condensates in the framework of QCD sum rules it is known that the vacuum density of strange-antistrange quark pairs is comparable to the density of u and d quarks. It is natural to assume that the polarized constituent quark can contain an $\bar{s}s$ pair with the vacuum quantum numbers that is in $^3P_0$ state. Hence, in the polarized nucleon, the spin of $\bar{s}$ will be antiparallel to the valence quark spin, $S_z(\bar{s}) = -1/2$ for $S_z(q_v) = 1/2$. In Ref. [9] we have considered the case (in the following referred to as **A**) when the angular momentum projection of $\bar{s}s$ pair, $L_z(\bar{s}s) = +1$ ($S_z(\bar{s}s) = -1$). In this case any s quark in the target fragment should have *negative* longitudinal polarization, so that the longitudinal Λ polarization should also be *negative*, see Fig. 1.

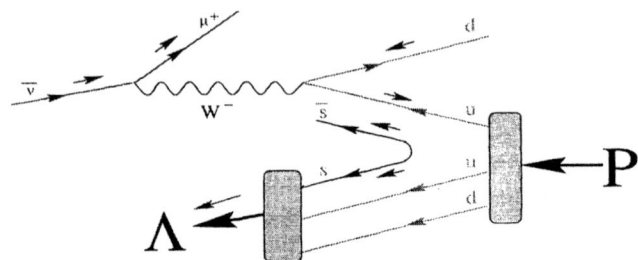

FIGURE 1. *Dominant diagram for Λ production in the target fragmentation region due to scattering on a valence u-quark. Each small arrow represent the longitudinal polarization of the corresponding particle.*

In the quark-parton model of deep inelastic ν or $\bar{\nu}$ scattering, the net longitudinal polarization P_s of the remnant s quark, is given by

$$P_s = \frac{\sum_q C_{sq} N_q - \sum_{\bar{q}} C_{s\bar{q}} N_{\bar{q}}}{N_q + N_{\bar{q}}}, \qquad (5)$$

where N_q ($N_{\bar{q}}$) is the total number of events in which a quark (antiquark) is struck, and C_{sq} is the spin-correlation coefficient. The antiquarks contribute with a negative sign because their charged-current weak interactions are righthanded. The final Λ polarization $P_\Lambda = D_F P_s$, where D_F is a dilution factor which describes the spin transfer during hadronization.

Let us consider also the scenario (**B**) when both projections $L_z(\bar{s}s) = +1$ and $L_z(\bar{s}s) = 0$ of the $\bar{s}s$ pair angular momentum contribute with equal probabilities. The $L_z(\bar{s}s) = 0$ means neglecting the transverse motion of the $\bar{s}s$ pair [10].

The correlation of the remnant s-quark polarization with that of any other struck sea quark ($q_{sea} \neq \bar{s}$) depends on whether they come from the same parent constituent quark. If yes, which might be the dominant case, then a strong spin-correlation is expected (case **a**). If no, the correlation should be reduced (case **b**) Then for the spin-correlation coefficients we have

$$\mathbf{A}: \quad C_{sq_{val}} = -1, \ C_{s\bar{s}} = 1$$

$$\begin{align}\mathbf{B}: \quad & C_{sq_{val}} = -1,\ C_{s\bar{s}} = 1 \\ \mathbf{Aa}: \quad & C_{sq_{sea}} = 1, \\ \mathbf{Ba}: \quad & C_{sq_{sea}} = \frac{1}{9}, \\ \mathbf{b}: \quad & C_{sq_{sea}} = 0.\end{align} \qquad (6)$$

The results for the remnant s-quark polarization and the predictions of **NQM** are presented in the Table 1 together with the measured Λ polarization.

TABLE 1. Λ polarization in the TFR of (anti)neutrino SIDIS.

Experiment (Reaction)	Data	Aa	Ab	Ba	Bb	NQM
WA21 [1] ($\nu_\mu - p$)	-0.29 ± 0.18	-0.51	-0.75	-0.22	-0.25	-0.47
WA21 [1] ($\bar{\nu}_\mu - p$)	-0.38 ± 0.18	-0.85	-0.92	-0.30	-0.31	0.03
WA59 [2] ($\bar{\nu}_\mu - Ne$)	-0.63 ± 0.13	-0.82	-0.91	-0.29	-0.30	-0.22
E632 [3] ($\nu_\mu - Ne$)	-0.43 ± 0.20	-0.70	-0.84	-0.27	-0.27	-0.22
NOMAD [4] ($\nu_\mu - C$)	-0.21 ± 0.04	-0.59	-0.80	-0.24	-0.27	-0.22
NOMAD [4] ($\nu_\mu - $"$p$")	-0.29 ± 0.06	-0.54	-0.77	-0.23	-0.26	-0.47
NOMAD [4] ($\nu_\mu - $"$n$")	-0.16 ± 0.05	-0.61	-0.81	-0.25	-0.27	0.03

DISCUSSION AND CONCLUSIONS

The data on longitudinal Λ polarization from NOMAD experiment [4] have the best statistical accuracy, and we will base the conclusions mainly on comparisons with this data. As one can see from Table 1

- The predictions of NQM are in a surprisingly good agreement with the NOMAD data on the iso-scalar (mainly Carbon) target, but contradict to the NOMAD 'p" and 'n" and WA21 data.

- The meson cloud model predicts zero polarization and is in contradiction with all data.

- The best description of the NOMAD data is achieved in the polarized intrinsic strangeness model with scenarios **Ba** and **Bb** provided that $D_F \approx 1$. If one allows $D_F < 1$ then the scenarios **Aa** and **Ab** can also provide description of all data.

It is possible to distinguish between the scenarios **A** and **B** of the last model by measuring the $\bar{\Lambda}$ polarization in the TFR of the neutrino SIDIS. One should expect that $P_{\bar{\Lambda}} \approx P_\Lambda$ in the case **A** and $P_{\bar{\Lambda}} \approx 3P_\Lambda$ in the case **B**.

The considered models predict different polarizations for intrinsic s- $\bar{s}$-quark sea:

- NQM: $\Delta s = \Delta \bar{s} = 0$.

- Meson cloud model [11]: $\Delta \bar{s} \approx 0$, $\Delta s < 0$.
- Intrinsic strangeness model **A**: $\Delta s \approx \Delta \bar{s} < 0$.
- Intrinsic strangeness model **B**: $\Delta s \approx 1/3 \Delta \bar{s} < 0$.

In principle, this predictions can be tested by measuring the asymmetries of strange particle production in the current fragmentation region of SIDIS or of (anti)neutrino DIS on polarized target.

I would like to mention that models considered here are only first attempts to describe the new phenomenon – the polarization transfer from lepton to the final hadron in the TFR of SIDIS. More theoretical work is devoted to spin transfer from a polarized quark to polarized Λ (see, for example [12] and references therein). The different models are able to describe the existing LEP data.

For the TFR it is also interesting to take into account the $SU(6)$ symmetry breaking and try to improve NQM as it was done in [12]. In the meson cloud model one can expect that the contributions from higher possible fluctuations with vector meson $K^{+*}\Lambda$ will lead to nonzero Λ polarization. However, the estimates of ref. [11] show that the relative probability of this state is small as compared with $K^+\Lambda$ (less than 10%). Moreover they predict a positively-correlated s-quark spin in polarized proton.

ACKNOWLEDGMENTS

I would like to thank J. Ellis, A. Khodjamirian and M. Sapozhnikov for discussions. I am grateful to the Organizing Committee of the Second EPIC Workshop for support and hospitality.

REFERENCES

1. J.T. Jones et al, (WA21 Collaboration), *Z. Phys.* **C28** (1987) 23.
2. S. Willocq et al, (WA59 Collaboration), *Z. Phys.* **C53** (1992) 207.
3. D. DeProspo et al, (E632 Collaboration), *Phys. Rev.* **D50** (1994) 6691.
4. P. Astier et al, (NOMAD Collaboration), Preprint CERN-EP/2000-111
5. I.I. Bigi, *Nuovo. Cim.* **41A** (1977) 43, *ibid* 581.
6. W. Melnitchouk and A.W. Thomas, *Z. Phys.* **A353** (1996) 311.
7. D. Ashery and H.J. Lipkin, *Phys. Lett.* **B469** (1999) 263.
8. J. Ellis, M. Karliner, D.E. Kharzeev and M.G. Sapozhnikov, *Nucl. Phys.* **A673** 263 (2000) 256.
9. J. Ellis, D.E. Kharzeev and A. Kotzinian, *Z. Phys.* **C69** (1996) 467.
10. R. Carlitz and M. Kislinger, *Phys. Rev.* **D2** (1970) 336.
11. S.J. Brodsky and B.-Q. Ma, *Phys. Lett.* **B381** (1996) 317.
12. C. Boros, J.T. Londergan and A.W. Thomas, *Phys. Rev.* **D61** (2000) 014007; *ibid* **D62** (2000) 014021.

Polarized Heavy Flavor Production in Next-to-Leading Order QCD

Marco Stratmann

Institute for Theoretical Physics, University of Regensburg, D-93040 Regensburg, Germany

Abstract. We present a calculation of the NLO QCD corrections to heavy flavor production with longitudinally polarized beams and briefly discuss the main sources of theoretical uncertainties. We apply our results to study the spin asymmetry for total charm quark production which will soon be used by the COMPASS experiment to measure Δg. The results are also relevant for future experiments like EPIC or eRHIC.

MOTIVATION

Despite significant progress in the field of spin-dependent DIS, the polarized gluon density Δg still remains almost completely unconstrained by presently available DIS data, see Fig. 1. An important role plays here the lack of any direct constraint on Δg from other processes or from sum rules. Upcoming spin experiments will thus put a special emphasis on more exclusive measurements to further complete our understanding of the spin structure of the nucleon. In this context heavy quark ($Q = c, b$) production is considered to be one of the cleanest probes for Δg because it is dominantly driven by photon-gluon (gluon-gluon) fusion, $\gamma g\,(gg) \to Q\bar{Q}$, in case of photo(hadro)-production. However, LO estimates of these processes are rather unreliable. Apart from a sizable dependence on the renormalization and factorization scales, μ_r and μ_f, respectively, we already know from corresponding unpolarized NLO calculations [3] that the NLO corrections are large in certain regions of phase space. In addition, the 'clean picture' of, e.g., γg fusion is perhaps obscured by new light quark induced NLO subprocesses $\gamma q \to Q\bar{Q}q$. Therefore the knowledge of the polarized NLO corrections is mandatory for a meaningful extraction of Δg in the future. In what follows we will mainly focus on photoproduction where more details can be found in [4]. Complete NLO results for the case of hadroproduction are expected to be available soon [5].

TECHNICAL FRAMEWORK

The NLO QCD corrections to $\gamma g \to Q\bar{Q}$ consist of three parts: (i) the one-loop virtual corrections, (ii) the real corrections, $\gamma g \to Q\bar{Q}g$, and (iii) the new

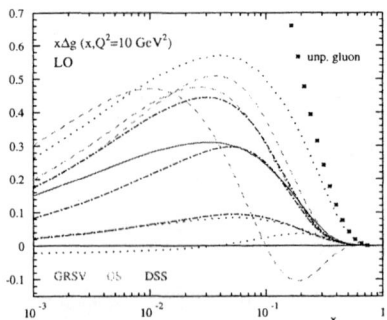

FIGURE 1. Different gluon densities [1,2], all compatible with present DIS data.

genuine NLO production mechanism $\gamma q (\bar{q}) \to Q\bar{Q}q (\bar{q})$. In the calculation of (i)-(iii) one encounters UV, IR, and mass (M) singularities which are removed by renormalization, in the sum of (i) and (ii), and by factorization, respectively. To make all these singularities manifest we choose to work in the framework of n-dimensional regularization. The required polarized squared matrix elements

$$\Delta |M|^2 = \frac{1}{2}\left[|M|^2(++) - |M|^2(+-)\right] , \qquad (1)$$

where the $\pm$ denote the helicities of the incoming particles, are obtained by projecting onto the helicity states of the bosons (photons or gluons) and quarks using the $\epsilon_{\mu\nu\rho\sigma}$ tensor and the γ_5 matrix, respectively. Our results [4] agree with a recent calculation [6] using dimensional reduction, and by taking the sum instead of the difference in (1) we also agree with the known unpolarized results in [3].

The presence of γ_5 and $\epsilon_{\mu\nu\rho\sigma}$ in the polarized calculation introduces some complications since these objects have no unique continuation to $n \neq 4$. In the HVBM prescription, which we use, the usual n-dimensional scalar products $k\cdot p$ are accompanied by their $(n-4)$-dimensional subspace counterparts $\widehat{k\cdot p}$ ('hat momenta'). These terms deserve special attention when performing the $2 \to 3$ phase space integrations and they can contribute to the final expressions.

Finally, it should be recalled that in the factorization procedure for (iii) one has to introduce the parton content of the polarized *photon* [7] where no data exist so far. A scheme independent result in $\mathcal{O}(\alpha_s^2\alpha)$ can thus only be obtained for the sum of the 'direct' *and* 'resolved' photon contributions. However, for small $\sqrt{S}$ the resolved contribution is estimated to be negligible [8], see Tab. 1 below.

RESULTS AND PHENOMENOLOGICAL ASPECTS

The total photon-parton cross section in NLO can be expressed in terms of scaling functions ($i = g, q, \bar{q}$; for simplicity we choose $\mu_r = \mu_f$)

$$\Delta\hat{\sigma}_{\gamma i}(s, m^2, \mu_f) = \frac{\alpha\alpha_s}{m^2}\left[\Delta f_{\gamma i}^{(0)}(\eta) + 4\pi\alpha_s\left\{\Delta f_{\gamma i}^{(1)}(\eta) + \Delta \bar{f}_{\gamma i}^{(1)}(\eta)\ln\frac{\mu_f^2}{m^2}\right\}\right] , \qquad (2)$$

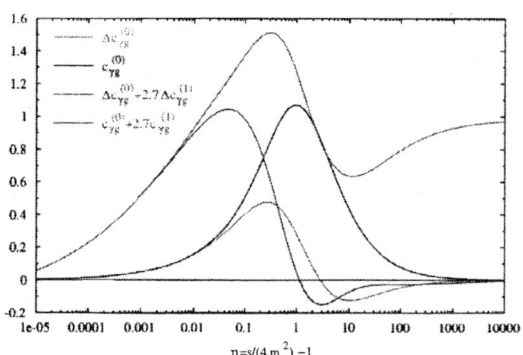

FIGURE 2. Polarized and unpolarized scaling functions for γg fusion in LO and NLO.

where $\Delta f_{\gamma i}^{(0)}$ and $\Delta f_{\gamma i}^{(1)}$, $\Delta \bar{f}_{\gamma i}^{(1)}$ stand for the LO and NLO corrections, respectively, and $\eta \equiv s/4m^2 - 1$. The scaling functions can be further decomposed depending on the electric charge of the heavy and light quarks, e_Q and e_q, respectively:

$$\Delta f_{\gamma g}(\eta) = e_Q^2 \Delta c_{\gamma g}(\eta) \,, \quad \Delta f_{\gamma q}(\eta) = e_Q^2 \Delta c_{\gamma q}(\eta) + e_q^2 \Delta d_{\gamma q}(\eta) \quad (3)$$

with corresponding expressions for the $\Delta \bar{f}_{\gamma i}$.

In Fig. 1 we compare the polarized and unpolarized scaling functions for γg fusion in LO QCD and for the full NLO ($\overline{\mathrm{MS}}$) combination $\Delta c_{\gamma g}^{(0)} + 4\pi \alpha_s \Delta c_{\gamma g}^{(1)}$ using a typical value for α_s relevant for charm production[1]. In the threshold region $\eta \to 0$ one observes that in LO and NLO $\Delta c_{\gamma g} \to c_{\gamma g}$. The NLO corrections on the parton level are large in this region due to the 'Coulomb singularity' which leads to a non-zero cross section at threshold in NLO. A further enhancement in NLO is due to large logarithms $\propto \beta \ln^2 8\beta^2$ stemming from soft gluon radiation as $\beta^2 \equiv 1 - 4m^2/s \to 0$. However, these logarithms can be resummed to all orders, as was recently demonstrated, e.g., in the case of g_1^c in polarized DIS [9]. At high energies, $\eta \to \infty$, $\Delta c_{\gamma g}^{(1)}$ and $c_{\gamma g}^{(1)}$ behave rather differently. $c_{\gamma g}^{(1)}$ approaches a plateau due to t-channel gluon exchange, but such an enhancement is absent for $\Delta c_{\gamma g}^{(1)}$ because it drops out in the difference of the two helicity combinations in Eq.(1).

In general the NLO corrections are large, however, one has to keep in mind that the coefficients have to be convoluted with the parton densities Δf to yield a physical, i.e., measurable, cross section. As a function of the photon-proton c.m.s. energy $S_{\gamma p}$ the total hadronic heavy flavor photoproduction cross section reads

$$\Delta \sigma_{\gamma p}^Q(S_{\gamma p}, m^2, \mu_f) = \sum_{f=q,\bar{q},g} \int_{4m^2/S_{\gamma p}}^{1} dx \, \Delta \hat{\sigma}_{\gamma f}(xS_{\gamma p}, m^2, \mu_f) \, \Delta f(x, \mu_f^2) \,. \quad (4)$$

It is important to notice that the threshold region in Fig. 1 is enhanced by the gluon density at small values of x, while the coefficient functions at the largest

---

[1] The numerically less important quark coefficients can be found in [4].

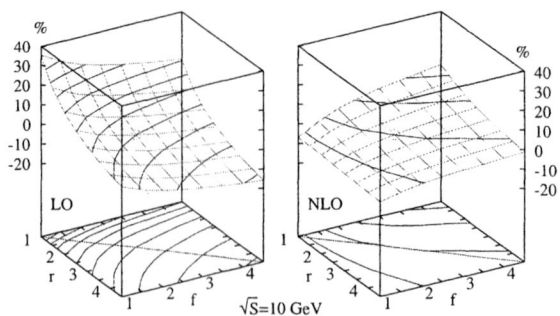

FIGURE 3. Scale dependence of $\Delta\sigma^c_{\gamma p}$ at $\sqrt{S_{\gamma p}} = 10\,\text{GeV}$ in LO and NLO (see text).

possible η are convoluted with parton densities at large x where they tend to zero. The importance of the threshold region was also observed, e.g., for F_2^c in DIS [10].

In Fig. 3 we study the scale dependence of $\Delta\sigma^c_{\gamma p}$ by varying μ_f and μ_r independently in steps of the charm mass, $m_c^2 \leq \mu_{r,f}^2 \leq 4m_c^2$, using the GRSV densities [1] and $m_c = 1.5\,\text{GeV}$. As can be easily read off from the contour lines at the base of the plot (given in steps of 5%), the NLO results are much more stable against variations of $\mu_{r,f}$. Actually the dominant theoretical uncertainty for charm production at COMPASS energies stems from the unknown value of m_c as can be inferred from Fig. 4 where we plot the spin asymmetry $A^c_{\gamma p} \equiv \Delta\sigma^c_{\gamma p}/\sigma^c_{\gamma p}$ using different gluon densities Δg [1,2] to demonstrate the sensitivity of COMPASS[2] to Δg. It should be noted that we have imposed at cut $p_T^c < 1.2\,\text{GeV}$ to enhance $A^c_{\gamma p}$. The NLO corrections are sizable but under control, and the 'background' from γq induced processes and resolved photons is small for energies accessible at COMPASS and EPIC [8] but not necessarily at higher $S_{\gamma p}$, see Tab. 1.

TABLE 1. Direct and resolved-γ contributions to $\Delta\sigma^c_{\gamma p}$.

$\sqrt{S_{\gamma p}}$ [GeV]	GRSV Δg		'large Δg'		GSC Δg	
	direct [nb]	resolved [nb]	direct [nb]	resolved [nb]	direct [nb]	resolved [nb]
10	10.2	-0.72	23.0	-0.63	-3.48	-0.70
20	13.9	-0.29	23.2	0.33	19.6	-0.70
100	-4.76	2.09	-9.71	4.00	0.58	2.27

Unfortunately, COMPASS can measure $\Delta g/g$ from $A^c_{\gamma p}$ only for a single value of x of about 0.15. A further complication arises from the fact that also the *unpolarized* gluon density g is not very well known for $x > 0.1$ [12] which makes it difficult to extract Δg from a measurement of $\Delta g/g$ without an improved knowledge of g. A measurement of the unpolarized charm cross section by COMPASS would be extremely helpful in this respect and, perhaps, would reduce the uncertainty on m_c as well. RHIC will provide first information about the shape of Δg from various

---

[2] The error bar indicates the expected statistical accuracy of COMPASS [11].

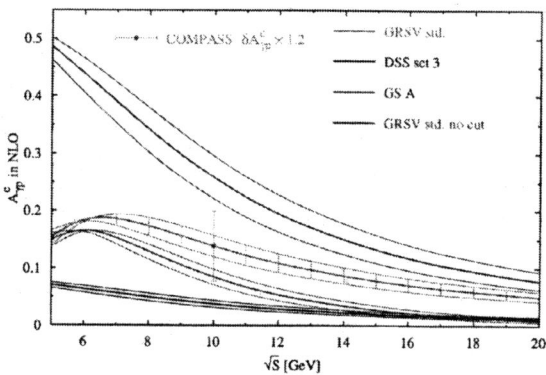

FIGURE 4. $A^c_{\gamma p}$ in NLO QCD for different gluon densities. The bands indicate variations of the charm mass in the range $1.4 \leq m_c \leq 1.6\,\text{GeV}$.

processes such as the hadroproduction of charm and bottom. In the latter case one has to resolve the long-standing discrepancy between b production data from the Tevatron and NLO QCD before one can reliably extract Δg from corresponding polarized measurements. However, recent b-jet data from DØ [13] agree much better with NLO QCD, and hopefully run II will help to diminish this discrepancy.

ACKNOWLEDGMENTS

It is a pleasure to thank Ingo Bojak for a pleasant collaboration.

REFERENCES

1. M. Glück et al., *Phys. Rev.* **D53**, 4775 (1996).
2. T. Gehrmann and W.J. Stirling, *Phys. Rev.* **D53**, 6100 (1996); D. de Florian et al., *Phys. Rev.* **D57**, 5803 (1998); *Phys. Rev.* **D62**, 094025 (2000).
3. R.K. Ellis and P. Nason, *Nucl. Phys.* **B312**, 551 (1989); J. Smith and W.L. van Neerven, *Nucl. Phys.* **B374**, 36 (1992).
4. I. Bojak and M. Stratmann, *Phys. Lett.* **B433**, 411 (1998); *Nucl. Phys.* **B540**, 345 (1999); *ibid.* **B569**, 694 (2000) (E); I. Bojak, Ph.D. Thesis, *hep-ph/0005120*.
5. I. Bojak and M. Stratmann, work in progress.
6. A.P. Contogouris et al., *Phys. Lett.* **B482**, 93 (2000); *hep-ph/0007050*.
7. See, e.g., M. Stratmann and W. Vogelsang, *Phys. Lett.* **B386**, 370 (1996).
8. M. Stratmann and W. Vogelsang, *Z. Phys.* **C74**, 641 (1997).
9. T.O. Eynck and S. Moch, *hep-ph/0008108*.
10. A. Vogt, *hep-ph/9601352* (in proc. of the HERA workshop, Durham, UK, 1995).
11. G. Baum et al., COMPASS collab., CERN/SPSLC-96-14.
12. W. Vogelsang and A. Vogt, *Nucl. Phys.* **B453**, 334 (1995); J. Huston et al., *Phys. Rev.* **D58**, 114034 (1998); A.D. Martin et al., *Eur. Phys. J.* **C4**, 463 (1998).
13. B. Abbott et al., DØ collab., *hep-ex/0008021*.

Soft Diffraction at EPIC Scales

M.A. Pichowsky

Nuclear Theory Center, Indiana University, Bloomington, IN 47405, USA.

Abstract. Experiment has provided a nearly complete phenomenological description of the soft Pomeron in recent decades. Yet, little is understood of the quark and gluon dynamics underlying soft-Pomeron phenomena. Why haven't canonical experiments provided additional insights into this issue? Can spin observables shed light on soft diffraction?

In recent years, we have made exciting advancements towards understanding the properties of the "hard Pomeron," believed to be closely related to the better-known "soft Pomeron" from Regge theory. Some of the excitement surrounding the discovery of a new, hard Pomeron, is the hope that its study would provide additional insight into the nature of the soft Pomeron, which has long remained mysterious. However, differences between the soft and hard Pomerons suggest that their relation may be very tenuous; the question remains of how to learn more about the soft Pomeron. Studies of diffractive, strong-interaction physics have come a long way in the last few decades. Surprisingly, after years of experiments we still lack a satisfactory picture of the structure of the soft Pomeron in terms of QCD. Its deceptively simple and universal behavior is the main reason for the present lack of understanding. The mystery would be resolved by understanding how the non-perturbative quark and gluon dynamics result in the simple, universal behavior observed in soft-Pomeron phenomena. Unfortunately, the measurement of canonical observables may not provide additional details of the quark and gluon dynamics underlying soft diffraction.

The Pomeron trajectory was originally introduced in Regge theory to account for the high-energy behavior of a large class of scattering processes that could not otherwise be described using the well-established hadron trajectories. The Pomeron trajectory

$$\alpha_P(t) = \alpha_0 + \alpha_1 t, \tag{1}$$

with $\alpha_0 = 1.08$ and $\alpha_1 = 0.25$ GeV^{-2}, provides the correct energy dependence for inclusive and exclusive processes, like πp, Kp and pp scattering, as well as vector-meson electroproduction. With most of the hadrons already accounted for by the usual Regge trajectories being predominantly comprised of valence quarks, it is possible that the Pomeron is associated with gluonic degrees of freedom. Hence, the soft Pomeron may be due to the exchange of gluonic bound states (hybrids or glueballs). From Eq. (1), the lightest of such states would be an isoscalar $J^{PC} = 2^{++}$ meson of mass $m_{2^{++}} = 1.92$ GeV. The $f_2^{++}(2010)$ meson, with mass $2.011^{+0.06}_{-0.08}$ MeV, may be such a gluon state [1].

A unique feature of QCD is that of all Regge trajectories, only the Pomeron trajectory intercept $\alpha_0 > 1$. Thus, the soft Pomeron (and gluon exchanges) dominate over all other interactions for diffractive processes at high energies, providing a unique oportunity to explore the dynamics of non-perturbative gluon exchange. If the soft Pomeron dominates over all other interactions, why don't we know more about it? The reason is that diffractive processes depend on deceptively few scales!

The short-ranged, strong interactions of QCD entail that the two most probable outcomes of high-energy hadron scattering are: 1) the beam and target completely miss each other, or 2) the beam and target collide, breaking apart into many hadrons. Such a process is *diffractive* and well-described by assuming the two hadrons are perfectly absorbing disks – the black-disk model. The resulting diffractive cross section is given by cylindrical Bessel functions, and for small-t behaves like

$$\frac{d\sigma}{dt} = A\, e^{b\,t}, \qquad (2)$$

where A is a t-independent constant, and b is the *impact parameter*, given by $b = R^2/4$ with R being the radius of the hadrons. For unpolarized diffractive hadron-hadron scattering, Regge theory gives,

$$\frac{d\sigma}{dt} \propto (s/s_0)^{2(\alpha_0-1)+2\alpha_1 t} F_B^2(t)\, F_T^2(t), \qquad (3)$$

where s and t are the usual Mandlestam variables, s_0 is an arbitrary energy scale, and $F_B(t)$ and $F_T(t)$ are hadron form factors representing the sizes of the beam and target. Thus, small-t diffractive processes are well-parametrized by Eq. (2) with

$$A(s) = A(s_0)(s/s_0)^{2\alpha_0-2}, \qquad (4)$$

$$b(s) = \frac{1}{4}(R_B^2(s_0) + R_T^2(s_0)) + 2\alpha_1 \ln(s/s_0), \qquad (5)$$

where R_B and R_T are *diffractive radii* of the hadrons.

The universal behavior of the soft Pomeron entails that the relations in Eqs. (2)–(5) provide a concise account of most of what can be measured in diffractive processes, which are described in terms of the Pomeron trajectory $\alpha_P(t)$ and the sizes of the hadrons R_B and R_T. Models of hadron structure may provide predictions of the diffractive radii R_B and R_T, thereby eliminating them as unknown parameters, but an explanation of how the underlying *gluon dynamics* result in the Pomeron trajectory $\alpha_P(t)$ remains a mystery.

The relevant degrees of freedom and interactions for ρ-meson electroproduction is shown in Fig. 1. Above CM energies of $W \approx 10$ GeV (regions IV–VI), ρ-meson

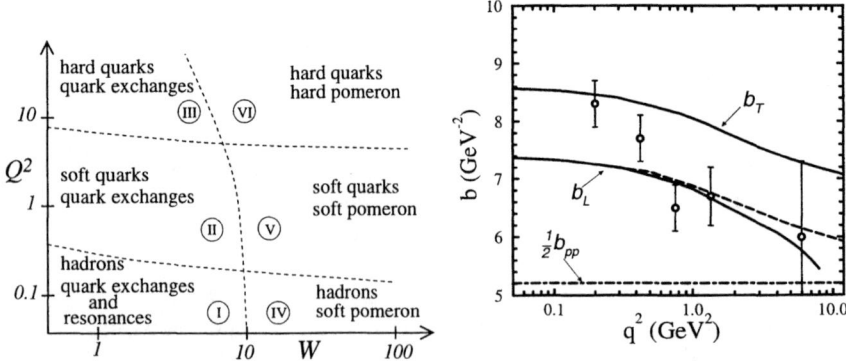

FIGURE 1. (Left) Relevant degrees of freedom and interactions of ρ-meson electroproduction.
FIGURE 2. (Right) Q^2 dependence of electroproduction impact parameters for transversely (b_T) and longitudinally (b_L) polarized ρ mesons at $W = 13$ GeV, with data from Ref. [2]. The dot-dashed line $\frac{1}{2}b_{pp}$ represents the size of the proton target.

electroproduction is diffractive. For small Q^2 (region IV), this diffractive process is like any other in that its behavior is completely determined by the Pomeron trajectory, the target-proton size R_p, and the ρ-meson size R_ρ according to Eqs. (2)–(5).

Increasing Q^2 allows one to probe smaller scales within the ρ meson. Using the model of Ref. [3] to calculate the Q^2-dependence of the impact parameter $b(W^2, Q^2)$, shown as solid curves for transversely and longitudinally polarized photons in Fig. 2, one can estimate the length-scales probed by electroproduction at various values of Q^2. The distances from each of the curves b_L or b_T to the horizontal dot-dashed line (representing the proton-target size) represents the length-scales probed within the longitudinally- or transversely-polarized ρ meson, respectively. One observes from Fig. 2, that for small values of Q^2, most of the produced ρ mesons are transversely polarized, while for large Q^2, most are longitudinally polarized; this is in accordance with s-channel helicity conservation. As a result, the low-Q^2 data are found on the curve for b_T, while large-Q^2 data are found on the curve for b_L. The difference between the solid curve b_L and the dashed curve (calculated by eliminating small length-scales or "hard components" in the ρ-meson wavefunction, provides a measure of the importance of perturbative QCD. The increasing importance of perturbative QCD may be indicative of the emergence of hard-Pomeron exchange (and, consequently our passing from region V into region VI). Hence, values of Q^2 below a few GeV^2 should be employed to ensure that the process is dominated by soft-Pomeron exchange.

With the study of soft diffraction necessarily limited to regions IV and V, it appears that little insight into the quark and gluon dynamics underlying soft-Pomeron

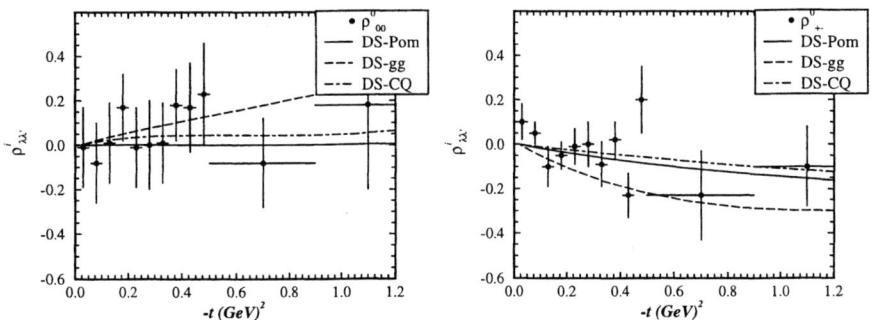

FIGURE 3. Spin-density matrix elements ρ_{00}^0 and ρ_{+-}^0 for ϕ-meson photoproduction with data from Ref. [4].

exchange would be gained from future experiments unless new observables could be found that are particularly sensitive to such dynamics. Polarization experiments may provide such observables! As an example of this, the model of Ref. [3] is used to calculate spin-density matrix elements of ϕ-meson photoproduction. In Fig. 3, the results are shown for the Pomeron-exchange interaction with Dyson-Schwinger confined-quark propagators (solid curves), the same interaction with constituent-quark propagators (dot-dashed curves), and a two-gluon exchange interaction with Dyson-Schwinger propagators (dashed curves). One observes that these spin observables are more sensitive to differences between the Pomeron-exchange and two-gluon exchange interactions employed than they are to the dynamics of the quark propagators. Such preliminary studies give us an indication that spin observables may be used to probe quark and gluon dynamics underlying diffraction in QCD.

In this article, I have tried to convince the reader that unpolarized observables are primarily determined by the diffractive sizes of hadrons and the well-known Pomeron trajectory $\alpha_P(t)$. As a result, our best hope to understand the origin of the soft Pomeron in QCD requires the measurement of spin observables in vector-meson electroproduction.

Acknowledgments. This work is supported by the U.S. Department of Energy under contract DE-FG02-87ER40365 and the National Science Foundation under contract PHY97-22076.

REFERENCES

1. D.E. Groom, et al., Eur. Phys. J. **C15**, 1 (2000).
2. E665 Collaboration, M. Adams et al., Z. Phys. **C74**, 237 (1997).
3. M.A. Pichowsky and T-S.H. Lee, Phys. Rev. **D56**, 1644 (1997).
4. Barber et al., Z. Phys. **C12**, 1 (1982).

The strange sea inside the nucleon

V. Barone

*Università "A. Avogadro", 15100 Alessandria, Italy
and INFN, Sezione di Torino, 10125 Torino, Italy*

C. Pascaud and F. Zomer

LAL, IN2P3-CNRS and Université de Paris-Sud, 91898 Orsay, France

Abstract. We present the results of a new global analysis of DIS data, characterized by an enlarged neutrino and antineutrino data set. Special emphasis is given to the strange sector. The strange sea distribution is determined independently of the non-strange sea. The possibility of a charge asymmetry, $s(x) \neq \bar{s}(x)$, is tested.

I INTRODUCTION

Neutrino DIS measurements play a relatively minor role in most of the present-day global fits. Since the determination of the strange sea density relies completely on charged-current DIS, $s(x)$ turns out to be quite poorly known. Recently we tried to improve the situation by performing a new analysis of DIS data with an enlarged neutrino and antineutrino data set [1]. This enabled us to provide an accurate determination of $s(x)$ in the framework of a global fit. We also tested the hypothesis of the charge asymmetry of the strange sea. In the following we shall briefly outline the results of our analysis focusing in particular on the strange sector.

II NEUTRINO MEASUREMENTS AND GLOBAL FITS

To start with, let us see why so far only a small part of the information accumulated in $\nu, \bar{\nu}$ DIS experiments has been exploited in global fits.

The old data (BEBC, CDHS, CDHSW) cannot be immediately used because the radiative corrections are either incomplete or not applied at all and/or the bin center corrections are not performed. On the other hand, recent published data (CCFR) are, so to speak, analyzed 'too much'. The cross sections are not available and only structure functions are provided. The latter come from a preanalysis, which includes assumptions on $R = \sigma_L/\sigma_T$, nuclear effects, slow rescaling for charm etc. This can lead to problems of consistency with the QCD analysis eventually performed within the global fits.

The most sophisticated global parametrizations on the market (CTEQ [2], MRST [3]) use only the F_2^ν and xF_3^ν structure functions from CCFR. Two difficulties arise. First of all, the CCFR structure functions are not compatible at low x with the charged-lepton structure functions. In particular, below $x \simeq 0.1$, there is a clear discrepancy between the CCFR F_2^ν and the F_2^μ measured by NMC. This discrepancy might be due to a different nuclear shadowing in νDIS and μDIS and to a different longitudinal-to-transverse cross section ratio in charged-current and neutral-current DIS. The second problem arising from the use of CCFR structure functions (especially, of F_2^ν) is that they tend to favor an unrealistically large value of α_s. The MRST global fit clearly shows that the function $\chi^2(\alpha_s)$ has no minimum below $\alpha_s(M_Z^2) = 0.123$.

The difficulty of including neutrino data in global fits reflects itself in an uncertain knowledge of $s(x)$.

III THE DETERMINATION OF $S(X)$

There are two ways to extract the strange sea distribution from DIS data: *i)* by a direct determination; *ii)* by a global fit (like the other flavor densities).

Direct determination means that the strange-charm sector of charged-current DIS is selected either by looking at particolar signatures in the final state or by properly combining the inclusive structure functions.

The opposite-sign dimuon production in $\nu, \bar{\nu}$ DIS probes the strangeness in the proton. In principle this is a very effective way to determine $s(x)$. However the analysis of this process is affected by many uncertainties (charm fragmentation, acceptance correction, etc.). The dimuon determination has been performed by CCFR [4]. Their data sample consists of about 5,000 ν events and 1,000 $\bar{\nu}$ events. The strange-to-non-strange momentum ratio κ found in the NLO CCFR analysis is $\kappa = 0.48$ at $Q^2 = 20$ GeV^2.

The determination of $s(x)$ within a global fit is made difficult by the lack of large and reliable neutrino and antineutrino data sets. Both MRST and CTEQ are unable to fit $s(x)$ independently of the non-strange sea. Thus they borrow the value $\kappa \simeq 0.5$ from the CCFR analysis and constrain $s(x)$ as $s(x) + \bar{s}(x) = 0.5[\bar{u}(x)+\bar{d}(x)]$. The results of the fits for the strange distribution are clearly biased by this constraint and the $s(x)$ found by MRST and CTEQ depends ultimately on the CCFR dimuon measurement.

IV A NEW GLOBAL ANALYSIS AND THE EXTRACTION OF $S(X)$

We performed a new global analysis of DIS data at the NLO level in QCD. The main features of this analysis are (for details see Ref. 1): 1) The analysis is based on a large $\nu, \bar{\nu}$ data set, which includes all available ν, $\bar{\nu}$ cross section data (BEBC, CDHS, CDHSW). We also fit charged-lepton data (BCDMS, NMC, H1) and Drell-Yan data (E605, NA51, E866). For the sake of consistency, the CCFR structure function data, coming from a preanalysis different from ours, are not included. 2) The data have been properly reanalyzed. Bin center corrections, electroweak radiative corrections, corrections for nuclear and isoscalarity effects have been applied. 3) Error correlations have been taken into account. 4) A massive factorization scheme is used: we chose the Fixed Flavor Number Scheme, which is known to be more stable in the moderate Q^2 range of the data. 5) The kinematic cuts imposed are: $Q^2 \geq 3.5$ GeV^2, $W^2 \geq 10$ GeV^2 (in this region higher-twist effects are negligible). As for the CDHSW data, we excluded the controversial region $x < 0.1$.

Due to the abundance of neutrino data, we are able to fit the strange sea *independently* from the non-strange sea. Moreover, the good balance between ν and $\bar{\nu}$ measurements in the high-statistics CDHSW data allows us to test the charge asymmetry in the strange sea: $s(x) \neq \bar{s}(x)$ (see next section).

In our main fit, called fit1, the parton densities are parametrized at the input scale $Q_0^2 = 4$ GeV^2 imposing as usual $s = \bar{s}$, but without any strong constraint between $s, \bar{s}$ and $\bar{u}, \bar{d}$. We will present the results for the strange sector, referring the reader to Ref. 1 for a full account of the fit. In Fig. 1(left) we show the strange

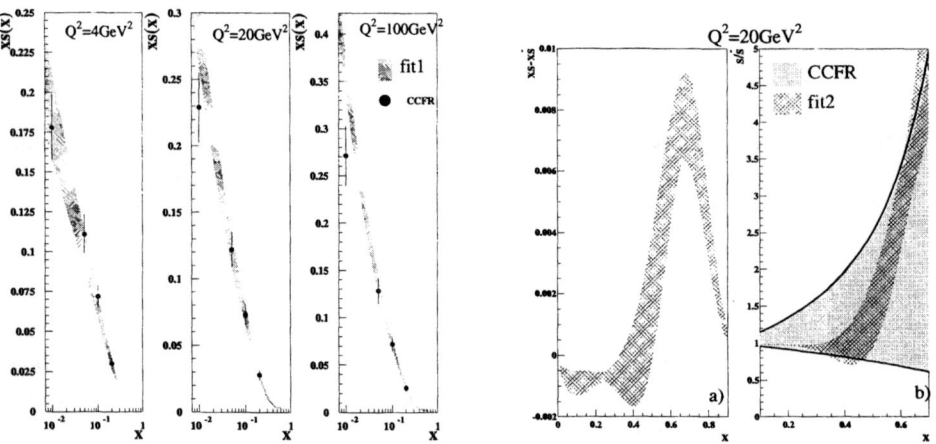

FIGURE 1. Left plot: strange distribution in fit1 at various Q^2 values. Right plot: strange vs. antistrange distribution in fit2.

distribution at different Q^2 values. The results of the CCFR dimuon determination [4] is shown for comparison. The meaning of the error bands is explained in detail in Ref. 1. They correspond to an increase of the χ^2 by one unit and do not take into account the uncertainties related to the choice of the functional form of the distributions. Note that although the CCFR points seem to be in good agreement with our curves, the strange-to-non-strange ratio we find is quite different from CCFR's: $\kappa = 0.67$ at $Q^2 = 20$ GeV^2, to be compared with the CCFR value 0.48 at the same scale. We also performed a modified fit, called `fit1b`, imposing, as it is done by CTEQ and MRST, the condition $s + \bar{s} = 0.5(\bar{u} + \bar{d})$, motivated by the CCFR result on κ. It turns out that `fit1b` is definitely worse (χ^2(`fit1`)=2431, χ^2(`fit1b`)=2493 and χ^2(`fit2`)=2405 for 2657 data points). We found that it is especially the $\bar{\nu}$ data which favor `fit1` with respect to `fit1b`. Incidentally, we notice that no discrepancy whatsoever emerges in our fit between neutrino and charged-lepton data. We checked that the fit worsens if the CCFR structure functions data are taken into account.

A Test of the charge asymmetry in the strange sea

Is $s(x)$ different in shape from $\bar{s}(x)$? In order to answer this question by fitting neutrino and antineutrino data one needs a good balance between ν and $\bar{\nu}$ statistics. This is the case of our data set. A charge asymmetry in the strange sea is not forbidden by first principles (clearly, as the nucleon has no net strangeness, one must have $\int dx\, (s - \bar{s}) = 1$), and is actually expected in the framework of the intrinsic sea theory of Brodsky et al. [5]. Intrinsic $q\bar{q}$ pairs have a relatively long lifetime and arrange themselves into higher Fock states of the proton $|uudq\bar{q}\ldots\rangle$. By minimizing the kinetic energy on the light-cone one finds that the larger the mass of the intrinsic quark the higher its average momentum. Thus the intrinsic sea tends to occupy the large x region. In the specific case of the strange sea, the $s\bar{s}$ pairs give rise to $N \to \Lambda K$ fluctuations [6]. In Ref. 7 it was shown, by simple chiral symmetry arguments, that one should expect $\langle x_s \rangle > \langle x_{\bar{s}} \rangle$.

In order to test the charge asymmetry of the strange sea, we released the constraint $s = \bar{s}$ and performed another fit, `fit2`, looking for a possible difference between $s(x)$ and $\bar{s}(x)$. In Fig. 1(right) we plot $xs(x) - x\bar{s}(x)$ and $s(x)/\bar{s}(x)$ at $Q^2 = 20$ GeV^2. The strange distribution turns out to be harder than the antistrange one, in agreement with the expectation of the intrinsic sea theory. In Fig. 2 we show the difference Δ_ν between ν and $\bar{\nu}$ differential cross sections which is directly sensitive to $s - \bar{s}$: `fit2` is favored at large x with respect to `fit1`. One can see in Table 1 that the minimum χ^2 of `fit2` is 25 units smaller than the χ^2 of `fit1`, with an overall number of 2657 data points. It is clear that new high-statistics ν and $\bar{\nu}$ data would allow to increase the significance of the result and to draw a more definite conclusion.

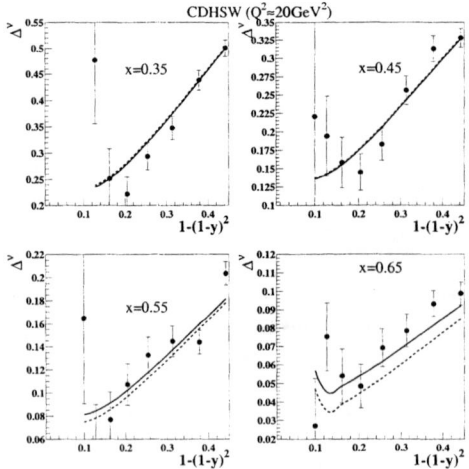

FIGURE 2. Difference between the ν and $\bar{\nu}$ differential cross sections. The solid line corresponds to `fit2`, the dashed line to `fit1`.

V CONCLUSION

We have shown how the full use of available $\nu, \bar{\nu}$ cross sections provides important information on the flavor structure of the nucleon, in particular on the strange distribution. A proper analysis of the forthcoming neutrino data (CHORUS and NuTeV) will certainly improve our knowledge of $s(x)$ and allow a more conclusive test of the charge asymmetry of the strange sea.

REFERENCES

1. V. Barone, C. Pascaud and F. Zomer, *Eur. Phys. J.* C **12**, 243 (2000).
2. H.L. Lai et al., CTEQ Collaboration, *Eur. Phys. J.* C **12**, 375 (2000).
3. A.D. Martin, R.G. Roberts, W.J. Stirling and R.S. Thorne, *Eur. Phys. J.* C **4**, 463 (1998).
4. A.O. Bazarko et al., CCFR Collaboration, *Z. Phys.* C **65**, 189 (1995).
5. S.J. Brodsky, C. Peterson and N. Sakai, *Phys. Rev.* D **23**, 2745 (1981).
6. A.I. Signal and A.W. Thomas, *Phys. Lett.* B **191**, 205 (1987).
   S.J. Brodsky and B.-Q. Ma, *Phys. Lett.* B **381**, 317 (1996).
7. M. Burkardt and B.J. Warr, *Phys. Rev.* D **45**, 958 (1992).
8. See the transparencies of Bodek's talk at
   http://moriond.in2p3.fr/QCD00/transparencies/6_friday/am/bodek
9. M.G. Aivazis, F.I. Olness and W.-K. Tung, *Phys. Rev. Lett.* **65** (1990) 2339.
10. V. Barone et al., *Phys. Lett.* B **268** (1991) 279; *Z. Phys.* C **70** (1996) 83.
11. V. Barone, U. D'Alesio and M. Genovese, *Phys. Lett.* B **357** (1995) 435.

e-A PHYSICS

QCD in eRHIC and EPIC

Jamal Jalilian-Marian

Physics Department, University of Arizona, Tucson, AZ 85716 USA

Abstract.
We give a brief review of QCD effects in eRHIC and EPIC. We show that perturbative QCD predicts a new state of matter, the Color Glass Condensate, which determines the outcome of high energy collisions. We show how this state can be detected and discuss its implications for high energy heavy ion collisions. We discuss how the proposed new experiments such as eRHIC and EPIC can help detect and understand the properties of the Color Glass Condensate.

INTRODUCTION

QCD is now commonly accepted as the theory of strong interactions. The agreement of theoretical predictions with experimental data is excellent. However, QCD is a very complicated theory and despite its success in describing the experimental date, is not theoretically well understood. A major issue in QCD which is yet to be understood is the behavior of cross sections at high energies. Perturbative QCD predicts a fast growth of cross sections with energy due to the sharp and unlimited rise of partonic distribution functions with energy (x_{bj}). This unlimited growth would eventually violate unitarity as expressed by the Froissart bound. This unlimited growth is expected to be slowed by high parton density effects at high energies which would restore unitarity.

At high energies (small x_{bj}) gluons are the most abundant parton species in the wave function of a proton/nucleus. As one goes to higher and higher energies, the gluon number density in the wave function increases and gluons will start to spatially overlap and interact. In [1] it was shown that the interactions of gluons in the wave function lead to saturation, or more precisely, a slow down in the growth of the gluon number density. In other words, the gluon number density per unit rapidity per unit area $\frac{1}{\pi R^2}\frac{dN}{d^2 k_t}$ grows like $\ln k_t^2$ at small k_t rather than the usual perturbative $1/k_t^2$ growth. At large k_t where the gluon self-interactions are negligible, the growth of the gluon number density matches with the perturbative one. One can define a new scale, the so-called Saturation scale Q_s^2, where the perturbative and non-perturbative regions match. It is trivial to show that most gluons in the wave function of a hadron or nucleus at high energies will have momenta of order Q_s.

Therefore, most gluons in the wave function of hadron or nucleus in the high energy limit condense into a state with momentum Q_s. As shown in [1] the mathematical properties of this state are analogous to those of a spin glass. Since the gluons carry color, this state is called a Color Glass Condensate. It is characterized by its momentum scale Q_s the precise value of which depends on energy and the atomic number since it is caused by the non-linearities of the gluon fields, the strength of which depends on the energy and atomic number. However, the existence of this state at high energies is independent of the hadron or nucleus considered. This has very interesting and important consequences. It means that at high energies, the cross sections are determined by the Color Glass Condensate alone and are therefore universal. Furthermore, the initial conditions of a high energy collision such as relativistic heavy ion collisions at RHIC and LHC are determined by the properties of the Color Glass Condensate. If the saturation scale Q_s is large enough compared to Λ_{QCD}, then one can use weak coupling $(\alpha_s(Q_s^2) << 1)$ methods to investigate the properties of the Color Glass Condensate.

THE SATURATION SCALE Q_S

The best way to measure the saturation scale Q_s is the measurement of hadronic/nuclear structure functions in DIS experiments. For momenta $Q^2 >> Q_s^2$ non-linear QCD effects which would give rise to the saturation phenomenon are negligible and one expects that the structure function F_2 and its derivative $dF_2/d\ln Q^2 \sim xG(x,Q^2)$ change very fast with x, Q^2 (almost exponentially) as predicted by pQCD. However, for momenta which are less than the saturation scale, the behavior of the structure function and its derivative is dramatically different. Specifically, one expects $dF_2/d\ln Q^2 \sim Q^2 \pi R^2$ in the saturation region.

Alternatively, plotting $dF_2/d\ln Q^2$ vs. x and Q^2 as suggested first by Caldwell, one would see an eventual leveling off and decrease of the derivative of F_2 as x and Q^2 decrease. The momentum Q_s^2 where $dF_2/d\ln Q^2$ levels off will depend sensitively on the impact parameter and the atomic number A of the target. There was a great deal of excitement when it was realized that the HERA data exhibit such a behavior. However, it turned out that one could fit all the data using the standard DGLAP evolution equations which would indicate we are far from the saturation region. It should be emphasized, however, that F_2 and its derivative are fully inclusive quantities and are not well constrained.

A more exclusive quantity is the diffractive structure function F^{diff}. This is a better quantity in which to look for saturation effects. There is much less flexibility in fitting the experimental data on diffraction and therefore F^{diff} and its log derivative will be more sensitive to saturation effects. Another venue is the vector meson production at high energies. There already exist saturation based models which reproduce the experimental data on diffraction and vector meson production quite well while the Reggie based models fail. This is a strong indication that saturation effects are already present in the data but are masked by large uncertainties of the data points.

It should be emphasized that saturation effects can be present in both hadrons and nuclei. However, since parton number densities are typically enhanced by $A^{1/3}$, saturation effects in nuclei will show up at lower energies than hadrons. Therefore, nuclei are a better environment in which to measure the saturation scale. High parton densities which lead to saturation effects also lead to the breakdown of the standard collinear factorization theorems in pQCD. Since calculations of cross sections in QCD is based on some factorization theorem, one needs to invent a new formalism to perform calculations in the saturation region. In [1] an effective action and Wilson renormalization group approach was proposed and developed which enables one to perform calculations in the saturation region. This new formalism also leads to a new evolution equation which is the generalization of the known pQCD evolution equations such as DLA DGLAP and BFKL. Using the new evolution equation, one can calculate the amount of nuclear shadowing of gluons generated by QCD radiation [2].

eRHIC AND EPIC

A high energy (polarized) electron- (polarized) ion collider such as the proposed (EPIC) eRHIC will go a long way in establishing the non-linear QCD effects which lead to a new state of matter, the Color Glass Condensate. The proposed experiments will significantly expand our knowledge of the (polarized) inclusive nuclear structure functions as well as the diffractive ones. The kinematic region in x and Q^2 covered will be much larger than any previous experiment. It is expected that the non-linear QCD effects which lead to a new state of matter will manifest themselves in the measurements of the inclusive and diffractive structure functions in the kinematic region covered by the proposed experiments. Also, at a given x, the value of Q^2 will be larger than that of fixed target experiments. This will lead to smaller uncertainties in theoretical predictions. Alternatively, at fixed Q^2, one will reach smaller values of x where interesting and novel phenomena are expected to occur. Furthermore, the experimental uncertainties of the structure function measurements will be much less than the uncertainties of the current measurements. Another crucial advantage of the proposed experiments will be the significantly wider Q^2 coverage at fixed x. This would enable us to measure the (polarized) gluon distribution functions much more precisely and will help establish whether the nuclear shadowing of quarks and gluons is a leading or higher twist effect which will discriminate between pQCD and Regge based theoretical models of nuclear shadowing. One will also be able to measure the longitudinal structure function at eRHIC directly for the first time due to the tunable beam energy.

SUMMARY

A high energy electron-nucleus collider is urgently needed. In addition to discovering and exploring a new state of matter, the Color Glass Condensate, a high energy electron-nucleus collider will vastly improve the precision of the structure function measurements as well as widely expanding the kinematic region in x and Q^2 where these structure functions can be measured for the first time.

ACKNOWLEDGMENTS

I would like to thank the organizers of the EPIC workshop for the invitation and the opportunity to give a talk.

REFERENCES

1. L. McLerran and R. Venugopalan, *Phys. Rev.* **D49**, 335 (1994); **D49**, 2233 (1994); A. Ayala, J. Jalilian-Marian, L. McLerran and R. Venugopalan, *Phys. Rev.* **D52**, 2935 (1995); **D53**, 458 (1996); J. Jalilian-Marian, A. Kovner, L. McLerran and H. Weigert, *Phys. Rev.* **D55**, 5414 (1997); J. Jalilian-Marian, A. Kovner, A. Leonidov and H. Weigert, *Nucl. Phys.* **B504**, 415 (1997); *Phys. Rev.* **D59**, 014014 (1999); **D59**, 014015 (1999); **D59**, 034007 (1999); A. Kovner and J.G. Milhano, hep-ph/9904420.
2. J. Jalilian-Marian and X.N. Wang, *Phys. Rev.* **D60**, 054016 (1999), hep-ph/0005071.

QCD at High Parton Density

Yuri V. Kovchegov[1]

Department of Physics, University of Washington, Box 351560
Seattle, WA 98195-1560

Abstract. We discuss the physics of quarks and gluons in the nuclear wave function at very high energies. We show that the density of gluons and quarks in the nuclear structure functions can become very large as the energy of the nucleus increases. This leads to parton recombination in the nuclear wave function, which compensates the parton branching to keep the parton density from exceeding the maximum physically possible value. The effect is known as parton saturation. We derive an equation governing the structure functions in the saturation regime and present a solution of this equation. We will argue that eRHIC collider would be an ideal machine to study the strong gluonic fields associated with the saturation regime in the nucleus.

INTRODUCTION

One of the most fundamental problems in the physics of strong interactions is the problem of understanding QCD at high energies. The physics of high energy QCD is relevant to many scattering processes, such as proton–proton collisions (pp), deep inelastic scattering on a proton or nucleus (ep and eA), nuclear collisions (AA). The energies of the modern AA accelerators are beginning to reach up to hundreds of GeV per nucleon at Relativistic Heavy Ion Collider (RHIC) at BNL and at Large Hadron Collider (LHC) which is being constructed at CERN. To be able to properly interpret the data to be provided by these machines it is vital to clarify the underlying dynamics of strong interactions.

QCD is the only non-Abelian gauge field theory with such interesting feature as confinement that we can test experimentally. Unfortunately many fundamental properties of QCD have not been understood theoretically and have not been tested experimentally yet. One of such features of QCD is the dynamics of strong interactions at high energies. It has been conjectured [1] that it leads to creation of very strong gluonic fields in the nuclear and hadronic wave functions. These strong fields are freed in the nuclear collisions producing gluonic matter with the field strength $F_{\mu\nu} \sim 1/g$ at the early stages of the collisions. Subsequent interactions

---

[1] This work has been sponsored in part by the U.S. Department of Energy under Grant No. DE-FG03-97ER41014.

may equilibrate this gluonic matter, possibly leading to the state commonly known quark–gluon plasma (QGP). However, since the maximum possible in QCD gluonic field strength has been reached at the very early stages of the collision these final state interactions can not make it stronger. That way the strong gluonic fields will be created independent of the possible subsequent QGP formation. Final state rescatterings can only modify the distribution of particles in this gluonic matter, without increasing the field strength. Thus it is very interesting and important to study the properties of this gluonic matter, which could be observed already in the wave function of a single nucleus. The proposed electron–heavy ion collider (eRHIC) is the machine which is perfectly suitable to explore the strong field physics.

The internal structure of hadrons and nuclei are usually described in terms of the quark and gluon distribution functions, $xq(x, Q^2)$ and $xG(x, Q^2)$. However, the observable quantities are the cross sections of DIS, which, in turn, are related to the structure functions $F_1(x, Q^2)$ and $F_2(x, Q^2)$. Proton's structure functions have been extensively studied at Hadron Electron Ring Accelerator (HERA) at DESY. The majority of the data shows that the proton's structure functions at large Q^2 and not very small x can be described by the DGLAP evolution equation, which is a linear equation. It is convenient to present different properties of hadronic and nuclear structure functions in terms of a "phase diagram" in Q^2 and $\ln 1/x$ plane (see Fig. 1). The DGLAP physics corresponds to the lower right section of the plane, where Q^2 is large and x is not too small. As one goes towards smaller x in the same region of Q^2 the hadronic and nuclear structure functions rise. Most of the data in that region can be explained in terms of either small-x limit of DGLAP equation or, alternatively and more interestingly, by the BFKL equation, which is also a linear equation but could be responsible for evolving the system towards a higher gluonic density regime. However, the DIS cross sections can not rise forever. This, for instance, would violate unitarity. That means that at some very small x the hadronic and nuclear structure function have to undergo a significant qualitative change of behavior, becoming a much slower varying functions of x. The slow down of the growth of hadronic structure functions is usually associated with non-linear effects in the quark and gluon dynamics, such as parton recombination, which eventually balances the parton splitting process. Therefore the partons in the hadronic or nuclear wave function reach the state of *saturation*. The region of saturation of the structure functions is represented in yellow in Fig. 1. The transverse momentum scale Q^2 corresponding to the transition to the saturation region is different for different values of Bjorken x, increasing with decreasing x. We will denote this transition line by $Q_s(x)$. There has recently been obtained new experimental data at HERA which might contain evidence of saturation transition in DIS on a proton at $x \approx 10^{-3} - 10^{-4}$ and $Q^2 \approx 2 - 4 GeV^2$. There are, however, different possible explanations of that data.

The saturation transition line varies as one goes from DIS on a proton to DIS on a nucleus. Approximately we can argue that all the momenta in the system scale by powers of the atomic number A. The saturation transition scale, $Q_s^2(x)$,

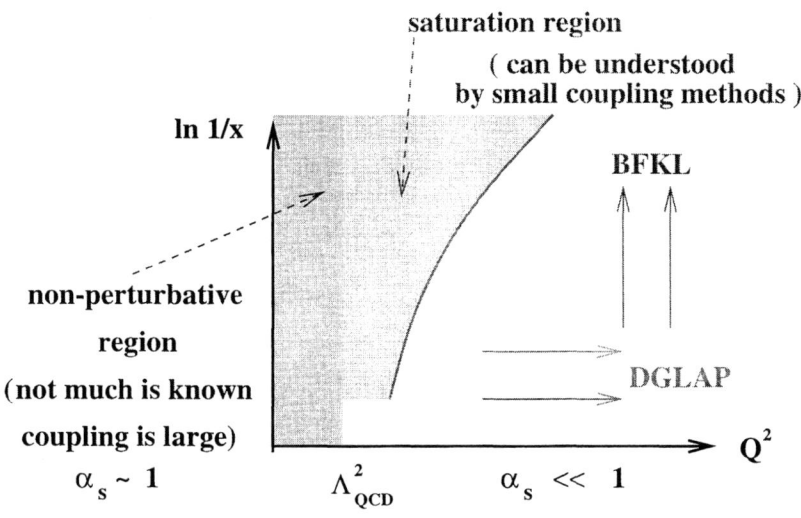

FIGURE 1. "Phase diagram" of high energy QCD.

changes as $Q_s^2(x) \to A^{1/3} Q_s^2(x)$. That means that the whole saturation curve in Fig. 1 would be shifted towards higher momenta (to the right in Fig. 1) by a factor of say 5 or 6 for a heavy (Au or Pb) nucleus. The effect of saturation in DIS on a heavy ion would be much more pronounced, and will be accessible at much larger values of x than in DIS on the proton for the same value of Q^2. That means one needs much lower energies in eA colliders as compared to ep colliders to achieve the same density of partons. DIS on a heavy ion would allow us to observe saturation and the non-linear effects of high gluonic density associated with it in the region of large Q^2, which would also allow the use of perturbation theory in the theoretical calculations, i. e. theoretical understanding of the process would be much more feasible than in ep collisions.

EVOLUTION EQUATION DESCRIBING THE SATURATION REGION

A non–linear evolution equation describing the nuclear F_2 structure function can be written in the large N_c limit using Mueller's dipole model [2]. The F_2 structure function can be written as a convolution of the wave function of the virtual photon splitting in the quark–antiquark pair $\Phi(\mathbf{x}_{01}, z)$ and the cross section (or forward amplitude) of the $q\bar{q}$ pair interacting with the target nucleus $N(\mathbf{x}_{01}, \mathbf{b}_0, Y)$:

$$F_2(x, Q^2) = \frac{Q^2}{4\pi^2 \alpha_{EM}} \int \frac{d^2 \mathbf{x}_{01} dz}{2\pi} \, \Phi(\mathbf{x}_{01}, z) \, d^2 b_0 \, N(\mathbf{x}_{01}, \mathbf{b}_0, Y). \tag{1}$$

The forward amplitude $N(\mathbf{x}_{01}, \mathbf{b}_0, Y)$ obeys the following evolution equation which was derived in [3]

$$N(\mathbf{x}_{01}, \mathbf{b}_0, Y) = -\gamma(\mathbf{x}_{01}, \mathbf{b}_0) \exp\left[-\frac{4\alpha C_F}{\pi} \ln\left(\frac{x_{01}}{\rho}\right) Y\right] +$$

$$+ \frac{\alpha C_F}{\pi^2} \int_0^Y dy \exp\left[-\frac{4\alpha C_F}{\pi} \ln\left(\frac{x_{01}}{\rho}\right)(Y-y)\right] \times$$

$$\times \int_\rho d^2 x_2 \frac{x_{01}^2}{x_{02}^2 x_{12}^2} [2\, N(\mathbf{x}_{02}, \mathbf{b}_0 + \frac{1}{2}\mathbf{x}_{12}, y)$$

$$- N(\mathbf{x}_{02}, \mathbf{b}_0 + \frac{1}{2}\mathbf{x}_{12}, y)\, N(\mathbf{x}_{12}, \mathbf{b}_0 + \frac{1}{2}\mathbf{x}_{02}, y)], \tag{2}$$

with $\gamma(\mathbf{x}_{01}, \mathbf{b}_0)$ the propagator of a dipole of size $\mathbf{x}_{01}$ at the impact parameter $\mathbf{b}_0$ through the target nucleus or hadron, which was taken to be of Glauber form

$$\gamma(\mathbf{x}_{01}, \mathbf{b}_0) = \exp\left[-\frac{\alpha \pi^2}{2 N_c S_\perp} x_{01}^2 AxG(x, 1/x_{01}^2)\right] - 1. \tag{3}$$

here $S_\perp$ is the transverse area of the hadron or nucleus, A is the atomic number of the nucleus, and xG is the gluon distribution of the nucleons in the nucleus, which was taken at the two gluon (lowest in α) level. Y is the rapidity variable. Eq. (3) is written for a cylindrical hadron or nucleus. In the large N_c limit $C_F \approx N_c/2$. In Eq. (2) ρ is an ultraviolet regulator, which will not present in any physical quantities such as cross sections and structure functions.

The linear term on the right hand side of Eq. (2) corresponds to the well-known BFKL equation [4]. The quadratic term in Eq. (2) is responsible for the parton recombination and due to the minus sign in front it introduces damping in the cross section N. That way our equation is similar to GLR-MQ equation [5], except that our equation has been rigorously derived in the leading logarithmic approximation, whereas GLR-MQ was only proven in the double logarithmic approximation. Similar equation has been obtained by Balitsky for the correlators of Wilson lines [6].

The solution of Eq. (2) for the case of a nuclear target was found in [7]. The resulting behavior of DIS cross section as functions of energy is shown in Fig. 2. For not very high energies F_2 structure function (and the total DIS cross section) grows as a power of the center of mass energy of the $\gamma^* A$ system s

$$F_2(x, Q^2) \sim s^{\alpha_P - 1}, \tag{4}$$

which corresponds to the single BFKL pomeron exchange ($\alpha_P - 1 = 4\alpha N_c \ln 2/\pi$). As energy s increases to asymptotically high values the growth of the cross section slows down

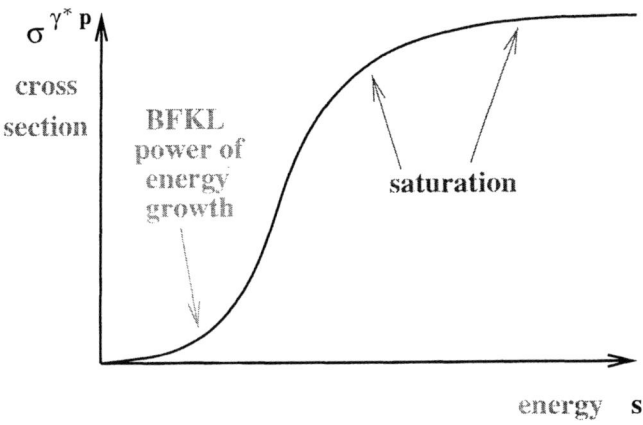

FIGURE 2. DIS cross sections as predicted by Eq. (2).

$$F_2(x, Q^2) \approx (Q^2 R^2 N_c/3\pi^2)(\alpha_P - 1)\ln s. \tag{5}$$

The slow (logarithmic) dependence of the structure functions on the center of mass energy s given by Eq. (5) corresponds to the saturation regime. Finally, the solution of Eq. (2) allows us to derive an expression for the border of the saturation region $Q_s(x)$, as shown in Fig. 1. As was estimated in [7]

$$Q_s(x) \sim \Lambda A^{1/3} (1/x)^{\alpha_P - 1}. \tag{6}$$

An interesting feature of Eq. (6) is that the saturation scale grows as $A^{1/3}$ with the atomic number A and not as $A^{1/6}$ as would be predicted by usual multiple rescattering models. This makes observation of saturation phenomena much more feasible at the eRHIC collider.

REFERENCES

1. L. McLerran and R. Venugopalan, Phys. Rev. D **49**, 2233 (1994); 3352 (1994); D **50**, 2225 (1994).
2. A.H. Mueller, Nucl. Phys. **B415**, 373 (1994); A.H. Mueller and B. Patel, Nucl. Phys. **B425**, 471 (1994); A.H. Mueller, Nucl. Phys. **B437**, 107 (1995).
3. Yu. V. Kovchegov, Phys. Rev. D **60**, 034008 (1999).
4. E.A. Kuraev, L.N. Lipatov and V.S. Fadin, *Sov. Phys. JETP* **45**, 199 (1978); Ya.Ya. Balitsky and L.N. Lipatov, *Sov. J. Nucl. Phys.* **28**, 22 (1978).
5. L.V. Gribov, E.M. Levin, and M.G. Ryskin, Nucl. Phys. **B188**, 555 (1981); Phys. Reports **100**, 1 (1983); A.H. Mueller, J.-W. Qiu, Nucl. Phys. **B268**, 427 (1986).
6. I. Balitsky, Nucl. Phys. **B463**, 99 (1996).
7. Yu. V. Kovchegov, Phys. Rev. D **61**, 074018 (2000); E. Levin and K. Tuchin, Nucl. Phys. **B573**, 833 (2000); M. Braun, Eur. Phys. J. **C16**, 337 (2000).

QCD Instantons in the Soft Pomeron

Yuri V. Kovchegov[1]

Department of Physics, University of Washington, Box 351560
Seattle, WA 98195-1560

Abstract. We propose a model of soft pomeron in which the interaction between the hadrons is mediated by a gluon ladder with the QCD instanton induced interactions at the vertices. Thus we propose a new type of instanton-induced interactions that leads to the rising with energy cross sections $\sigma \sim s^\Delta$ of Regge type (the Pomeron). We calculate the intercept of the pomeron Δ and derive the Pomeron trajectory.

INTRODUCTION

This talk is based on the work done in collaboration with Dima Kharzeev and Genya Levin [1].

At the high energy electron-nucleus collisions proposed at eRHIC a very high density of partons in the nuclear wave function will be created and probed by interactions with virtual photon coming from electron. The corresponding high field strength of the gluon field parametrically will be of the order $F_{\mu\nu} \sim 1/g$, where $g \ll 1$ is the strong coupling constant. The field strength of order $1/g$ is the maximum obtainable field strength in QCD. If one tries to increase gluonic fields even further QCD vacuum would break up creating particle-antiparticle pairs which would introduce color charge screening insuring that the field would not increase any further. Creation of such a strong gluonic field in eA collisions makes one wonder whether the instanton-induced effects may become important in this process. After all the field strength of the QCD instantons is also parametrically of the order $1/g$. Here we will argue that instantons may play an important role in the high energy scattering processes which have been studied in accelerators around the world for many years, such as deep inelastic scattering (DIS) and proton-proton collisions.

Donnachie and Landshoff [2] parameterized a variety of deep inelastic and proton-proton scattering data by a relatively simple model with two types of pomeron interactions. One type of pomeron in their model has an intercept of 0.435 and is

---

[1] This work has been sponsored in part by the U.S. Department of Energy under Grant No. DE-FG03-97ER41014.

important at large (hard) transverse momentum scales $Q^2 > 10\,\text{GeV}^2$ [2]. Correspondingly it could be called "hard" pomeron. Possibly the hard pomeron can be obtained in QCD by summation of gluon ladder diagrams with perturbative vertices. This is the idea behind the BFKL pomeron [3] and is outside of the scope of this work. Another type of pomeron interaction in Donnachie and Landshoff model [2] comes with the intercept of 0.08 and is important at relatively soft transverse momentum scales $Q^2 \sim 1\,\text{GeV}^2$. It also dominates the total cross section in proton-proton collisions. Correspondingly this kind of interaction has been named "soft" pomeron. Here we will concentrate on understanding this phenomenon.

Recently Kharzeev and Levin [4] proposed a theoretical model of the soft pomeron based on QCD low energy theorems which were formulated and proved in [5]. The low energy theorems imply that there exists a large momentum scale which is characteristic to QCD vacuum. The scale was estimated to be $M_0^2 \approx 4 \div 6\,\text{GeV}^2$ [5]. The largeness of the scale M_0 insures that one can use gluonic degrees of freedom in this problem. The essence of the model of Kharzeev and Levin [4] is that the soft pomeron is proposed to be a sum of ladder diagrams with gluon lines in the t-channel, and the vertices together with the s-channel exchanges being given by the correlators of pairs of QCD energy-momentum tensors, which are given by the low energy theorems of [5]. The resulting soft pomeron intercept is

$$\Delta = \frac{1}{48} \ln \frac{M_0^2}{m_\pi^2}, \qquad (1)$$

where m_π is the pion's mass. After plugging in the numbers into Eq. (1) the intercept comes out to be quite close to the experimental value of 0.08 [4].

The interesting feature of the soft pomeron model proposed in [4] is that the vertices in the ladder graph were given by some non-perturbative interactions resulting from low energy theorems. As one can convince oneself the low energy theorems could only be satisfied by a strong gluonic field, $A_\mu \sim 1/g$, which is hinting that there is some strong field constituting the QCD vacuum. This field was responsible for gluonic vertices in the model of [4]. Here we will elaborate on the idea of using some non-perturbative interactions in the pomeron lader. We will employ a relatively simple model of that strong vacuum gluonic field: we will assume that gluonic interactions are generated by an instanton field. The soft pomeron conjectured here is represented in Fig. 1. It is still a set of ladder diagrams, just like the BFKL pomeron [3]. The difference is in the vertices: in the BFKL pomeron the vertices are the perturbative vertices introducing factors of g, so that each rung of the BFKL pomeron parametrically yields us with a factor

$$\alpha \ln s, \qquad (2)$$

where α is the strong coupling and s is the center of mass energy, leading to the BFKL pomeron intercept being proportional to α. Alternatively, the vertices of the pomeron in Fig. 1 are induced by instantons, with instantons and anti-instantons generating multi-gluon interactions. Each instanton brings in a factor of $\exp(-2\pi/\alpha)$, such that each rung of the ladder brings in a factor of

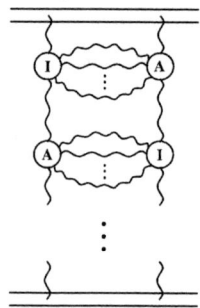

FIGURE 1. The structure of soft pomeron as conjectured in the paper. Reprinted from [1] with permission from Elsevier Science.

$$\exp\left(-\frac{4\pi}{\alpha}\right) \ln s. \qquad (3)$$

with the pomerons intercept proportional to $\exp(-4\pi/\alpha)$. The BFKL pomeron resums all powers of the parameter given in Eq. (2) and our pomeron of Fig. 1 resums all the power of the parameter in Eq. (3). We make an assumption that the scale of the coupling constant is set by the typical size of QCD instantons ($\alpha = \alpha(1/\rho_0)$) and that this typical size ρ_0 is small enough for the coupling to be small. This assumption seems to be justified by lattice data [6]. If the coupling is small than the parameter $\exp(-4\pi/\alpha) \ll 1$ and since at high energies $\ln s \gg 1$ then $\exp(-4\pi/\alpha) \ln s \sim 1$ and resummation of this parameter is justified.

Another difference between our pomeron of Fig. 1 and the BFKL pomeron is that in our ladder there is more than one gluon produced in the s-channel in each rung. In fact there could be as many gluons produced as physically allowed by the $2 \to n$ process mediated by an instanton in QCD. (For simplicity we first consider pure gluodynamics.) Here we run into a serious difficulty: to be able to calculate the $2 \to n$ transition amplitude through an instanton one needs to know the so-called "holy grail" function, which has been known only at not very high energies [7]. To make an approximate estimate we assume that the $2 \to n$ is given by the tree level diagrams for center of mass energies below the sphaleron energy and is zero for energies above the sphaleron energy. For arguments justifying this assumption see [1]. Sphaleron energy in QCD is given by

$$E_{sph} = \frac{3\pi}{4\,\alpha\,\rho_0}, \qquad (4)$$

which for $\rho_0 = (700\text{MeV})^{-1}$ [6] we estimate to be $E_{sph} = 2.4 \div 2.6$ GeV. In the results presented below we had to average over instanton sizes. There we make another crude assumption based on lattice data [6]: we assume that the instanton size distribution falls off rapidly for $\rho > \rho_0$, which allowed us to use ρ_0 as the upper (infrared) cutoff of the distribution function $n(\rho)$ [1].

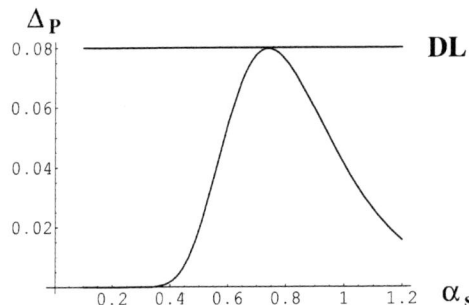

FIGURE 2. Intercept of our soft pomeron as a function of α for $E_{sph} = 2.4\,\text{GeV}$, $\rho_0 = 0.3fm$ and the virtual corrections giving a factor of $\delta = 0.31$ suppression. The horizontal line corresponds to the phenomenological pomeron intercept [2]. Reprinted from [1] with permission from Elsevier Science.

After a lengthy calculation, the details of which we are going to skip here, we obtain the following expression for the soft pomeron intercept in gluodynamics

$$\Delta_{soft} = \frac{\pi I_0 d^2}{(N_c^2 - 1)^2} \left(\frac{2\pi}{\alpha}\right)^{4N_c} e^{-\frac{4\pi}{\alpha}} \frac{(4\pi)^6}{\alpha^3} \frac{1}{81} \frac{E_{sph}^2 \rho_0^2}{6 e^2}$$

$$\times \left[{}_2F_4\left(\frac{9}{2}, \frac{9}{2}; 1, 1, \frac{11}{2}, \frac{11}{2}; \frac{\pi E_{sph}^4 \rho_0^4}{6 \alpha e^2}\right) - 1\right], \qquad (5)$$

with $I_0 \approx 0.014$ and $d \approx 1.5\,10^{-3}$ [8]. Inclusion of the quark contribution modifies the expression for the soft pomeron intercept

$$\Delta_P = \delta \left[\prod_{q=u,d,s,\ldots} 1.3 \left(m_q \rho_0 - \frac{2\pi^2}{3} \langle 0|\bar{q}q|0\rangle \rho_0^3\right)\right]^2 \Delta_{soft}, \qquad (6)$$

where the product in Eq. (6) goes over the quark flavors, m_q's are quark masses and the value of the quark vacuum condensate was taken to be $\langle 0|\bar{q}q|0\rangle = -(0.25\,\text{GeV})^3$ for all flavors [1]. The factor δ in Eq. (6) is due to the virtual corrections, which are not shown in Fig. 1 and were not included into Eq. (5). Unfortunately we could not perform a rigorous inclusion of virtual corrections into Eq. (5) and have just estimated them to decrease the value of the intercept by a factor of $2 \div 3$.

The intercept of Eqs. (5) and (6) is plotted in Fig. 2 as a function of α for $E_{sph} = 2.4\,\text{GeV}$, $\rho_0 = 0.3fm$ and $\delta = 0.31$. Since at the scale of $\rho_0 = 0.3fm$ higher order perturbative corrections may significantly change α we therefore have a certain freedom in the value of α and that is why we have used it as a variable in Fig. 2. The horizontal line in Fig. 2 corresponds to the Donnachie and Landshoff

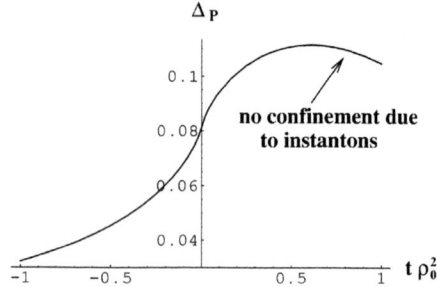

FIGURE 3. Soft pomeron's trajectory. t is measured in the units of ρ_0^2. Reprinted from [1] with permission from Elsevier Science.

value of the intercept 0.08. Due to the uncertainty in the world's knowledge of QCD instanton parameters, such as E_{sph} and ρ_0 we have an ambiguity in the value of the intercept in Fig. 2. However, for the set of values that we have chosen we can obtain an intercept close to phenomenological for $\alpha = 0.7 \div 0.8$.

Finally, we have also calculated the soft pomeron trajectory, which is plotted in Fig. 3. The corresponding slope is given in terms of the intercept $\Delta = 0.08$ by [1]

$$\alpha' = \Delta \frac{\pi \rho_0^2}{256 I_0} \ln \frac{1}{q \rho_0} \approx 0.16 \,\text{GeV}^{-2}, \qquad (7)$$

which is close to the experimental value of $\alpha'_{exp} = 0.25 \,\text{GeV}^{-2}$.

Therefore we have constructed a model of the soft pomeron based on the instanton interactions, which gives values for the pomeron's intercept and trajectory close to the experimental ones. Further work is needed to better quantify the problem.

REFERENCES

1. D. E. Kharzeev, Yu. V. Kovchegov, E. M. Levin, Nucl. Phys. **A690**, 621 (2001).
2. A. Donnachie, P. V. Landshoff, Phys. Lett. **B296**, 227 (1992); Phys. Lett. **B437**, 408 (1998) and references therein.
3. E.A. Kuraev, L.N. Lipatov and V.S. Fadin, *Sov. Phys. JETP* **45**, 199 (1978); Ya.Ya. Balitsky and L.N. Lipatov, *Sov. J. Nucl. Phys.* **28**, 22 (1978).
4. D. Kharzeev and E. Levin, Nucl. Phys. **B 578**, 351 (2000).
5. V. A. Novikov, M. A. Shifman, A. I. Vainshtein, and V. I. Zakharov, Nucl. Phys. **B191**, 301 (1981).
6. A. Ringwald, F. Schrempp, Phys. Lett. **B459**,249 (1999) and references therein.
7. For a review, see e.g. V.A. Rubakov and M.E. Shaposhnikov, Sov. Phys. Usp. **39** (1996) 461; hep-ph/9603208.
8. T. Schäfer, E. V. Shuryak, Rev. Mod. Phys. **70**, 323 (1998).

Diffractive Electroproduction of Vector Mesons from Light Nuclei

Leonid L. Frankfurt[1], Misak M. Sargsian[2] and Mark I. Strikman[3]

[1] *Tel Aviv University, 69978, Tel Aviv, Israel*
[2] *Florida International University, Miami, Florida 33199*
[3] *Pennsylvania State University, University Park, Pennsylvania 16802*

Abstract. Coherent electroproduction of vector mesons from light nuclei is considered in the kinematics of deep inelastic scattering. Special emphasis is given to the $-t \sim 0.8 - 1$ GeV^2 region where the cross section is dominated by the interaction of the $q\bar{q}$ configuration in the γ^* with two nucleons. This kinematics provides the unique possibilities to study the onset of Color Transparency with increasing Q^2 as well as its gradual disappearance at very small x, i.e., perturbative Color Opacity. It also allows to discriminate between leading twist and higher twist mechanisms of gluon shadowing. We demonstrate that the discussed kinematics of Electron Polarized Ion Collider well suited for such investigations and emphasize advantages of using polarized deuterons.

INTRODUCTION

Recent theoretical analyses [1–5] demonstrated the existence of the number of "two-body" hard exclusive processes off nucleons and nuclei which can be calculated legitimately within perturbative QCD (pQCD) based one the formulation of new type of partonic distributions (generalized parton distributions). The main requirement for application of pQCD is the onset of the factorization regime. For diffractive electroproduction of vector mesons by longitudinally polarized photons ($\gamma_L^* + p \to V + p$) the onset of this regime seems to be observed already for $Q^2 \geq 5$ GeV^2, for the recent review see [6].

In the hard diffractive limit of small x and large Q^2 vector meson production in the target rest frame is essentially a three stage process. First, at a distance $l_c \sim \frac{1}{2m_N x}$ before the target, the γ_L^* transforms into a small transverse size $q\bar{q}$ pair where $b_{q\bar{q}} \equiv r_{qt} - r_{\bar{q}t} \approx 3/Q$, i.e., $b_{q\bar{q}}(Q^2 \sim 10 \text{ GeV}^2) \approx 0.4$ fm. Then, the small $q\bar{q}$ pair interacts with the target with an amplitude [7]:

$$A(q\bar{q}T)\mid_{t=0} = \frac{Q^2 \pi^2}{3x}\left[b^2 \alpha_s(Q^2) \cdot \left(i - \frac{\pi}{2}\frac{d}{\ln x}\right) x G_T(x, Q^2)\right], \quad (1)$$

where $G_T(x, Q^2)$ is the target's gluon density (we neglect here small and calculable difference between the gluon density and the skewed gluon density). The third stage, the transformation $q\bar{q} \to V$ occurs long after the target at distances $\sim 2q_0/m_V^2$. The use of completeness over the diffractively produced states allows to express the result in terms of bare parton distributions within the target and the vector meson similar to DIS processes [1]: $A^{\gamma^*T \to VT} = \psi^{\gamma^* \to q\bar{q}} \otimes A(q\bar{q}T) \otimes \psi^{q\bar{q} \to V}$, where $\psi^{\gamma^* \to q\bar{q}}$ is the wave function of the $\gamma^* \to q\bar{q}$ transition, $\psi^{q\bar{q} \to V}$ is the $q\bar{q}$ component of the V's wave function. $A(q\bar{q}T)$ describes the scattering of the $q\bar{q}$ off the target T with a cross section given by Eq.(1).

Two key features of Eq.(1) are a dipole like decrease of interaction with he size of $q\bar{q}$ system and proportionality of the cross section to the gluon distribution of the target $G(x, Q^2)$, which leads to a fast growth of the cross section with energy.

The use of nuclei allows to check the prediction of QCD (Eq.(1)) that the interaction cross section of a small $q\bar{q}$ with a nucleon is indeed small, i.e., Color Transparency, and whether it can reach (due to the increase of the gluon density at small x) values comparable to those for the interaction of light hadrons (pions), i.e., perturbative Color Opacity. One possible strategy is to study coherent vector meson production off heavy enough nuclei at $t \sim 0$. Another possibility, which we consider here, is to use the process $\gamma^* + A \to V + A$ with the lightest nuclei (A=2,3,4) for $V = \rho, \rho', \omega, \phi...$ at $x \leq 10^{-2}$, and to focus on the region of comparatively large $|t|$, where **the interaction with two nucleons dominates**. We restrict ourselves to the coherent channel because this case can be singled out experimentally in an unambiguous way. There is clear experimental evidence for the dominance of the rescattering diagrams in the coherent production off the deuteron at $|t| \geq 0.6$ GeV^2 (see Ref. [8] and references therein). Due to the quadrupole contribution, the diffractive minimum is filled up for deuteron target. However possibilities to investigate the polarized deuteron target will allow to significantly suppress the contribution from single scattering term, making the cross section to be sensitive solely to the mechanism of $q\bar{q}$ rescattering off nucleons. Also the use of other targets as ^3He and ^4He, for which the diffractive minimum occurs at $-t \sim 0.2$ GeV^2 allows to check the similar predictions at considerably smaller $|t|$.

CROSS SECTION

At small x and large Q^2, where the average $b_{q\bar{q}}$ is small, there exist two mechanisms for the double rescattering of the $q\bar{q}$ pair off the target nucleons. The leading twist mechanism has the same structure as the leading twist contribution to the gluon shadowing and it is shown in Fig.1. It is expressed through the total shadowed gluon density in the target and therefore through the diffractive parton densities in the nucleon [9,10]. At very small x the leading twist mechanism leads to a large gluon shadowing which reflects an observation that in QCD as a result of gluon emissions a small configuration can evolve to a large size configuration. As a result for sufficiently small x the double interaction will decrease only logarithmically with increase of Q^2 [2,9,10].

The leading twist shadowing and hence the leading twist double interaction competes with the higher twist effects of the eikonal type where two two-gluon ladders

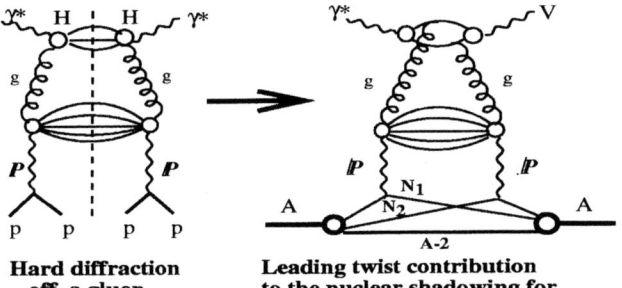

FIGURE 1. Hard diffraction off a gluon. Leading twist contribution to the nuclear shadowing for coherent vector meson production.

are attached to the $q\bar{q}$ pair, see e.g. [3].

In the following analysis we will mostly focus on the region of relatively large $x \geq 0.01$ and $Q^2 \geq few\ GeV^2$ where the leading twist shadowing effects are small while eikonal-type contributions are sufficiently large. We also estimate the leading twist contribution for small x.

In the b-space - eikonal approximation [11] the amplitude for coherent production of a vector meson off the deuteron is (suppressing spin indices):

$$\frac{d\sigma^{\gamma_L^* ^2 H \to V^2 H}}{dt} = \frac{1}{16\pi}\left|2f^{(1)}(t)S_d(\frac{q}{2},-\frac{q_-}{2}) + \frac{i}{2}f^{(2)}\left(\frac{t}{2}\right)\int \frac{d^2k'_\perp}{(2\pi)^2}S_d(k'_\perp,-k'_-)e^{-Bk'^2}\right|^2 \quad (2)$$

where $t = q_-q_+ - q_\perp^2$ and $S_d(k) = F_c(q) + \left(3\frac{(Jq)^2}{q^2} - 1\right)F_Q(q)/\sqrt{2}$. F_C and F_Q represent the deuteron's charge and quadrupole form factors. The $q_- = q_0 - q_3 = \frac{Q^2+m_V^2-t}{2\nu}$ and $k'_- = k'_0 - k'_3 = \frac{Q^2+m_V^2+t}{2\nu}$ appear in the argument of of the deuteron structure function as a result of accounting for the target recoil which is essential for large values of t [12,13]. The similar expression can be written for coherent vector meson production from ^3He and ^4He where for later case no quadrupole form factor will contribute since the target is spherical. For ^3He and ^4He targets we restrict ourselves by the single and double scattering amplitudes only since multi-scattering amplitudes violate energy-momentum conservation due to the production of multi-particle states. Their contribution should be zero (S. Mandelstam cancellation) within the approximation when only a $q\bar{q}$ pair is considered.

In Eqs.(2), the (1) and (2) - fold scattering amplitudes are defined as:

$$f^{(n)} \sim \int d^2b\, \psi_{\gamma^*}^L(b)\psi_V(b)\left[\sigma_{q\bar{q}N}(b,x)\right]^n exp(Bt/n). \quad (3)$$

Here $\sigma_{q\bar{q}}$ is given by Eq.(1). Because of the small size of $q\bar{q}$, the slope of the elementary amplitude, $B \approx 2.5\ GeV^{-2}$ [6], is mainly determined by the nucleon's two-gluon form factor. In the leading twist

$$f^{(2)} \sim \int d^2b\, \psi_{\gamma^*}^L(b)\psi_V(b)\sigma_{q\bar{q}N}(b,x)\sigma_{eff}(b,x)exp(B't/2), \quad (4)$$

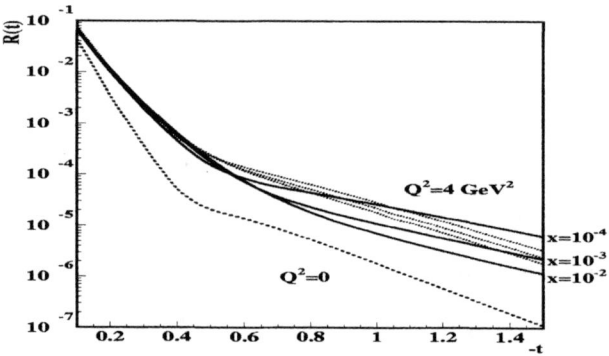

FIGURE 2. *The $-t$ dependence of the ration R as defined in the text, for different x at fixed Q^2. Dashed curve - VMD prediction for $Q^2 = 0$, dotted - VMD for $Q^2 = 4 GeV^2$. Solid curves - calculation discussed in the text.*

where $\sigma_{eff}(x, b)$ is expressed through the diffractive gluon density [9], and $B' = 5 - 7 GeV^{-2}$ is the slope the gluon induced hard diffraction.

ONSET OF COLOR TRANSPARENCY (CT)

The onset of CT leads to a rather nontrivial dependence of the coherent production cross section on x, t and Q^2. To estimate those effects in the eikonal approximation, we analyze the ratio $R(t) = \frac{d\sigma^{\gamma_L^*}}{dt}(t)/\frac{d\sigma^{\gamma_L^*}}{dt}(t = 0)$. It follows from Eqs.(2-3) that at $-t \geq -t_0 \sim 0.5$ GeV^2 (> 0.2 GeV^2 for $A = 3, 4$), when the coherent cross sections are dominated by the rescattering terms, $R(|t| > |t_0|) \sim x^2 G_N^2(x, Q^2) e^{Bt}/Q^4$. Therefore, one should expect a strong decrease of R with increasing Q^2. At the same time, for fixed Q^2, one expects a fast increase of the cross section with decreasing x. This increase is restricted by a unitarity condition [3]. An observation of a slowing down of such an increase at small x would be a clear signature of the onset of color opacity.

In Fig.1 we present $R(t)$ for fixed Q^2 and different x, calculated for coherent scattering off ^2H. We show also in this figure the expectation of the Vector Dominance model (VMD) in which a ρ-meson is produced in the first interaction and then scatters off the nucleons with a cross section similar to the πN cross section. As follows from this figure the region of $x \approx 10^{-2}$ accessible for kinematics of EPIC one may expect the significant drop of the double scattering term as compared to the prediction of VMD - this corresponds to the observation of color transparency. This region of x corresponds to the moderate densities of gluon distribution and the increase of the $xG_t(x, Q^2)$ in Eq.1 does not overcome the smallness of dipole $q\bar{q} - N$ interaction. It is interesting to note that further decrease of x to 10^{-4} will result to the increase of the double scattering term as compared to the ordinary hadronic interaction (such an increase eventually will be saturated due to unitarity). This situation will correspond to the phenomenon of color opacity.

The leading twist mechanism of double rescattering is likely to dominate for $x \leq 10^{-2}$ and lead to a gross enhancement of the cross section at $-t \sim 0.8 GeV^2$ as compared to the estimate of Fig.2. For example, the expected enhancement for $Q^2 = 4 GeV^2$ and $x \sim 10^{-3}$ is by a factor ~ 10. Another distinctive feature of the leading twist mechanism is a relatively weak x dependence of the shape of the differential distribution.

Note that in the case of a ^3He and ^4He targets, one expects a clean diffractive minimum which is very sensitive to the amount of rescattering in Eq.(3). The investigation of the depth of the diffraction minima would allow to check another prediction of eikonal approximation, namely the large relative phase of the impulse approximation and the rescattering amplitude, corresponding to $Re(f^{(2)})/Im(f^{(2)}) \sim 0.3 - 0.4$.

Possibilities to explore the polarized targets opens new dimension in studies color transparency/opacity phenomena. The uniqueness of the vector meson production from transverse polarized deuteron, is that for certain values of $-t$ ($\approx 0.4 GeV^2$) the single scattering term completely cancels out and the cross section is defined only by double scattering term [13]. Additionally, the asymmetry measured for tensor polarized deuteron targets changes the sign at $-t_0 \approx 0.5 GeV^2$. The implication of color transparency in this case will be significant shift of the $-t_0$ toward the larger values [13] ($-t \approx 0.7 GeV^2$ for $Q^2 = 5\ Gev^2$).

To summarize, the study of the coherent production of vector mesons in DIS will provide unique opportunities to observe onset of color transparency as a function of Q^2 for the range of $x \geq 0.01$ which can be probed at EPIC. At higher energies one would be able to investigate onset of color opacity and distinguish between the leading twist and the higher twist mechanisms of the Color Opacity.

REFERENCES

1. S. J. Brodsky L.L. Frankfurt, J.F. Gunion, A.H. Mueller and M.I. Strikman, Phys. Rev. **D50**, 3134 (1994).
2. H. Abramowicz, L.L. Frankfurt and M.I. Strikman, SLAC Summer Inst. 1994, 539.
3. L. L. Frankfurt, W. Koepf and M. I. Strikman, Phys. Rev. **D54**, 3194 (1996).
4. J. C. Collins, L. Frankfurt and M. Strikman, Phys. Rev. **D56**, 2982 (1997).
5. L. L. Frankfurt, M. F. McDermott and M. Strikman, JHEP **9902**, 002 (1999)
6. H. Abramowicz and A. Caldwell, Rev. Mod. Phys. **71**, 1275 (1999).
7. B. Blattel, G. Baym, L.L. Frankfurt, M. I. Strikman Phys. Rev. Lett. **70**, 896 (1993).
8. T.H. Bauer, R.D. Spital, D.R. Yennie, F.M. Pipkin, Rev. Mod. Phys. **50**, 261 (1978).
9. L. Frankfurt and M. Strikman, Eur. Phys. J. **A5**, 293 (1999).
10. L. Frankfurt, V. Guzey and M. Strikman, hep-ph/0010248.
11. H. Cheng and T. T. Wu, "Expanding Protons: Scattering at High-Energies," *Cambridge, USA: MIT-Pr. (1987) 285p.*
12. L. Frankfurt, W. Koepf, J. Mutzbauer, G. Piller, M. Sargsian and M. Strikman, Nucl. Phys. **A622**, 511 (1997).
13. L. Frankfurt, G. Piller, M. Sargsian and M. Strikman, Eur. Phys. J. **A2**, 301 (1998).

Physics of p-A Collisions and Its Relation to e-A Collisions

Jen-Chieh Peng

Los Alamos National Laboratory
Los Alamos, New Mexico 87545

INTRODUCTION

The advent of the Relativistic Heavy Ion Collider (RHIC), whose primary goal is to investigate matter at extraordinary temperature and energy density via nucleus-nucleus collisions, will also offer a unique opportunity to study some fundamental questions in strong interactions via proton-nucleus (p-A) collisions. It is important to point out that RHIC, for the first time, will also allow us to study p-A interaction in a collider mode. At RHIC, the center-of-mass energy reached in p-A collision is roughly an order of magnitude higher than any existing fixed-target experiments. Moreover, the large-acceptance collider detectors are capable of measuring many particles produced in the p-A collisions simultaneously, which could provide qualitatively new information not accessible in fixed-target experiments.

It has long been recognized that p-A collisions serve an important role in understanding nucleus-nucleus (A-A) collisions. In addition to their connection to A-A physics, p-A collisions are also closely related to electron-nucleus (e-A) collisions. Many outstanding questions in hadron physics can be addressed in e-A as well as in p-A collisions. These questions include:

- What is the quark and gluon content of nuclei? How is it different from that in a free nucleon?

- Do parton densities saturate at small x? Can they be described by the conventional DGLAP evolution equations at the small-x region?

- What are the parton distributions of mesons? What are the roles of mesons in describing the properties of nucleon and nuclei?

To probe the parton structures of nuclei, Deep-Inelastic Scattering (DIS) has been used as a powerful tool in e-A collisions. In p-A collisions, an analogous tool is the Drell-Yan process. A unique feature of the Drell-Yan process is that it probes the antiquark distributions in the target nucleus. This is complementary to the DIS which probes the sum of quark and antiquark distributions. Indeed, the Drell-Yan data have been used to determine the up and down sea quark distributions in the proton in various global parton distribution function parametrizations.

In this paper, we discuss two physics topics which can be studied in the near future at RHIC in p-A collisions. These two topics are selected to illustrate the complimentarity between p-A and e-A collisions.

PARTON DISTRIBUTIONS AT SMALL X

One of the major physics goals at an e-A collider is to study the saturation effects of parton distributions at small x via the nuclear dependence measurements. A suppression of the Drell-Yan yields from heavy nuclear targets is observed in E772/E866 at small x_2 [1]. This is consistent with the shadowing effect observed in DIS. In fact, E772/E866 provide the only experimental evidence for antiquark shadowing. The reach of small x_2 in E772/E866 is limited by the mass cut ($M \geq 4$ GeV) and by the relatively small $\sqrt{s}$ ($\sqrt{s} = 39$ GeV).

P-A collisions at RHIC clearly offer the exciting opportunity to extend the study of shadowing to smaller x. Figure 1 compares the E772 Fe/D Drell-Yan ratio data with what could be obtained at PHENIX in a two-month run. The PHENIX detector consists of a forward and a backward muon arm for dimuon detection, as well as a barrel detector capable of detecting e^+e^- pairs. The forward and

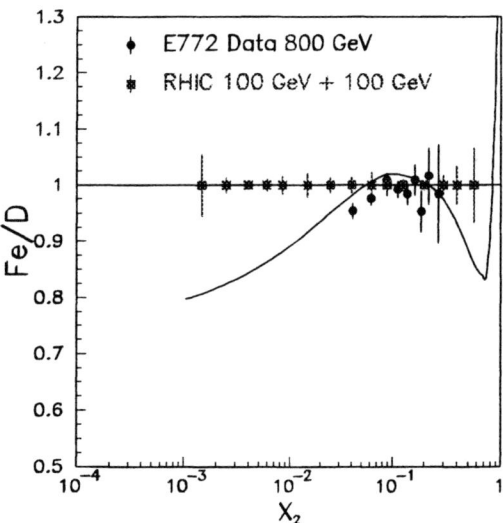

FIGURE 1. Projected statistical accuracy for Fe/D Drell-Yan cross section ratios as a function of x_2 in a 2-month run at PHENIX. The E772 data are also shown. The solid curve is the EKS98 parametrization for parton distributions in nuclei.

backward muon arms and the barrel detector cover different kinematic regions. Figure 1 shows that the coverage in x_2 will be significantly increased not only for the small-x_2 but also for the large-x_2 region. Note that the reach at small x at RHIC is comparable to that in eRHIC. However, the corresponding Q^2 at RHIC is significantly greater than at eRHIC (roughly a factor of 10 for a given x).

The flavor asymmetry between the $\bar{u}$ and $\bar{d}$ antiquark distributions in the proton has been recently observed by the E866 experiment [2] and the Hermes experiment [3]. RHIC offers the opportunity to measure $\bar{d}/\bar{u}$ asymmetry to very small x. Such information is important for an accurate determination of the integral of $\bar{d} - \bar{u}$, as well as for a better understanding of the origins for flavor asymmetry. The statistical accuracy for measuring $\sigma^{pd}/2\sigma^{pp}$ in a two-month PHENIX run is shown in Figure 2. Also shown in Figure 2 are the data from E866. The lowest x_2 reachable at RHIC is around 10^{-3}, an order of magnitude lower than in E866.

The gluon distribution in nuclei is practically unknown. At RHIC, the gluon content of the nucleus could be measured using a variety of tools including direct-photon production, photon-jet production, di-jet production, and heavy-quark production. These results will provide new information on gluon content of nuclei, which is one of the physics goals at a future e-A collider.

In contrast to the Drell-Yan process, large nuclear effects are found in the hadronic production of J/Ψ, Ψ', and Υ [4]. A detailed study of the J/Ψ and

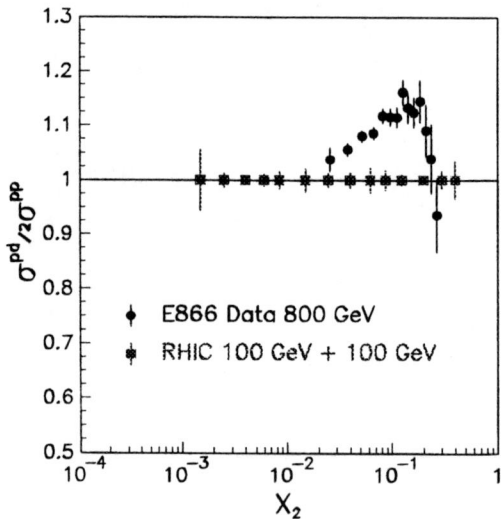

FIGURE 2. The solid circles are the $\sigma^{pd}/2\sigma^{pp}$ data from E866. The solid squares show the kinematic coverage in x_2 and the expected statistical accuracy for a two-month p+p plus p+d run. Both the dimuon and dielectron detections have been taken into account.

Ψ' A-dependence as a function of p_T and x_F was recently reported by E866 [5]. P-A collisions at RHIC would offer two unique opportunities for further clarifying the mechanisms for nuclear effects in quarkonium production. First, RHIC can reach much smaller values of x_2 ($x_2 \sim 5 \times 10^{-4}$). Second, the negative x_F region, not covered by fixed-target experiments, could be well studied at RHIC.

HARD PROCESSES TAGGED BY LEADING PARTICLES

The success of the meson-cloud model in explaining the $\bar{d}/\bar{u}$ asymmetry [6] suggests a novel technique to study meson substructures without using a meson beam. The idea is that the meson cloud in the nucleon could be considered as a virtual target to be probed by various hard processes. Recently at the HERA e-p collider, meson structure functions were measured in a hard diffractive process, where forward-going neutrons or protons were tagged in coincidence with the DIS events [7]. Analogous measurements could be made at RHIC for p-p collisions and in e-p collisions at a future collider [8]. In particular, a Drell-Yan pair in coincidence with a forward-going neutron or proton could provide information on the antiquark

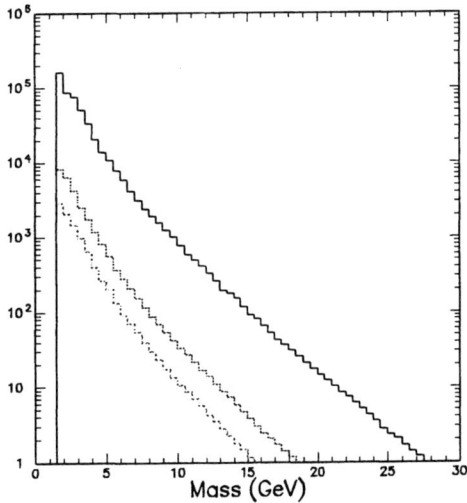

FIGURE 3. The solid curve corresponds to dimuon event distributions in one muon arm for a one-month $p+p$ PHENIX run at $\sqrt{s} = 200$ GeV. The dotted and dashed curves correspond to the yields for $p+p \to n + \mu^+\mu^- + X$ with $Y > 0.6$ and $Y > 0.8$, respectively, where Y is the fraction of proton beam momentum carried by the neutron.

distributions in pions at small x. The large-acceptance detectors and the collider environment at RHIC are ideal for such measurements.

Assuming factorization, one could write down the following expression for tagged Drell-Yan cross section:

$$d\sigma^{DY}/dy\, dm\, dx_F(p+p \to N + \mu^+\mu^- + x)$$
$$= d\sigma^{DY}/dm\, dx_F(p+M \to \mu^+\mu^- + x) \cdot f_{MB}(y). \quad (1)$$

$f_{MB}(y)$ is the probablity for proton to fluctuate into a meson-baryon pair, where the baryon carries a fraction y of the initial proton momentum. The underlying picture of the tagged Drell-Yan process is as follows. A proton fluctuates into a baryon plus a meson ($p \to n + \pi^+$, $p \to \Lambda + K^+$, for example) and the other proton beam interacts with the meson, producing a lepton pair. The remaining baryon moves roughly along the initial proton beam direction and could be detected at forward angles. A simulation for the $p + p \to n + \mu^+\mu^- + X$ for the PHENIX detector, where the muon pairs are detected in the muon arms and the neutrons are detected by a small-angle calorimeter (the Zero Degree Calorimeter [9]), has been carried out and the result is shown in Figure 3.

Drell-Yan experiment tagged by forward-going baryons at RHIC would provide a direct test of the meson-cloud model. One could also conceive tagging forward-going Δ or Λ. The Λ-tagging is of special interest since it can shed light on the strange-quark contents of the proton. Another extension is the measurement of double-helicity asymmetry, A_{LL}, in neutron-tagged $\vec{p} + \vec{p}$ Drell-Yan process. If the dominant underlying process is $\vec{p} + \pi$ interaction, A_{LL} is expected to be zero. Doubly tagged Drell-Yan process, which involves meson-meson interaction, could also be contemplated. Finally, it is natural to extend tagged Drell-Yan measurement to other hard processes such as J/Ψ production and di-jet production.

REFERENCES

1. Alde D. A. et al., *Phys. Rev. Lett.* **64**, 2479 (1990); Vasiliev M. A. et al., *Phys Rev. Lett.* **83**, 2304 (1999).
2. Hawker E. A. et al., *Phys. Rev. Lett.* **80**, 3715 (1998); Peng J. C. et al., *Phys Rev.* **D58** 092004 (1998).
3. Ackerstaff K. et al., *Phys. Rev. Lett.* **81**, 5519 (1998).
4. McGaughey P. L., Moss J. M. and Peng J. C., *Annu. Rev. Nucl. Part. Sci.* **49**, 217 (1999).
5. Leitch M. J. et al., *Phys. Rev. Lett.* **84**, 3256 (2000).
6. For recent reviews, see Kumano S., *Phys. Rept.* **303**, 183 (1998); Peng J. C. and Garvey G. T., hep-ph/9912370 (1999).
7. Adloff C. et al., *Eur. Phys. J.* **C6**, 587 (1999).
8. Holt R., Talk presented at this Workshop (2000).
9. Adler C. et al., nucl-ex/0008005 (2000).

MACHINE

Conceptual Design Study of the Electron-Proton Storage Ring Collider with Polarized Beams

I.A. Koop, M.S. Lorestelev, I.N. Nesterenko, A.V. Otboev,
V.V. Parkhomchuk, E.A. Perevedentsev, V.B. Reva, V.G. Sahmovsky,
D.N. Satilov, P.Yu. Shatunov, Yu. M. Shatunov and A.N. Skrinsky

Budker Institute of Nuclear Physics
Novosibirsk, 630090, Russia

R. Milner and C. Tschalaer

MIT-Bates Linear Accelerator Center
Middleton, MA 01949, USA

Abstract: This report presents a preliminary feasibility study of the electron-proton collider with polarized electron and proton beams with c.m. energy in the range of 15 to 30 GeV.

INTRODUCTION

A ring-ring machine realization of the electron-proton collider with longitudinally polarized particles was investigated in the feasibility study presented here. The objectives of this study are mainly to address the issues of achieving high luminosity and high polarization of both colliding beams. Electron cooling of the proton beam is essential for the suppression of emittance enhancement, which originates due to intra-beam scattering as well as from beam-beam effects. Technical aspects of the facility are the subject of future investigations but we have tried to use realistic constrains based on existing parameters of the beams and equipment.

Design Overview

Primary performance goals for the collider, based on physics motivations and requirements, are as follows:

- to achieve a peak luminosity of $1 \cdot 10^{33}$ cm^{-2}s^{-1};
- to operate in the energy range $E_{c.m.}$=15 to 30 GeV with an energy asymmetry of about 1 to 4 which corresponds to an electron beam energy of E_e =3.5 to 7 GeV and a proton beam energy of E_p=16 to 32 GeV;

- to arrange the longitudinal polarization of electrons and protons in two interaction regions to be $P \geq 0.5$ with adequate polarization life time.

To meet all these requirements we found that the key features of the design have to be:
- round beams;
- low $\beta_x = \beta_z$ values at the interaction point;
- head-on collisions;
- multibunch operation;
- electron cooling of the proton beam at the experiment energy;
- separation of colliding beams by the transverse magnetic field
- preservation of proton beam polarization during acceleration by implementing a Siberian Snake scheme in the lattice design;
- possibility to use high field polarizing wigglers to increase the self-polarization rate of the electron beam.

Luminosity Considerations

Achieving the extremely high luminosity value of $1 \cdot 10^{33}$ cm^{-2}s^{-1} is the main challenge and needs special consideration.

Beam Sizes and number of bunches

The nominal luminosity is given by the equation

$$L = F_{rev} \frac{n_b N_e N_p}{4\pi \sigma^{*2}} \quad (1)$$

where F_{rev} is the revolution frequency, n_b is the number of bunches in each beam, σ^* is the round beam size at collision and N_e and N_p are the single bunch populations for electrons and protons.

The collision frequency, which in fact is equal to $F_{coll} = F_{rev} n_b$, is determined by the bunch spacing D_{bb}, which in turn is dictated by the RF wave length. We choose $\lambda_{RF} = 1.52$ m ($F_{RF} = 197$ MHz). Then the collision frequency in each interaction point is also equal to 197 MHz, if we fill every bucket.

Single bunch populations N_e and N_p are limited by many reasons; beam-beam effects, different kinds of instabilities etc. For electrons the most severe intensity threshold is set by the head-tail transverse mode-coupling instability, that limits the one bunch population. A modern accelerator experience (for instance, in both B-factories and LEP collider), tells us, that $N_e = 3 \cdot 10^{10}$ is more or less a safe number. The proton bunch population is admitted to $N_p = 1 \cdot 10^{11}$, which is based on BNL and FNAL experimental results.

To achieve the required luminosity of $L = 1 \cdot 10^{33}$ cm^{-2}s^{-1} the beam size at the IP should be $\sigma^* = 68$ μm together with the other parameters fixed above.

A list of the main collider parameters at maximum energy is presented in Table 1.

Table 1.

	Units	Electron ring	Proton ring
Circumference	m	1387.94	1387.35
Energy	GeV	7	32
Arc radius	m	108.50	108.50
Bending radius	m	63.53	63.53
Number of bunches		913	913
Bunch spacing	m	1.52	1.52
Bunch population		$3 \cdot 10^{10}$	$1 \cdot 10^{11}$
Beam currents	A	0.95	3.16
Energy losses/turn	MeV	3.6	
Total radiated power	MW	3.42	
Beam emittances, $\varepsilon_{x,z}$	$\mu m \cdot mrad$	46	46
Beta function at IP	cm	10	10
Beam size at IP, $\sigma^*_{x,z}$	μm	68	68
Bunch length, σ_l	cm	10	10
Beam-beam parameter		0.035	0.0023
Lasslett tune shift			0.036
Luminosity	$cm^{-2}s^{-1}$	$1.0 \cdot 10^{33}$	

Beam-beam interactions

Let us discuss briefly the advantages and drawbacks of the flat and round beam cross section geometries.

Some obvious advantages of the flat colliding scheme include much lower beta functions in the quads near the IP and readiness of orbit separation in the first parasitic crossing point. On the other hand, a flatter proton beam will result in an increase of the intra-beam scattering rate and the luminosity life time will become shorter.

The evident advantage of round colliding beams (RCB) is that with the fixed particle density, the tune shift from the opposite bunch becomes twice as small as the tune shift in the case of flat colliding beams. Besides, the linear beam-beam tuneshift in the round beams becomes independent of the longitudinal position in the bunch, thereby weakening the action of synchro-betatron resonances.

The main feature of the RCB is rotational symmetry of the kick from the round opposite beam; complemented with the X-Z symmetry of the betatron transfer matrix between the collisions, it results in an additional integral of motion $M = xz' - zx'$, i.e. the longitudinal component of particle's angular momentum is conserved. Thus, the transverse motion becomes equivalent to a one-dimensional (1D) motion. The resulting elimination of all betatron coupling resonances is of crucial importance, since they are believed to cause the beam lifetime degradation and blow-up.

The above arguments in favor of RCB have been checked by computer simulation of the beam-beam effects in the VEPP-2M collider lattice, modified to the RCB option [1]. The main results of the simulations [2] are presented in the Figure 1, where the beam sizes are plotted versus the space charge parameter ξ. One can see that the beam blow-up for the round beam option is much weaker than what is simulated by the same code for flat colliding beams (dashed line). The simulations have also demonstrated the stability of theRCB against the "flip-flop" effect, similarly to conclusions using simple flip-flop models [3].

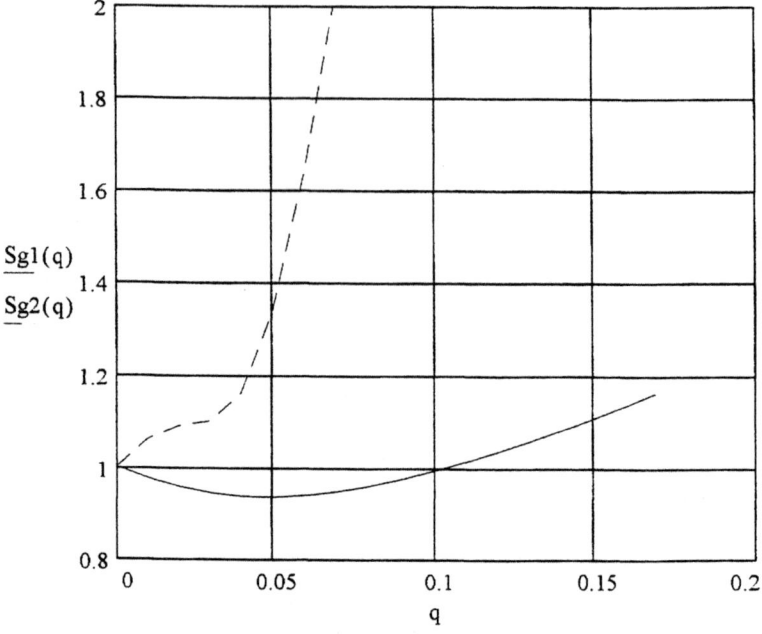

Figure 1. Variation of the weak beam size vs.the space charge parameter ξ(or q): solid curve for the round colliding beams, dashed curve for the flat beam option.

Taking into account also a higher luminosity life time in the case of the RCB, we make our choice in favor of round colliding beams geometry.

What does the round beam mean in practice?
- Small and equal β-functions $\beta_0 = \beta_x, \beta_y$ at the IP.
- Equal beam emittances ε_x, ε_y.
- Equal betatron tunes v_x, v_y and no betatron coupling in the arcs.
- Small and positive fractional tunes.

Unfortunately, a solenoidal beam focusing in the interaction straight at high energies is completely impractical, therefore achieving low and equal β-functions $\beta_0 = \beta_x, \beta_y$ at the IP should be realized by using triplets of quads. Nevertheless, requirements 1 and 3 can be more or less easily fulfilled. To make equal beam emittances ε_x, ε_y we need to introduce some small coupling of vertical and horizontal betatron motions. The best way is to use a small solenoidal field in the IP straight section, where beta functions are equal. A solenoidal field conserves the angular momentum M and, therefore, does not deteriorate the beam-beam effects. A betatron coupling of about 10^{-3} is adequate to have perfect round cross section beams at the IP.

Item 4 is also important for attainment of large values of the space charge parameter ξ_{max}. The area near the integer resonance is freer from the low-order nonlinear resonances.

Space charge parameters ξ_p and ξ_e for protons and electrons are determined by the expressions:

$$\xi_p = \frac{N_e r_p}{4\pi \gamma_p \beta_p \varepsilon}$$

$$\xi_e = \frac{N_p r_e}{4\pi \gamma_e \varepsilon} \qquad (2)$$

where r_p and r_e are the classical radii of the proton and the electron, γ_p and γ_e are the relativistic factors of the corresponding beams, $\beta_p = v_p/c$ is the proton beam velocity and ε is the round beam emittance.

For the beam parameters listed in table 1 one gets values ξ_p=0.0025 and ξ_e=0.045, which are typical values for proton-proton and electron-positron colliders.

Collider design

A layout of the designed collider is presented in Fig. 2.

Figure 2. Layout of e-p collider.

We have chosen the scheme of two intersecting in two points rings. Both rings have approximately equal circumferences - 1388 m. Each ring has two experimental straight sections, two technical straights and four identical arcs. The rings are separated vertically about 1 m outside the interaction areas.

The geometry of the interaction regions is dictated by the requirement of preserving the electron beam polarization. To minimize depolarization, an equilibrium spin direction of the electrons should be exactly vertical in the arcs. Thus, to provide the longitudinal polarization in the collision point we need to install two spin rotators on both sides of the each interaction area. Therefore, asymmetric orbit separation is preferable, because it cancels any spin rotation in the straight. We follow here closely to the design of the interaction region of the SLAC PEP-II B-factory, placing asymmetric splitting magnet on both sides of the IP. However, there are many other features, which arise from the need to accommodate the ± 90° spin rotators.

Electron ring

A spin rotation from the vertical direction in the arcs to a longitudinal one in the IPs will be performed in two steps: first, by a solenoidal spin rotator to horizontal plane and then by dipoles. The ± 90° spin rotator consists of two superconducting solenoids, each 3 m long, and with a field of about 6 T, see Figure 2. Between solenoids is placed a focusing structure, which cancels the betatron coupling and also creates the spin transparency. The solenoids are located in the drift spaces, where the velocity vector **v** has an angle ± 7.55° with respect to the collision axis, see Figure 2. After the solenoids, the spin precesses around the vertical magnetic field, becoming purely longitudinal at the IP, if the electron energy is E_e=5.25 GeV. The last bend (wiggling magnets) is divided also into several parts to create an achromat, see Figure 3. Two final magnets provide a proton and electron separation as close as possible to the IP.

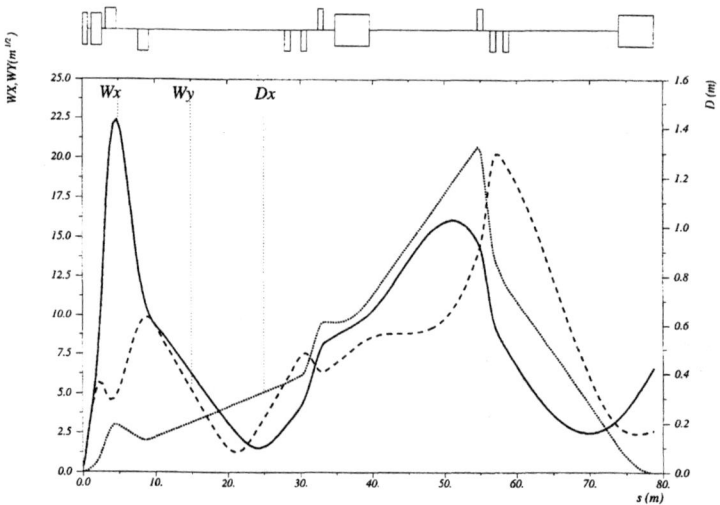

Figure 3. Lattice functions of the first half of the electron ring interaction region.

On the opposite side of the interaction straight, the spin is restored to the vertical direction by the negative spin rotator. As a result, the spin tune is undisturbed by the interaction region insertion, and the polarization behavior is essentially the same as without the spin rotators.

At energies other than 5.25 GeV, the polarization will be not purely longitudinal. At energies $E_e=3.5$ GeV and $E_e=7$ GeV, the spin at the IP is directed by $\pm 30°$ and the effective electron beam polarization will become about 86% compared with the value for 5.25 GeV. We would like to emphasize, that at arbitrary energy the spin is always restored to the vertical direction in the next arc, because of the zero total spin rotation over the interaction straight section.

The lattice of regular FODO cells in the arcs is shown in Figure 4. The phase advance is 60° in both planes. The total number of cells is 16 per one quadrant. Dispersion suppressors are made from the last three cells in the end of each arc. We use a scheme with one missing magnet in the second cell, see Figure 5. The arc ends by the special quad, which makes equal both, the vertical and the horizontal amplitude functions as well as their slope at the quad exit.

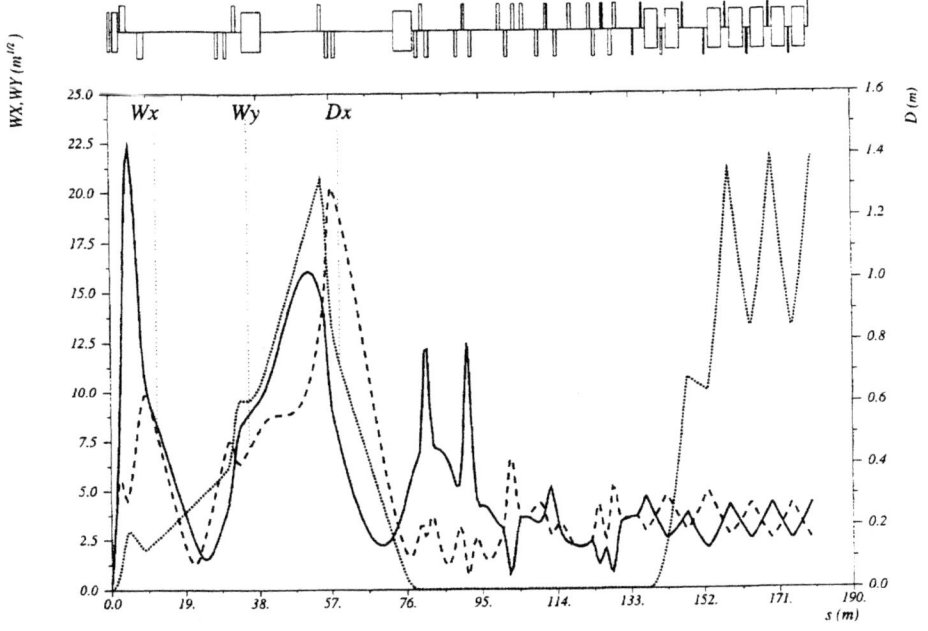

Figure 4. Lattice functions of the electron ring interaction region.

The other parts of the interaction region insertion serves for matching the IP lattice functions to those in the arcs.

With collisions at lower proton energy, say 15 GeV instead of 33 GeV, it is necessaru to compensate a decrease of the proton velocity by increasing the circumference of the electron ring. To do so we have foreseen two bypasses per each quadrant. One bypass increases the length of the electron orbit by 15.5 cm. It is possible to activate any number of bypasses and cover by this means a wide range of proton energies.

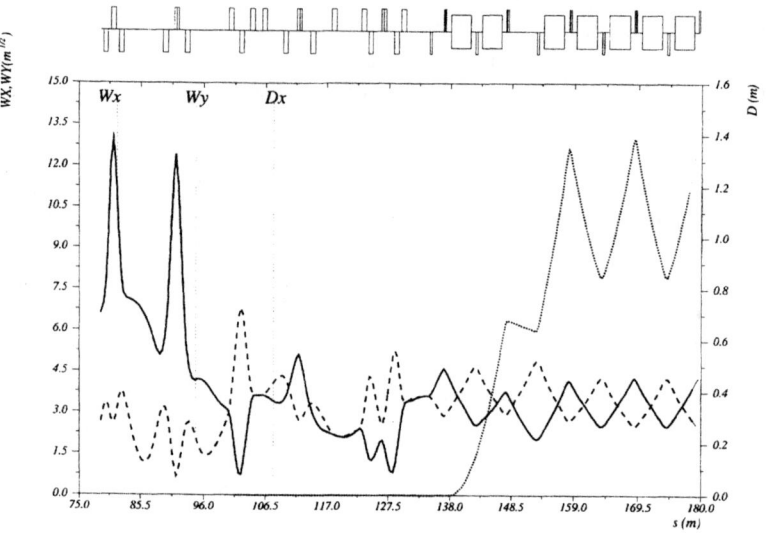

Figure 5. Lattice functions of the second half of the electron ring interaction region. This includes a vertical step and solenoids of the spin rotator.

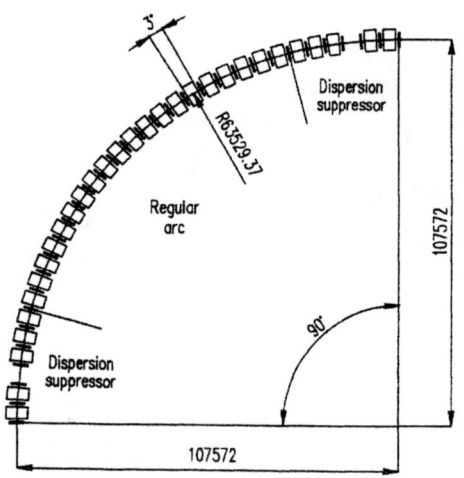

Figure 6. One arc of the electron ring.

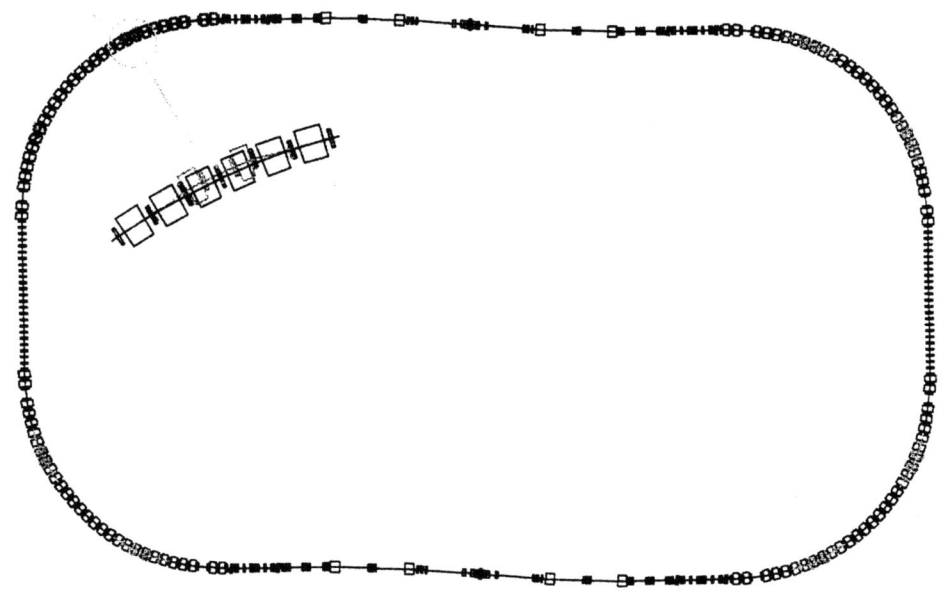

Figure 7. The electron ring with bypasses.

Proton ring

The maximum rigidity of the proton ring, which corresponds to the kinetic energy T_p=31.261 GeV, is 4.598 times higher than the maximum rigidity of the electron ring, and is equal to BR=107.362 T·m. The main dipoles in the arcs produce a magnetic field of 1.69 T and have a curvature radius of 63.53 m. Both rings have the same cell length: l=10.652 m. The only difference is in the length of the quads: the proton ring quads are twice as long: 0.8 m. The field gradient in these quads is about 25 T/m.

The optical functions of the proton ring are shown in Figure 8.

The design of the proton ring interaction region is also driven by the spin manipulation requirements. The straight begins with the spin rotator, which rotates the spin around the longitudinal axis by 90° (in the case using of the full Siberian Snake scheme) or somewhat less (if a partial snake scheme is used). The betatron coupling compensation scheme is the same, as it was assumed for the electron beam. The total longitudinal magnetic field integral of the full snake is equal to 120.768 T m.

An antisymmetric lattice after the spin rotator is needed to match the arc optics, to provide the low beta function at IP and also to organize horizontal orbit bumps, see Figs. 2, 9.

As we mention before the plane of the proton arcs is separated vertically from the electron ring. To bring the beams together, special vertical bridges are designed in the interaction regions between the spin rotators, see Figure 9.

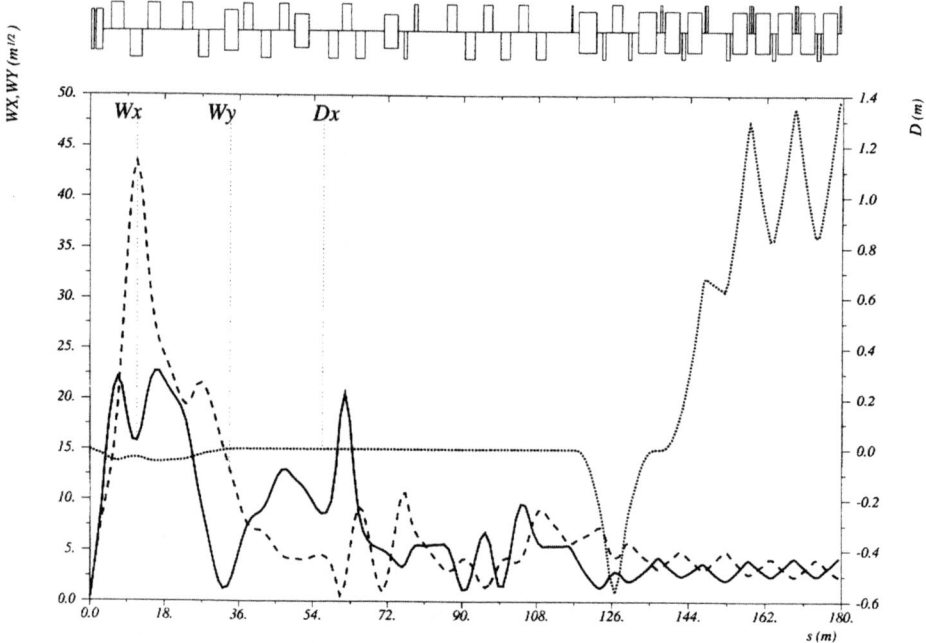

Figure 8. Lattice functions of the proton ring interaction

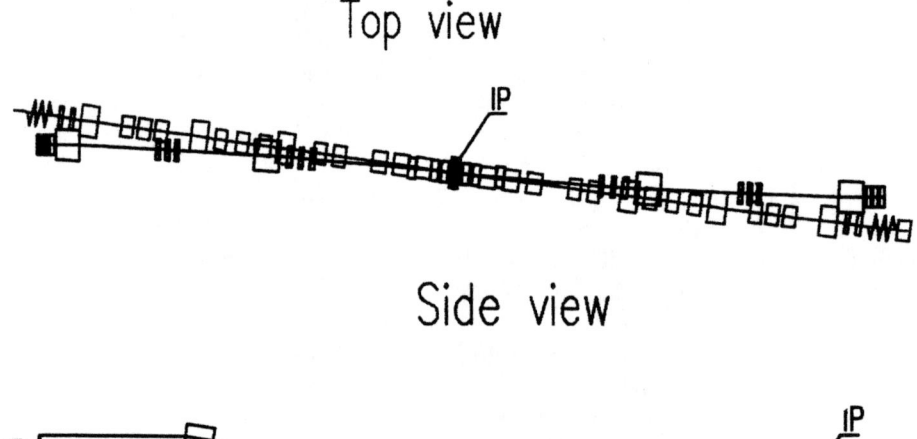

Figure 9. Top and side views of the left side of the interaction area.

POLARIZATION ISSUES

Polarization of electrons

The equilibrium polarization direction (vector **n**) is vertical in the main part of the ring and therefore one can expect a relatively low depolarization rate of the electron beam. Moreover, the Sokolov-Ternov polarization mechanism should provide sufficiently large beam polarization [7]. However, there are some other requirements, which should be satisfied by the insertion optics, where the spin vector **n** lies in the horizontal plane. To minimize the negative effect from a spin perturbation **w** over the whole straight section, we should fulfill the so called spin transparency condition, namely, that the integral of the perturbation through the insertion azimuth θ:

$$I = \int_{\theta_1}^{\theta_2} w\eta \, d\theta \qquad (3)$$

should be made zero or, at least, small. Here $\eta = \eta_1 - i\eta_2$ is a complex vector, which is constructed from the unity vectors η_1 and η_2, which in turn are the two orthogonal solutions of the spin motion equation for the equilibrium particle [8, 9]. The spin perturbation components are [10-12]:

$$w_x = v_0 z'' + K_x \frac{\Delta\gamma}{\gamma}$$

$$w_z = -v_0 x'' + K_z \frac{\Delta\gamma}{\gamma} \qquad (4)$$

$$w_x = K_y \frac{\Delta\gamma}{\gamma}$$

where $v_0 = \gamma a$ is the dimensionless spin tune; z'' and x'' are the second derivatives of the vertical or horizontal displacements over the azimuth θ; $K_{x,y,z}$ are respectively the normalized horizontal, longitudinal or vertical magnetic fields.

Careful analysis shows that the spin transparency condition for the interaction region straight section, which contains the spin rotators discussed above, can be fulfilled, and a small decrease of the equilibrium polarization degree is caused only by bending magnets in the straight between the two spin rotators.

We have identified a scheme for the focusing structure, which contains only regular quadrupoles inside the solenoidal spin rotator and cancels the betatron coupling as well as creates the spin transparency. The transfer matrices corresponding to a full insertion (from the first solenoid edge to the second solenoid edge) are:

$$T_x = \begin{pmatrix} 0 & -2r \\ (2r)^{-1} & 0 \end{pmatrix} \quad T_z = \begin{pmatrix} 0 & 2r \\ -(2r)^{-1} & 0 \end{pmatrix} \qquad (5)$$

Here r is a curvature radius in the solenoidal field B_y:

$$r = B\rho/B_y$$

In our proposal the condition (5) is fulfilled. Thus, all spin rotators as well as the horizontal wiggle between them, are spin transparent against betatron deviations. An inevitable spin-orbit coupling for the energy-off particles is excited by the first spin rotator and then compensated by the second one. As a result, the spin-orbit coupling vector $\mathbf{d} = \gamma\, \gamma \frac{\partial n}{\partial \gamma}$ is exactly equal to zero in the arcs, see Figure 10.

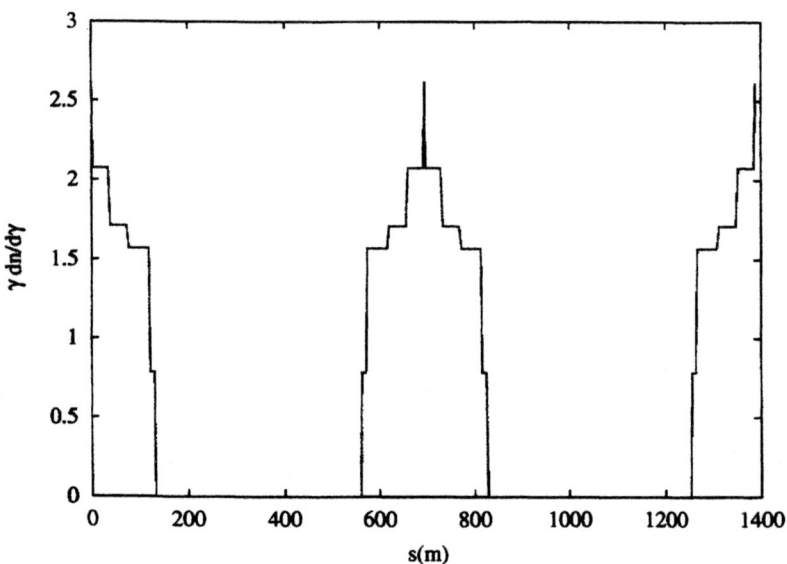

Figure 10. Spin-orbit coupling function along the electron ring for E=7 GeV.

The equilibrium polarization is given by the formula of Derbenev-Kondratenko [11]:

$$P_{eq} = \frac{8}{5\sqrt{3}} \frac{\alpha_-}{\alpha_+} = -\frac{8}{5\sqrt{3}} \frac{\left\langle |K_{x,z}|^3 b(n-d) \right\rangle}{\left\langle |K_{x,z}|^3 \left[1 - \frac{2}{9}(nv)^2 + \frac{11}{18}d^2\right]\right\rangle} \quad (6)$$

Due to the locality of the perturbation the spin-orbit coupling function does not have a resonance behavior. Loss of polarization from the maximum possible value of 92.4% is caused by the wiggling magnets.

Numerical calculations of the radiative polarization in the electron ring using the ASPIRRIN code [13] predict that the reduction in the polarization is of the order of 20%, see Figure 11.

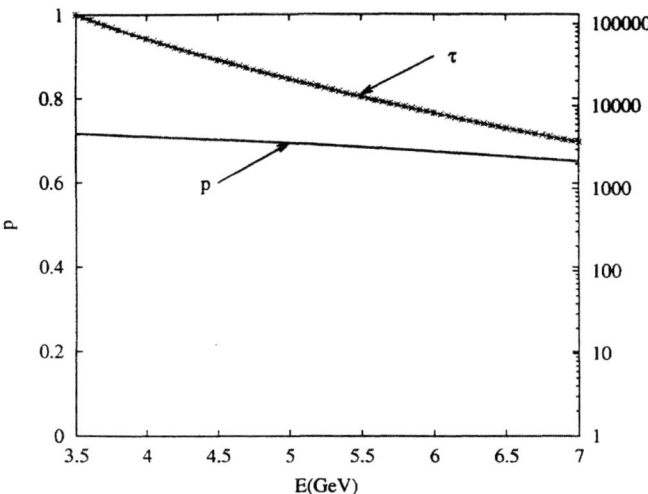

Figure 11. The equilibrium polarization and the spin relaxation time as functions of the electron energy.

These polarization losses can be reduced by a factor of 1.5-2 by higher fields in the arc dipoles (see Figure 12 or by installation of special polarizing wigglers. Also, it is reasonable in the final design to decrease the field value in the wiggling magnets which are responsible for the loss of polarization.

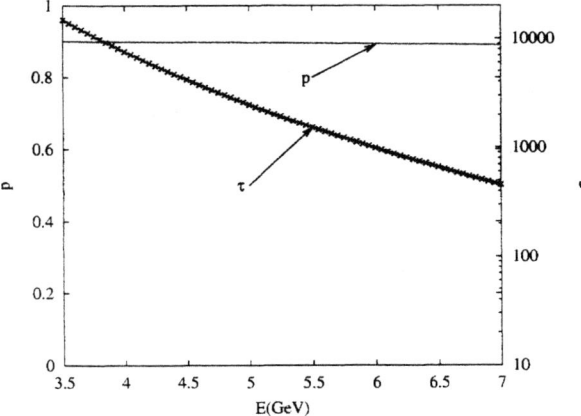

FIGURE 12. The equilibrium polarization and the spin relaxation time as functions of the electron energy
in the case of stronger main dipoles field (~ 1.2 T).

Polarization of protons

The realization of longitudinal proton polarization is based on the Siberian snake concept. It is well known that a single Siberian snake, which rotates the spin by 180° with respect to the velocity, always provides the longitudinal polarization in the opposite straight section. In addition, that, if a proton spin tune

$$v_p = \gamma_p \cdot 1.79284739$$

is an integer, then the equilibrium spin direction will be longitudinal also in the straight, where the snake is located. Thus, at the magic energies it is always possible to arrange the longitudinal polarization in both interaction regions.

Another purpose of using the snake is to eliminate all depolarizing resonances during the polarized protons acceleration from injection energy (2--3 GeV) to the experiment energy. With the full snake, a fractional part of the spin tune is equal to 0.5; hence all resonances, integer and intrinsic, are completely avoided, because betatron frequencies can exist far from this half-integer value.

The snake with the longitudinal axes of spin rotation can be configured in different ways. At energies of about 100 GeV/nucleon a more cost effective approach is to use spin rotators made of helical wigglers, as implemented in the RHIC design. But at energies below 30 Gev/nucleon the solenoidal design of the spin rotators is certainly preferable. This has a significant advantage, namely, that the closed orbit is not disturbed. With γ_p = 34.318 a full snake field integral is equal to 120 $T\,m$, only 1.67 times higher, than for the electron ring. In this proposal we divide a full snake in two parts, placed symmetrically on both sides of the IP. Each part consists of two solenoids, 5 m long and with a field value of 6 T. Between the solenoids is installed a FODO structure, which serves for the cancellation of betatron coupling.

It is possible to use a partial snake instead of full snake. Then the effective spin tune will be between 0.25 and 0.5, depending on the energy. Such a half-snake is sufficient for detuning from the integer and suppresses the intrinsic resonances by acceleration. Colliding experiments in such conditions should be performed only at magic energies, where the polarization will be exactly longitudinal in both IPs.

ELECTRON COOLING OF PROTONS

Without electron cooling of the proton beam, it's size will increase due to intrabeam scattering.

Intrabeam scattering

Following A. Piwinski's consideration of intrabeam scattering in the smooth focusing approximation [4, 5], which was slightly modified by one of the authors (I. A. Koop) for the case of round beam cross section geometry, the diffusion times of the momentum spread and of the emittance are:

$$\tau_\delta \approx \frac{N r_i^2 c}{4\pi \sigma_{x\beta}^2 \sigma_s \sigma_\delta^3 \beta^3 \gamma^2} \ln\left(\beta\gamma\sigma_{x'}\sqrt{\frac{2d}{r_i}}\right) f(x)$$

$$\tau_x \approx \frac{N r_i^2 c}{8\pi \sigma_{x\beta}^4 \sigma_s \sigma_\delta \beta^3 \gamma^2} \frac{\beta_x^2}{v_x^2} \ln\left(\beta\gamma\sigma_{x'}\sqrt{\frac{2d}{r_i}}\right)\left(1-\frac{v_x^2}{\gamma^2}\right)f(x) \qquad (7)$$

The diffusion times are defined as:

$$\tau_\delta = \left(\frac{1}{2\sigma_\delta^2}\frac{d\sigma_\delta^2}{dt}\right)^{-1}$$

$$\tau_x = \left(\frac{1}{2\varepsilon_x}\frac{d\varepsilon_x}{dt}\right)^{-1}$$

Other quantities in (7) have meanings: N is the number of particles per bunch; $r_i = Z^2 r_p/A$ is a classical radius of an ion; c is velocity of light; β and γ are the relativistic factors; d is the smallest beam size; $\sigma_{x\beta}$ is the betatron beam size; σ_δ is a relative momentum spread; σ_s is the longitudinal bunch length; β_x is the beta function in arcs and v_x is a betatron tune on arcs. The function $f(x)$ is determined by the expression:

$$f(x) = \left(1+\frac{3}{x-1}\right)\frac{\arctan\sqrt{x-1}}{\sqrt{x-1}} - \frac{3}{x-1} \qquad (8)$$

The argument of this function is:

$$x = \gamma^2 \sigma_{x'}^2 / \sigma_h^2$$

where the quantity σ_h represents the effective momentum spread of the scattering particles. The relation between σ_h and σ_δ is as follows:

$$\frac{1}{\sigma_h^2} = \frac{1}{\sigma_\delta^2} + \frac{D_x^2}{\sigma_{x\beta}^2} + \frac{D_x'^2}{\sigma_{x\beta}^2}$$

One should remember that σ_h is always smaller than σ_δ and this sometimes plays an important role.

In electron accelerators, the argument of the function $f(x)$ is typically very large due to large values of the relativistic factor γ. Then arctan (x) roughly is equal to π/2, and function $f(x)$ should be replaced by the function:

$$g(x) = \frac{\pi}{2\sqrt{x}} \qquad (9)$$

Below, the plot of the functions $f(x)$ and $g(x)$ is given.

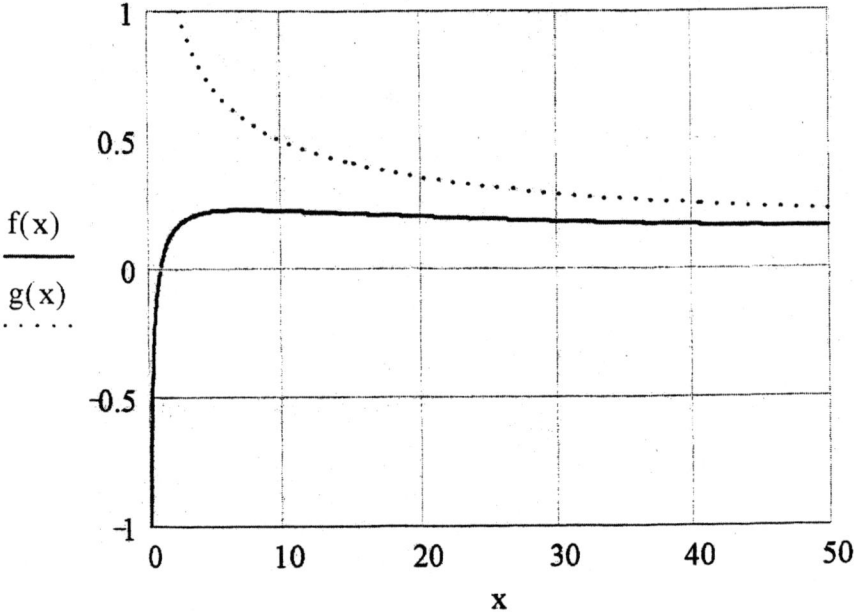

Figure 13. Plot of functions *f(x)* and *g(x)* is given.

Evaluation of the expressions (7) for the parameters, listed in the Table 1, gives the following numbers:

$$\tau_\delta = 3000\,\text{s} \qquad \tau_x = 85000\,\text{s}$$

Thus, the momentum spread grows quite rapidly, and electron cooling is clearly needed to suppress this process.

Electron cooling

According to the electron cooling concept we need an electron beam with the same velocity as protons, which is co-moving in the same direction. The cooling scheme may look as in the Figure 13.

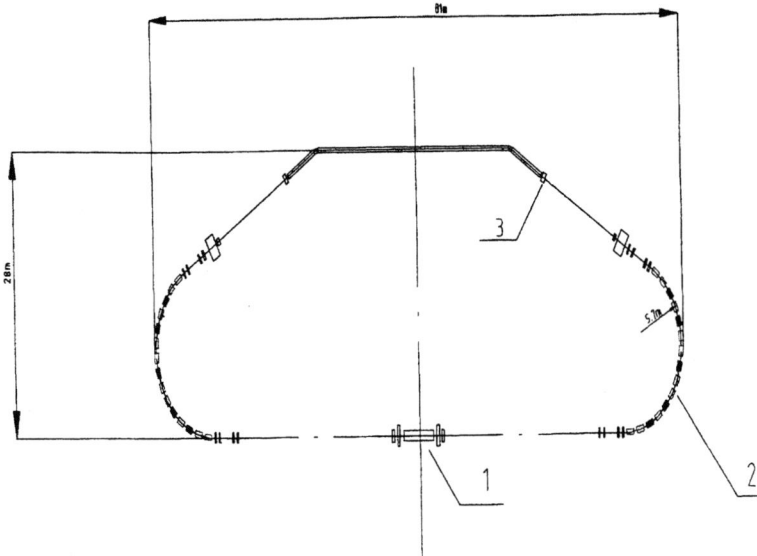

Figure 14. Schematic diagram of the electron cooling system. 1 --- main linac, 2 --- Chicane magnet system for bunching and debunching, 3 --- adapting elements to come in or come out of the cooling solenoid.

Implementation of electron cooling requires a cooling solenoid in the straight section of the proton ring, magnet optics for the chicane and transport channels, linac structure, RF for bunching and debunching, beam injector and a beam dump. The electron beam from the electron gun passes through the capture section and is accelerated in the main linac to full energy (8 to 16 MeV). At the chicane magnet structure, the length of the electron bunches is increased to 10-20 cm for optimization of the cooling process. After the chicane, the electron beam passes through the system of RF cavities to reduce the momentum spread to a value of about 10^{-4} and then enters the cooling solenoid. After the cooling section, the electron beam is returned to a second chicane before being modulated in energy (as necessary) for optimization of the bunch length for recovery of the energy in the main linac. The electron beam with the rest energy 0.5 - 1 MeV is dumped at the absorber. The simplest schematic diagram of a cooler is shown in Figure 13. The key elements at this figure are: 1) the main linac, 2) a chicane magnet system for bunching and debunching, 3) adapting elements to come in or out of the cooling solenoid.

The length of the cooling section is determined by a number of considerations:

- The magnetized cooling needs a long interaction time in the beam rest frame:
-

$$t' = \frac{l_{cool}}{\gamma \beta c} \text{ and it should be: } \omega_L t' \gg 1, \ \rho_{max} = v_{tr} t' \gg \rho_L$$

where ω_L is the Larmour electron frequency in the solenoidal magnetic field, ρ_{max} is the maximum value of the impact parameter, v_{tr} is the transverse velocity in the proton rest system frame and ρ_L is the Larmor radius of an electron in the longitudinal magnetic field.

- Increasing the cooling length decreases the needed electron beam current.
- For more effective cooling, it is better to have in the cooling section larger values of the beta functions, thus decreasing the transverse protons velocities. The cooling time is proportional to velocity in the third power and inversely proportional to the electron beam density. As a result, increasing β_{cool} leads to decrease in cooling time as $1/\sqrt{\beta_{cool}}$.
- The coherent damping decrement for fluctuations in the proton beam is proportional to $(t')^4$ and if parameters of the cooling are far away from the dangerous limits, is better to have a faster coherent cooling.

The magnetized electron beam has the advantage that simultaneously there are fulfilled two requirements: locally the beam is hot (this is important for suppressing recombination), but for cooling purposes it is effectively cold.

The cooling time can be expressed as [6]:

$$\tau_{cool} = \frac{v_{tr}^3}{4 r_e r_p c^4} \cdot \frac{\gamma_p^2}{L_C n_e \eta} \qquad (10)$$

where the Coulomb logarithm is expressed as:

$$L_C = \ln \frac{\left[\left(\frac{\beta_p^2 \gamma_p^3 \varepsilon}{2\pi r_e n_e \beta_{cool}}\right)^{-1/2} + \frac{\sqrt{\beta_{cool}}}{l_{cool}\sqrt{2\varepsilon}}\right]^{-1} + \frac{m_e c v_{tr}}{eB}}{\frac{m_e c v_{tr}}{eB} + \frac{r_e \beta_{cool}}{2\beta_p^2 \gamma_p^2 \varepsilon}}$$

Here n_e is the cooling beam density, B is the magnetic field, η is the fraction of circumference used for cooling.

Evaluation of the cooling time, using these formulae gives the following numbers:

$$\begin{aligned} l_{cool} &= 30 \quad m \\ \beta_{cool} &= 60 \quad m \\ n_e &= 1.5 \cdot 10^9 \quad cm^{-3} \\ \langle I_e \rangle &= 0.164 \quad A \\ B &= 2 \quad T \\ \tau_{cool} &= 3000 \quad s \end{aligned} \qquad (11)$$

Discussion and conclusion

In this paper, we have not discussed injection schemes for both rings. It is clear that a conservative scheme is to provide the injection of polarized beams at the experiment energy. However, the proton ring design gives the possibility to ramp the beam energy without loss in polarization. One full Siberian snake is sufficient to suppress all depolarizing resonances over the entire machine energy range.

The acceleration of the polarized electrons up to the top energy of 7 GeV is questionable. However, the radiative electron polarization can be enhanced considerably by applying special polarizing wigglers with high magnetic field (up to 10 T).

Finally we would like to remark that all values of machine and beams parameters which we have used for the design have been already achieved at other storage rings (see, for example B-factories status reports) with the exception of the relativistic electron cooling. However, this issue is under investigation at many laboratories (FNAL, BNL). So, we can conclude, that the e-p collider with luminosity $L = 1 \cdot 10^{33}$ $cm^{-2}s^{-1}$ in the energy range $E_{c.m.} = 15$ to 30 GeV looks quite realistic.

REFERENCES

A. N. Filippov et al., in Proceedings of the 15th International Conference on High Energy Accelerators, 1992, 1145.
1. Nesterenko et al., in Proceedings of PAC97, 1997, 1762.
A. V. Otboyev and E.A. Perevedentsev, in Proceedings of PAC99, New York, 1999.
2. Piwinski, "Beam losses and lifetime", CAS-CERN Accelerator School: 2nd general accelerator physics course, Gif-Sur-Yvette, France, proceedings CERN 85-19, 1984, 432.
3. Piwinski, "Intra-Beam Scattering", CAS-CERN Accelerator school, Geneva, 1991, 226.
4. V. V. Parkhomchuk, "New insight in the theory of electron cooling", NIM **A441**, 9 (2000).
A. A. Sokolov and I. M. Ternov, Sov. Phys. Doklady, **8**, 1203 (1964).
5. Ya. S. Derbenev, A.M. Kondratenko and A.N. Skrinsky, Sov. Phys. Doklady, **15**, 583 (1970).
6. Ya. S. Derbenev, A.M. Kondratenko and A.N. Skrinsky, Sov. Phys. JETP, **33**, 658 (1971).
7. Ya. S. Derbenev, A.M. Kondratenko, Sov. Phys. JETP **35**, 230 (1972).
8. Ya. S. Derbenev, A.M. Kondratenko, Sov. Phys. JETP, **37**, 968 (1973).
9. Ya. S. Derbenev, A.M. Kondratenko and A.N. Skrinsky, "Radiative polarization at ultra-high energies in Particle Accelerators", **9**, 247 (1979).
10. E. A. Perevedentsev, V.I. Ptitsyn and Yu.M. Shatunov, Proc. of 5th Int. Workshop on High Energy Spin Physics, Protvino, 1994, 281.

Electron and positron polarisation at HERA: Past and Future

D.P. Barber
for the HERA polarisation group

Deutsches Elektronen-Synchrotron (DESY), Hamburg, Germany

Abstract. Some aspects of radiative spin polarisation in the HERA electron/positron ring are presented.

INTRODUCTION

HERA is the $e^{\pm} - p$ collider at DESY in Hamburg. The e^+ or e^- beams run at about 27.5 GeV. Up to the end of 1997 the proton ring ran at 820 GeV. Since 1998 it has been running at 920 GeV. An integral part of the design of HERA was the provision of longitudinally spin polarised $e^{\pm}$ for the high energy physics experiments at the interaction points and the HERA $e^{\pm}$ ring is the first and only high energy $e^{\pm}$ storage ring at which longitudinal spin polarisation has been obtained.

Stored $e^{\pm}$ beams can become vertically polarised due to the emission of spin-flip synchrotron radiation — the Sokolov-Ternov (S–T) effect [1,2]. The maximum polarisation achievable is 92.4%, corresponding to a planar ring. To provide longitudinal polarisation at an interaction point (IP) the naturally occurring vertical polarisation in the arcs must be rotated into the longitudinal direction just before the IP and back to the vertical just after the IP using special magnet configurations called spin rotators.

Synchrotron radiation not only generates polarisation but can also cause depolarisation due to the stochastic excitation of the electron orbits by emission of synchrotron radiation photons [1–3]. The depolarisation can be strong in the presence of orbit distortions and is potentially very strong in the presence of spin rotators. Furthermore the ratio: (depolarisation rate/polarisation rate) increases strongly with energy. However, the depolarising effects associated with orbit distortion can be minimised by applying "harmonic closed orbit spin matching" [4,2] and the strong depolarisation associated with the rotators can, at least in principle, be combatted by optimising the optic using the technique of "strong synchrobeta spin matching" [5,2]. This has indeed proved to be the case: after a pair of dipole spin rotators was installed at the ends of the East interaction region in 1994 for the HERMES experiment, high longitudinal polarisation was obtained on the first attempt. Since then the HERMES experiment has been running regularly with longitudinal $e^{\pm}$ polarisation.

Since the details of the project and early performance including diagrams are available in [5,6] it is not necessary to repeat the details here. Instead, this article will review some special topics, provide an update on experience since 1994 and describe future plans. The reader is referred to [1–3] for extensive descriptions of the necessary basic theory.

OPTIMISATION AND TUNING

Depolarisation is particularly strong at the spin–orbit resonance condition: $\nu = k_0 + k_1 Q_1 + k_2 Q_2 + k_3 Q_3$ where the Q's are the tunes of the orbital modes, the k's are integers and ν is the amplitude dependent spin tune discussed in [7] in these Proceedings. At the typical energies of $e^{\pm}$ rings ν can be approximated by its value on the closed orbit, ν_0, [7] and this is usually approximately proportional to the machine energy.

So for high polarisation one should set ν_0 close to 1/2 so that it is optimally away from the fractional betatron tunes. Indeed, 27.5 GeV corresponds to $\nu_0 \approx$ 62.5. Then the optic should be strongly spin matched [2] while ensuring that it remains suitable for obtaining high luminosity. For example, the beam size at the interaction points must remain small and the tunes must remain acceptable — etc. This is done at the design stage and then for each new optic. The polarisation that can then be expected can be estimated using one of several computer codes [2]. Only the perfectly aligned ring should be considered! There is no point in trying to include the effects of misalignments — they are not only unknown but they would also totally confuse the matching procedure. A sketch illustrating the estimated polarisation before and after such spin matching can be found in [8]. It is recommended that the strong synchrobeta spin matching be done with the spin transfer matrix formalism since that involves mathematics closely related to that used in the various algorithms used to calculate the polarisation [2].

Once the machine is running, the orbit should be flattened as well as possible. Then, after the polarisation has reached equilibrium the energy should be scanned to see if more polarisation can be achieved. This ensures that one is not sitting at a synchrotron sideband of a parent betatron resonance.

The next step is to apply harmonic closed orbit spin matching by adjusting the specially provided orbit bumps [4,5,2]. This is done empirically by observing the polarisation as the orbit bumps are adjusted.

Once an acceptable level of polarisation has been obtained, a further energy scan can be tried as well as small adjustments of the orbital tunes.

BEAM–BEAM EFFECT

The $e^{\pm}$ polarisation at HERA can also be affected by the beam–beam (b–b) force occurring during collisions with the proton beam at the South and North interaction points (the H1 and ZEUS experiments) where, incidentally, the polarisation is vertical. Since the b–b force is very nonlinear it is very difficult to make analytical calculations of its effect on $e^{\pm}$ beams. It is even more difficult to make analytical estimates of the effect on the polarisation. However, the naive expectation is that the b–b force reduces the polarisation. For example, even if the b–b force were linear, the spin matches could be disturbed. Normally is it assumed that it is a good idea to reduce the b–b tune shift but as usual, there is no substitute for measurement and in 1996 even during collisions with 50 mA of protons, positron polarisations of up to 70 % were observed with the rotators running. Earlier, in 1995, we maintained 70 % for a run lasting ten hours. So, at least under *those* conditions, b–b forces had little influence.

Since a few proton bunches, which would normally be in collision with electrons (positrons), are by intent missing, not all electron (positron) bunches come to collision with protons. Towards the end of 1996 a second Compton polarimeter, built by HERMES, came into operation [9]. At about the same time we began to run with up to 100 mA of protons and then noticed that at the high currents it was difficult to attain more than about 60 %. Whereas the first Compton polarimeter measures vertical polarisation in the West area, the new polarimeter measures longitudinal polarisation directly close to HERMES and collects data much more quickly. Note that the *value* of the polarisation is the same all around the ring. This is confirmed by the two polarimeters. With the faster polarimeter it became possible to study the polarisation with sufficient precision on a bunch-to-bunch basis. Then contrary to intuition, it was seen that the colliding bunches sometimes have *more* polarisation than the non-colliding bunches [10,9]. We interpret this unexpected result as being due to the b–b tune shift: the colliding bunches and non–colliding bunches have different betatron tunes. So it is possible that the non–colliding bunches are close to a depolarising spin–orbit resonance (probably a synchrotron sideband resonance of a parent resonance [2]) and likely that the colliding bunches are not on such a resonance. Moreover, in the presence of the (non-linear) b–b effect the resonance structure can be very rich. This interpretation is supported by the fact that with slightly different machine tunes, there is either little difference between the colliding and non–colliding polarisations or the colliding bunches indeed have less polarisation than the non–colliding bunches.

One also sometimes observes that in the presence of b–b effect the rise time for the polarisation after injection is larger than that expected from standard radiative polarisation theory and that the polarisation level is sometimes relatively insensitive to the settings of the closed orbit harmonics of the harmonic closed orbit correction scheme. Clearly, it makes little sense to calibrate a polarimeter, by measuring the rise time [4], while the beam is in collision with protons.

The electron (positron) bunches in HERA come in three groups of sixty bunches with gaps between the groups. This causes dynamic beam loading of the rf cavity system so that the synchrotron tune can vary along a group with the result that electrons (positrons) at the beginning of a group can be closer to a depolarising resonance than those at the end (or vice versa). Thus we sometimes see a variation of the polarisation of the colliding bunches across a group.

In 2000 under normal running conditions with typically 90 mA or more of protons we had about 50 % to 60 % of equilibrium polarisation, averaged over the bunches. Figure 1 shows measurements with the vertical and longitudinal polarimeters of the polarisation in July 2000 with about 100 mA of protons. We hope to get more insight into the effect of b–b forces on polarisation from numerical tracking simulations.

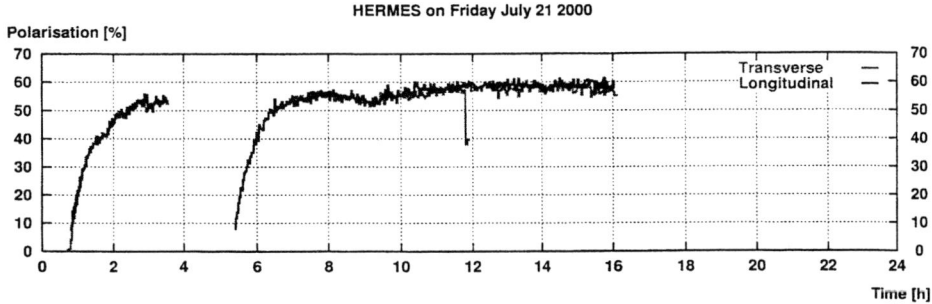

FIGURE 1. Longitudinal and vertical polarisation vs. time (hours) with positrons in collision with 100 mA of protons in North and South. The dip at 9.00 was due to adjustments to the closed orbit.

THE HERA LUMINOSITY UPGRADE

At the time of writing HERA is undergoing a major upgrade to increase the luminosity by a factor of about 4.7 [11]. This will be achieved by going to smaller beta functions at the interaction points and smaller horizontal $e^{\pm}$ emittance (by using an optic with 72 degree instead of 60 degree phase advances in the arcs and by shifting the rf frequency). The smaller beta functions require that the mini-beta combined function magnets are moved closer to the interaction points and that, in turn, entails removal of the "antisolenoids" used to counteract the effect of the H1 and ZEUS solenoids on the beam and the spin. The design orbit is then curved in the solenoids and coupling must be compensated with skew quadrupoles. Moreover, the b–b tune shifts will increase sharply [11]. H1 and ZEUS will also be equipped with spin rotators so that those experiments can run with longitudinal polarisation too. These modifications have major implications for $e^\pm$ polarisation and present nontrivial spin matching problems.

STRATEGY FOR ATTAINING HIGH RADIATIVE POLARISATION

In order to ensure attaining high longitudinal $e^\pm$ polarisation via the S–T effect at high energy (a few tens of GeV) in a storage ring, the lattice, the rotators, the optic etc, should be appropriately designed from the beginning. It is difficult to add polarisation as an after thought. Particular attention should be paid to good alignment control and beam position monitoring. For example, beam based calibration of monitor positions should be possible. The detector solenoids should be compensated locally with anti–solenoids. Very careful attention should be paid to polarimeter design so that synchrotron radiation and gas bremsstrahlung backgrounds are avoided. The feasibility of satisfactory strong synchrobeta spin matching should be demonstrated at the design stage and the resulting required number of independent quadrupole circuits should be provided.

At a few GeV the depolarisation effects tend to be much weaker than at (say) HERA energies but even for low energy rings calculations show that one should not be complacent.

REFERENCES

1. D.P. Barber et al., five articles in proc. ICFA workshop "Quantum Aspects of Beam Physics", Monterey, U.S.A., 1998, World Scientific (1999). DESY report 98-96, Los Alamos archive: physics/9901038 – physics/9901044.
2. D.P. Barber and G. Ripken, *Handbook of Accelerator Physics and Engineering*, Eds. A.W. Chao and M. Tigner, World Scientific (1999). DESY report 99-095 (1999), Los Alamos archive: physics/9907034.
3. K. Heinemann and D.P. Barber, *Nucl. Inst. Meth.* accepted for publication. DESY report 98-145 (1998), Los Alamos archive: physics/9901045.
4. D.P. Barber et al., *Nucl. Inst. Meth.* **A338** 166 (1994).
5. D.P. Barber et al., *Phys. Lett.* **343B** 436 (1995).
6. D.P. Barber, proc. 1995 Particle Accelerator Conference, Dallas, U.S.A., IEEE PAC 1995, 511 (1995).
7. D.P. Barber, G.H. Hoffstätter and M. Vogt in these Proceedings.
8. D.P. Barber, proc. EPAC-90, Nice, France 1990, Edition Frontières (1990).
9. M. Beckmann et al., DESY report 00-106 (2000), Los Alamos archive: physics/0009047.
10. D.P. Barber, proc. 13th International Symposium on High Energy Spin Physics (SPIN98), Protvino, Russia, 1998, World Scientific (1999).
11. G.H. Hoffstätter, "Future Possibilities for HERA", proc. EPAC-2000, Vienna, Austria, 2000, and DESY internal report M-00-04 (2000).

Polarized Electrons at Bates: Source to Storage Ring

Townsend Zwart, E. Booth, F. Casagrande, K. Dow, M. Farkhondeh,
W. Franklin, E. Ihloff, K. Jacobs, J. Matthews, R. Milner, T. Smith,
C. Tschalaer, E. Tsentalovich, W. Turchinetz, F. Wang

MIT – Bates Laboratory
Middleton, MA

Abstract. The MIT Bates 1 GeV electron scattering facility has recently completed a demanding set of parity violating experiments on the proton and the deuteron. These experiments required a polarized electron beam of unprecedented quality and high average currents. The facility has invested heavily in infrastructure for producing polarized beams and measuring beam polarization. This infrastructure includes a test beam set-up, a transmission polarimeter at 20 MeV, a laser back-scattering Compton polarimeter and a Siberian Snake. The polarized source group is also actively pursuing laser systems and photocathodes that could deliver high polarization, high peak current and moderate average current beams to meet the needs of the physics programs at Bates in the coming years.

INTRODUCTION

In 1999 the MIT Bates Accelerator Center completed the SAMPLE experiment (1). This experiment successfully extracted a limit on the contribution of strange quarks to the proton's magnetism by measuring a one ppm experimental asymmetry in the normalized yield of longitudinally polarized electrons on a 40 cm long liquid Hydrogen or Deuterium target. The SAMPLE experiment was completed in two three month runs in 1998 and 1999. To achieve the required 150 Coulombs of beam charge for each section of the experiment required a 40 uA electron beam of unprecedented stability and quality. To deliver 40 uA at the target it is necessary to operate with 120 uA at the polarized electron source, due to the 1/3 capture fraction of the front end RF system. The reliability of the beam delivery over the course of the two SAMPLE runs is shown in Fig.1.

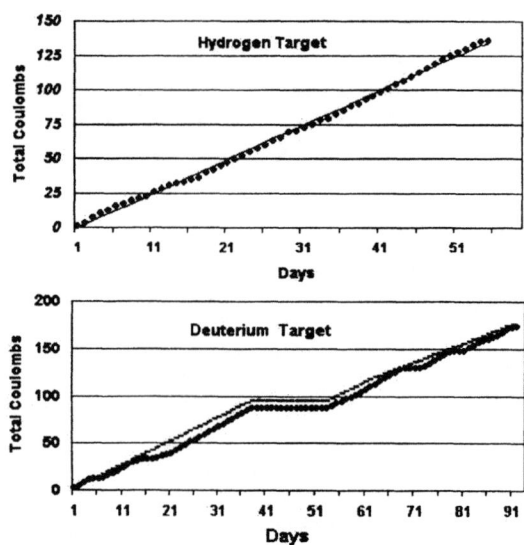

FIGURE 1. The top panel shows daily Coulombs delivered to the SAMPLE hydrogen target in 1998. The bottom panel shows daily Coulombs delivered to the SAMPLE deuterium target in 1999.

The SAMPLE experiment used a "bulk" GaAs photocathode which produces ~40% electron polarization and ~1% quantum efficiency (QE). In the past year Bates has invested in a Test Beam Set Up (TBSU) to actively pursue different photocathodes of higher polarization, ~75%, but unfortunately typically lower QE, ~0.05%. The TBSU shown in Fig.2 includes the standard Bates electron gun, a short beamline and a Mott scattering chamber to measure electron polarization. The TBSU has effectively allowed Bates to pursue a research and development program in photocathodes and laser systems without impacting the facility's experimental beam delivery, something that was previously impossible.

FIGURE 2. The test beam set up for polarized source development. The photocathode is located at top right above the ceramic insulator and the Mott polarimeter is at left at the end of the beamline.

Fig.3 shows both the quantum efficiency and electron polarization of a strained photocathode from the SPIRE corporation. This result is comparable to what has been achieved at other laboratories for these crystals. New laser systems have also been tested on the TBSU. The Bates facility obtained a 60 W CW fiber coupled diode laser at 850 nm from OPTO power corporation (2). By modulating the laser drive current at a 1% duty cycle the peak power has been increased to ~180 Watts and peak electron currents in excess of 60 mA have been obtained. The electron current as a function of applied laser power is shown in Fig.4.

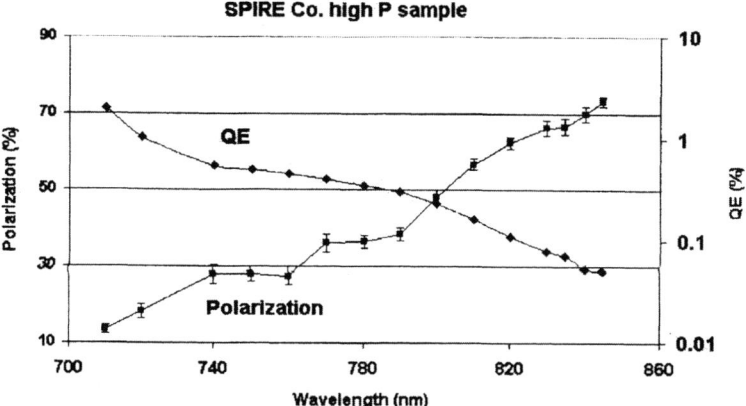

FIGURE 3. Polarization and Quantum efficiency are plotted as a function of laser wavelength. The operating point for this crystal would be near 850 nm.

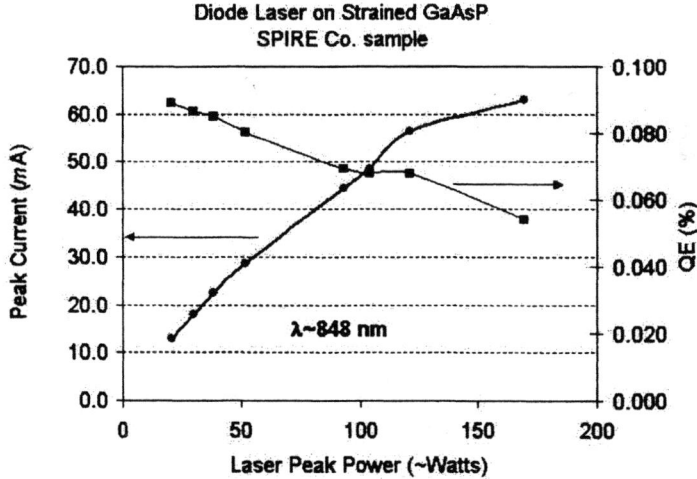

FIGURE 4. Peak current 4 and Quantum efficiency as a function of applied laser power. The full 11 mm diameter photocathode is illuminated.

The TBSU has very effectively allowed the polarized electron source group to prepare "high" polarization photocathodes and laser systems for both the upcoming stored beam BLAST physics program and the CW polarized extracted beam program.

The Bates facility has also recently installed a "transmission" polarimeter at the front end of the LINAC. This device is located where the electron beam energy is 20 MeV. The electrons are scattered in a thin BeO target and the resulting radiated photons are attenuated in a magnetized 15 cm long iron cylinder. An asymmetry is extracted by comparing the normalized photon yields downstream of the magnetized attenuator for positive and negative helicity electrons. This asymmetry is directly related to the electron polarization which is obtained from Moller polarimeters where the electron energy is typically 200 MeV or greater. The transmission polarimeter has proved very valuable for rapid relative measurements of the electron polarization at the front end of the accelerator. The calculated and measured analyzing power of the device is shown in Fig.5. Of particular note is the maximum of the analyzing power at 5 MeV, an operating point that could prove useful to other facilities.

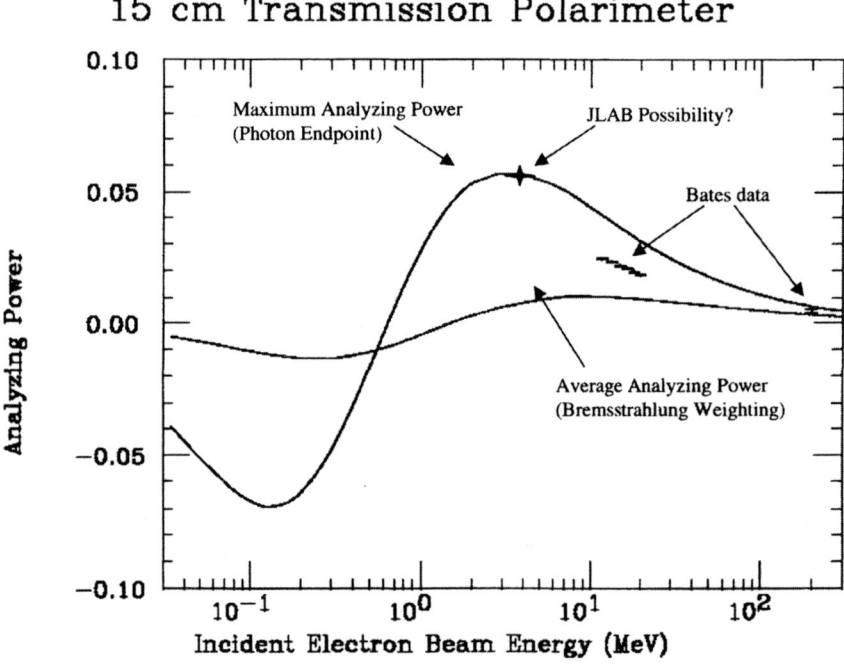

FIGURE 5. The endpoint analyzing power and weighted analyzing power are shown as a function of incident electron beam energy. Data from Bates is plotted. Also indicated is a possible operating point for a CW machine at 4 MeV.

THE BATES SOUTH HALL RING

In addition to the beamline where the SAMPLE experiment was performed Bates has constructed an electron storage/stretcher ring in it South experimental Hall (SHR.) The future physics program at Bates will rely heavily on this ring and a high current, ~100 mA, highly polarized electron beams. The SHR has been equipped with a Siberian Snake designed and constructed by the Budker Institute of Nuclear Physics in Novosibirsk Russia. (3) The Siberian Snake will maintain longitudinal electron polarization at the location of the BLAST detector for electron energies up to 1.1 GeV. The Siberian Snake has been tested and electron currents in excess of 80 mA have been stored with adequate lifetime for the BLAST physics program.

To measure the polarization of the stored electron beam Bates has constructed a laser back-scattering Compton polarimeter. This device is similar to one at the AmPS ring in the Netherlands (4). Initial results with this device are very promising. Fig.6 shows the back scattered photon spectrum when the laser is both blocked and unblocked. The measured backscattered Compton spectrum at 669 MeV shows the expected shape and has an excellent signal to noise ratio at 50 mA stored electron current. Fig.7 shows the asymmetry in the backscattered photon flux for opposite helicity states of the laser beam. Two curves are shown, corresponding to two different helicity states from the polarized electron source. Reversal of the electron helicity in the polarized source produces the expected sign change in the Compton polarimeter

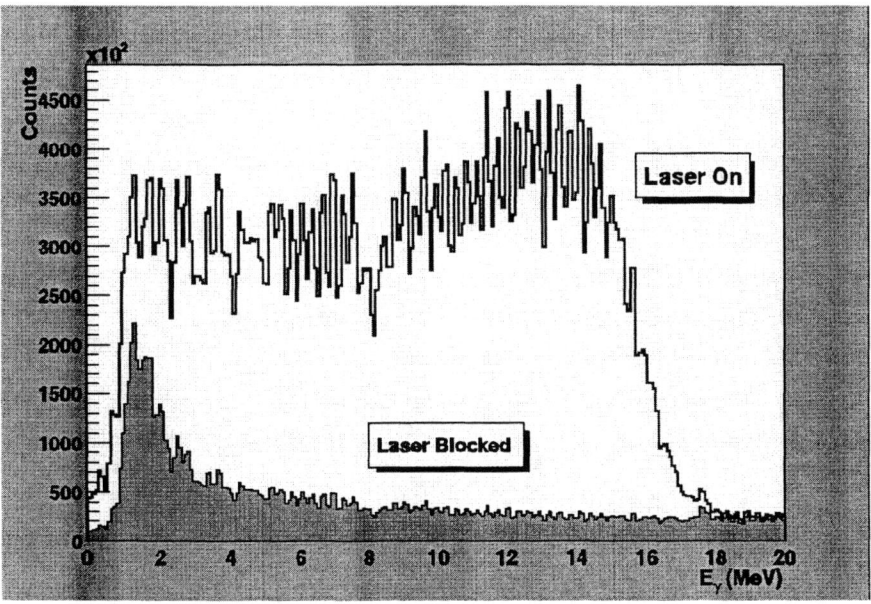

FIGURE 6. The backscattered Compton spectrum for laser blocked and unblocked at a current of ~5 mA and laser power of ~5 W. Note the excellent signal to noise ratio of five or better from 4 MeV to the Compton endpoint.

asymmetry. From the magnitude of the observed asymmetry we extract a preliminary measure of the electron polarization of 25%. Under these conditions the Siberian Snake preserved the magnitude of the polarization for times in excess of 30 minutes. This bodes well for the BLAST physics program which will require filling the ring at intervals of a few minutes.

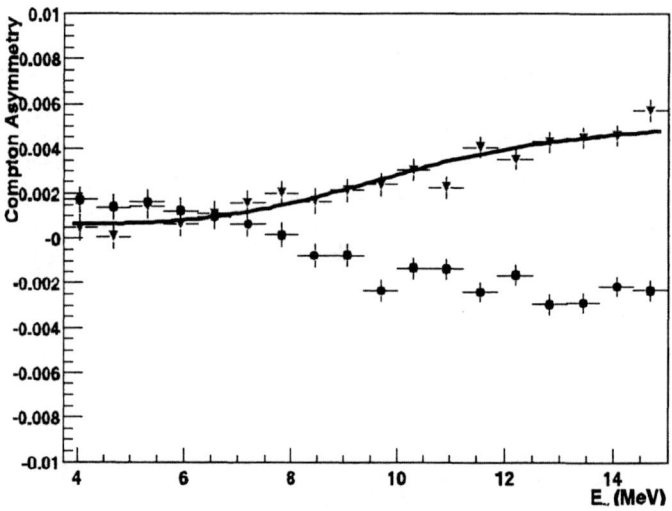

FIGURE 7. The asymmetry in the backscattered Compton spectrum is a function of photon energy. The triange and circles correspond to opposite electron helicities injected in the ring. The flip in the sign of the asymmetry for opposite electron helicities is clearly evident.

The SHR also provides an excellent venue to explore the behavior of stored polarized electrons at moderate energies. In collaboration with the Spin Physics Group at the University of Michigan (5) we have contstructed a RF spin flipper which should allow rapid reversal of the electron spin within a single fill. Initial tests of this device have been promising. This will provide important systematic checks on the data obtained in the BLAST detector.

At an energy of ~1 GeV the SHR will provide an excellent opportunity to test the dynamics of radiative polarization. The AmPS ring reported some preliminary data (6) but the SHR should have an opportunity to fully test the accuracy of the Derbenev-Kondratenko radiative polarization formula (7), including the elusive "kinetic" term.

SUMMARY

The Bates facility has successfully completed a demanding physics program using polarized electrons from bulk GaAs photocathodes. The construction of a Test Beam

Set Up has provided the infrastructure to prepare polarized electron beams which should be well matched to the anticipated physics program focussed on the stored polarized electron beams in the South Hall Ring.

REFERENCES

1. R. Hasty, *et. al.*, Science **290** (2000) 2117.
2. OPTO Power Corporation Model # OPC-BO60-mmm-FC
3. A Superconducting Spin Rotator for the MIT Bates SHR
I.   Passchier *et. al.*, Nucl Instr and Meth A **414**, 446 (1998)
4. Michigan Spin Physcis Flipping
I.   Passchier, Thesis
5. Ya.S. Derbenev, and A.M. Kondratenko, Soviet Phys. JETP, **37,** 6 (1973) 968.

Proton and electron polarisation in storage rings: some basic concepts

D.P. Barber[*], G.H. Hoffstätter[*] and M. Vogt[†]

[*]Deutsches Elektronen-Synchrotron (DESY), Hamburg, Germany
[†]University of New Mexico, Albuquerque, NM 87131, USA

Abstract. An introduction to some basic concepts central to a modern understanding of spin motion in storage rings is given.

INTRODUCTION

A proper understanding of spin–orbit resonance structure at high energy in storage rings can only be obtained with a correct definition of the "spin tune". This in turn requires establishing a proper coordinate system for "measuring" spin precession and that, in turn, requires the notion of the "invariant spin field". This paper shows how to embark on that approach. More comprehensive treatments can be found in [1–6].

SPIN–ORBIT DISTRIBUTIONS AND SPIN TUNE

Particle dynamics in storage rings is described in terms of six canonical coordinates $\vec{u} = (q_1, p_1, q_2, p_2, q_3, p_3)$ which could, for example, be $(x, p_x, y, p_y, \Delta t, \Delta E)$ where x, p_x, y, p_y describe transverse motion with respect to the curved periodic orbit and $\Delta t, \Delta E$ are the time delay relative to a synchronous particle and the deviation from the "design" energy. The independent variable is the distance along the ring s, ("the azimuth"). There is a corresponding classical Hamiltonian $h_{\text{orb}}(\vec{u}; s)$. In distorted rings $\vec{u}$ describes motion w.r.t. the resulting closed orbit.

We now make the idealisation that the beam can be described in terms of a smooth continuous density, $w(\vec{u}; s)$, which is a scalar function of $\vec{u}$ and the azimuth s. It is normalised to unity. In the absence of dissipation and noise (e.g. due to synchrotron radiation) and ignoring the effect of the tiny Stern–Gerlach forces on the orbital motion, w is constant along a phase space trajectory and obeys a relation of the Liouville type: $\frac{\partial w}{\partial s} + \sum_{k=1}^{3} \frac{dq_k}{ds} \frac{\partial w}{\partial q_k} + \frac{dp_k}{ds} \frac{\partial w}{\partial p_k} = 0$ which we write in terms of a Poisson bracket as $\partial w/\partial s = \{h_{\text{orb}}, w\}$. If the beam is stable, i.e. if w is the same from turn to turn, then it is 1–turn periodic in s and we write it as w_{st}

so that $w_{\rm st}(\vec{u};s) = w_{\rm st}(\vec{u};s+C)$, where C is the ring circumference: $w_{\rm st}(\vec{u};s)$ is a 1-turn periodic *scalar field* on $(\vec{u},s)$.

In the absence of spin flip, spin motion for electrons and protons moving in electric and magnetic fields is described by the T–BMT equation [1] $d\vec{S}/ds = \vec{\Omega} \times \vec{S}$ where $\vec{S}$ is the rest frame spin expectation value of the particle ("the spin") and $\vec{\Omega}$ depends on the electric and magnetic fields, the velocity and the energy so that it depends on $\vec{u}$ and s. Having assigned a phase space density to each point in phase space we now assign a polarisation $\vec{P}(\vec{u};s)$ to each point. $\vec{P}$ is the average over particles in an infinitisimal packet of phase space at $\vec{u}$ of the normalised spin expectation values $2\vec{S}/\hbar$. Since the T–BMT equation is linear in the spin and since the spins at $(\vec{u},s)$ all see the same $\vec{\Omega}(\vec{u};s)$, $\vec{P}(\vec{u};s)$ obeys the T–BMT equation $d\vec{P}/ds = \vec{\Omega}(\vec{u}(s);s) \times \vec{P}$ also. This can be rewritten as $\partial \vec{P}/\partial s = \{h_{\rm orb}, \vec{P}\} + \vec{\Omega}(\vec{u};s) \times \vec{P}$ in analogy with the Liouville equation for $w(\vec{u};s)$. It is assumed that $\vec{P}(\vec{u};s)$ is differentiable in all directions in phase space. The polarisation of the whole beam at azimuth s is $\vec{P}_{\rm av}(s) = \int d^6u \, w(\vec{u};s)\vec{P}(\vec{u};s)$. If the spin distribution is stable, i.e. if $\vec{P}(\vec{u};s)$ is the same from turn to turn, then $\vec{P}(\vec{u};s)$ not only obeys the T–BMT equation, but it is also 1-turn periodic in s and we write it as $\vec{P}_{\rm st}$ so that $\vec{P}_{\rm st}(\vec{u};s) = \vec{P}_{\rm st}(\vec{u};s+C)$. We denote the unit vector along $\vec{P}_{\rm st}(\vec{u};s)$ by $\hat{n}(\vec{u};s)$. Thus $\hat{n}(\vec{u};s)$ is a 1-turn periodic *vector field* on $(\vec{u},s)$ obeying the T–BMT equation. We call $\hat{n}(\vec{u};s)$ the *invariant spin field*. It can be visualised as a field of unit vectors in real space attached to every $\vec{u}$ at every s. Thus each $\vec{u}$ and s has its own unique $\hat{n}$ with the property that along particle orbits it obeys the T–BMT equation. For one turn $\hat{n}(\vec{M}(\vec{u};s);s) = R_{3\times 3}(\vec{u};s)\hat{n}(\vec{u};s)$ where $\vec{M}(\vec{u};s)$ is the new phase space vector after one turn starting at $\vec{u}$ and $R_{3\times 3}(\vec{u};s)$ is the corresponding spin transfer matrix. On the closed orbit $\hat{n}(\vec{u};s)$ becomes $\hat{n}(\vec{0};s)$ which we denote by $\hat{n}_0(s)$. Many authors make no clear distinction between $\hat{n}$ and $\hat{n}_0$ and many use the symbol $\hat{n}$ for $\hat{n}_0$. This causes confusion. Obviously $\hat{n}_0(s) = \hat{n}_0(s+C)$, i.e. $\hat{n}_0(s)$ is the 1-turn periodic solution of the T–BMT equation on the closed orbit. It is given by the real eigenvector of the 1-turn 3×3 spin transport matrix on the closed orbit.

Examples of the invariant spin field at 800 GeV for a HERA proton optic with a suitable arrangement of Siberian Snakes are shown in figure 1. In these particular simulations the protons only execute stable linear vertical betatron motion of fixed amplitude. The particle coordinates are not 1-turn periodic but at fixed azimuth they lie on a closed elliptical curve at positions depending on their vertical betatron phases. Likewise a spin at some $\vec{u}$ set parallel to $\hat{n}$ and tracked, is not 1-turn periodic but on tracking it turn to turn, it lies on the closed curve, parametrised by the orbital phase, of the field $\hat{n}$. Each picture shows the locus, on the surface of a sphere, of the tip of the $\hat{n}$ vector as the betatron phase varies at a point on the ring where $\hat{n}_0$ is vertical. The parameters are shown in the captions. An invariant emittance of 4π mm mrad corresponds to "1-σ". Both curves of figure 1 are invariant when tracked from turn to turn. Clearly, as the amplitude is increased, the invariant spin field becomes complicated. Near the spin–orbit resonances to be discussed below, the curves become very convoluted. For motion in one orbital plane, the loci on the sphere are closed as, for example, in figure 1. For motion in all three planes, the loci do not close in general although the field $\hat{n}$ is still an invariant of the 1-turn spin–orbit map. If the spins for an ensemble of particles distributed uniformly around the phase space ellipses for figure 1, are all set initially parallel to $\hat{n}_0$ and then tracked, the beam polarisation at that azimuth oscillates. If they are set parallel to $\hat{n}$, the beam polarisation at that azimuth is stationary. In general the maximum *stationary* beam polarisation that can be reached is $P_{\text{lim}}(s) = ||\int d^6u \, w_{\text{st}}(\vec{u};s)\hat{n}(\vec{u};s)||$. This can be calculated before carrying out simulations of particle acceleration and can give an impression of whether such a simulation would be worthwhile. P_{lim} for motion on vertical betatron ellipses is just given by the average of $\hat{n}$ over the betatron phase [3]. On the 64π mm mrad ellipse P_{lim} is much smaller than for the 4π mm mrad ellipse — it pays to devise ways to keep the spread of $\hat{n}$ small. Note that for $\vec{u} \neq \vec{0}$, the constraint $\hat{n}(\vec{u};s) = \hat{n}(\vec{u};s+C)$

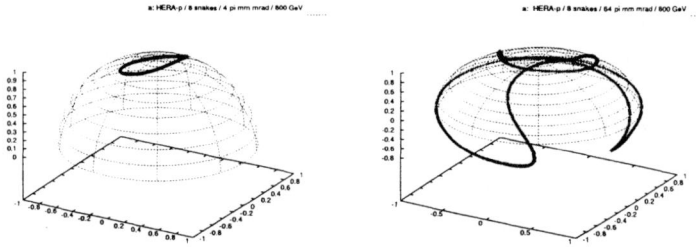

FIGURE 1. The field $\hat{n}$ in HERA–p calculated with SPRINT *on* the 4π mm mrad (left) and the 64π mm mrad (right) ellipses at 800 GeV.

for the invariant spin field is obviously *not* equivalent to the closure condition $\vec{N}(\vec{u};s) = R_{3\times3}(\vec{u};s)\vec{N}(\vec{u};s)$ since in general a spin at $\vec{u}$ set parallel to $\hat{n}(\vec{u};s)$ is not a "closed spin solution" but has a new direction after one turn. In general $\vec{N}$ does not obey the T–BMT equation everywhere along an orbit since the orbital motion is not 1-turn periodic. Furthermore, in contrast to the loci in figure 1, the loci of $\vec{N}(\vec{u};s)$ are *not* invariant beyond the first turn. In fact, the calculation of the real $\hat{n}(\vec{u};s)$ is computationally nontrivial and requires "stroboscopic averaging" using the code SPRINT [2], or Fourier analysis as in SODOM-II [7].

Although we have concentrated on protons and have introduced $\hat{n}$ via invariant spin distributions it was first motivated in another way, namely by Derbenev and Kondratenko [8,9], as a $\vec{u}$ and s dependent semiclassical spin quantisation axis for calculating radiative spin flip for electrons. That leads to the Derbenev–Kondratenko–Mane formula for the equilibrium electron polarisation in a storage ring [9]. That formula needs the vector $\partial \hat{n}/\partial \delta$ (where δ is the fractional electron energy deviation) at each $\vec{u}$ at azimuths s inside dipole magnets. So to calculate the attainable electron polarisation the correct definition of $\hat{n}$ is required. In fact $\hat{n}$ is not only central to the understanding of equilibrium spin distributions for noiseless and dissipationless motion, but it is also an essential starting point for perturbative calculations of various depolarisation effects [10].

Another key quantity in spin dynamics is the "spin tune". On the closed orbit this is defined as the number of precessions per turn of an arbitrary spin around $\hat{n}_0(s)$. We denote it by ν_0. For particles executing synchrobetatron motion the definition of spin tune is more subtle. Once the invariant spin field $\hat{n}(\vec{u};s)$ has been set up, two other unit vectors $\hat{n}_1(\vec{u};s)$ and $\hat{n}_2(\vec{u};s)$ are attached to all $(\vec{u},s)$ such that the sets $(\hat{n}_1, \hat{n}_2, \hat{n})$ form local orthonormal coordinate systems at all points in phase space at each s. Like $\hat{n}$, $\hat{n}_1$ and $\hat{n}_2$ are 1-turn periodic in s: $\hat{n}_i(\vec{u};s) = \hat{n}_i(\vec{u};s+C)$ ($i \in 1,2$). But unlike $\hat{n}$ they do not obey the T–BMT equation. If a spin $\vec{S}$ is followed along an orbit, the scalar product $\vec{S} \cdot \hat{n}$ of $\vec{S}$ and the local pre-established $\hat{n}$ is invariant since both vectors obey the T–BMT precession equation. Thus in the local pre-established $(\hat{n}_1, \hat{n}_2, \hat{n})$ coordinate system the motion of $\vec{S}$ is a precession around $\hat{n}$. Except for the uninteresting case of running on orbital resonance, the fields $\hat{n}_1(\vec{u};s)$ and $\hat{n}_2(\vec{u};s)$ can be chosen so that the rate of precession is constant and independent of the starting orbital phases [1–6]. The spin tune ν is the number of precessions per turn "measured" in this way. The spin tune depends only on the orbital amplitudes — a tune depending in some way on phases would hardly be a useful quantity since it would have to change as the phases advance! On the closed orbit ν reduces to ν_0 as required.

Spin motion is particularly strongly perturbed when the spin precession rate is near resonance with the orbital tunes: $\nu(J_1, J_2, J_3) = k_0 + k_1 Q_1 + k_2 Q_2 + k_3 Q_3$ where the Q's are the tunes of the orbital modes, the k's are integers and the J's are orbital amplitudes. Note that contrary to common practice we do not use ν_0 here. Indeed, that is the whole point of having a proper definition of spin tune as we now illustrate. Figure 2 (left) shows the dependence of the spin tune on orbital

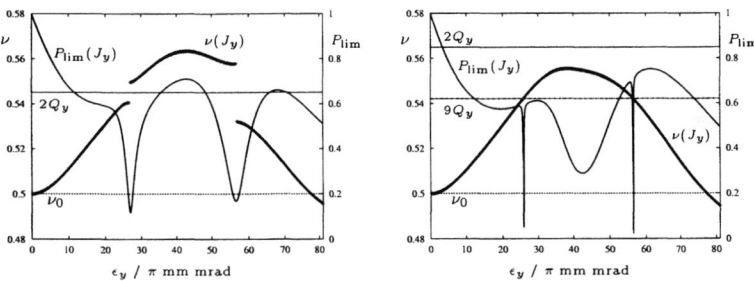

FIGURE 2. The amplitude dependent spin tune ν and P_{lim} on phase space ellipses with invariant vertical emittance ϵ_y as calculated with **SPRINT** for HERA–p at 805 GeV. Left: vertical tune $Q_y = 32.2725$, right: $Q_y = 32.2825$.

amplitude (= enclosed invariant emittance) for purely vertical betatron motion in HERA–p at 805 GeV with a suitable arrangement of snakes [3]. On the design orbit, i.e. at zero amplitude, ν is 1/2 as expected. But it deviates from 1/2 as the amplitude increases and at 27π mm mrad it jumps symmetrically across the resonant value $2Q_y$. After increasing further, ν then decreases and at an invariant emittance of 56π mm mrad it jumps back across the resonant value $2Q_y$. So ν never actually hits the resonant value but as one can see P_{lim} becomes small around the resonant amplitudes owing to the expected opening out of $\hat{n}$. Thus the behaviour of ν and P_{lim} are mutually consistent. Figure 2 (right) shows the behaviour of ν when Q_y is increased. The second order resonance can no longer be crossed but 9th order resonant behaviour occurs instead. These curves illustrate just how complicated spin motion can be at very high energy. Such phenomena could obviously not be seen without a properly defined spin tune. For example, a "fake spin tune" erroneously extracted from the complex eigenvalues of $R_{3\times 3}$ shows no correlation with dips in P_{lim}. That is no surprise since that "tune" depends on the orbital phase and is therefore unsuitable for describing long term spin–orbit coherence. With the properly defined ν, the changes in orbital tunes needed to avoid resonances can be properly estimated. Moreover, the size of a resonant jump in ν, $\Delta\nu$, for a high order resonance, is a measure of the strength of the resonance and it has been possible to parametrise polarisation loss, when varying machine parameters *dynamically* through such resonances, in terms of a generalised Froissart–Stora formula [5,6,11], containing $\Delta\nu$. Access to the properly defined ν also allows some misconceptions about depolarisation mechanisms to be swept away.

REFERENCES

1. D.P. Barber et al., five articles in proc. ICFA workshop "Quantum Aspects of Beam Physics", Monterey, U.S.A., 1998, World Scientific (1999). DESY report 98-96, Los Alamos archive: physics/9901038 – physics/9901044.
2. K. Heinemann and G.H. Hoffstätter, *Phys.Rev.* E **54**(4) 4240 (1996).
3. D.P. Barber, G.H. Hoffstätter and M. Vogt, proc. 13th International Symposium on High Energy Spin Physics (SPIN98), Protvino, Russia, 1998, World Scientific (1999).
4. G.H. Hoffstätter, M. Vogt and D.P. Barber, Phys. Rev. ST Accel. Beams **11**(2) 114001 (1999).
5. G.H. Hoffstätter, Habilitation Thesis, Technical University of Darmstadt (2000). Accepted for publication by Springer.
6. M. Vogt, Ph.D. Thesis, University of Hamburg (2000). To be published.
7. K. Yokoya, DESY report 99-006 (1999), Los Alamos archive: physics/9902068.
8. Ya. S. Derbenev and A. M. Kondratenko, *Sov.Phys. JETP.* **37** 968 (1973).
9. D.P. Barber and G. Ripken, *Handbook of Accelerator Physics and Engineering*, Eds. A.W. Chao and M. Tigner, World Scientific (1999). DESY report 99-095 (1999), Los Alamos archive: physics/9907034.
10. K. Heinemann and D.P. Barber, *Nucl. Inst. Meth.* accepted for publication. DESY report 98-145 (1998), Los Alamos archive: physics/9901045.
11. M. Froissart and R. Stora, *Nucl. Inst. Meth.* **7** 297 (1960).

Siberian Snakes and Spin-Flipping in Storage Rings

B.B. Blinov

Randall Laboratory of Physics, University of Michigan,
Ann Arbor, Michigan 48109-1120

Abstract. A Siberian snake, which is a 180° spin rotator, can be used to preserve electrons', protons' or light ions' beam polarization during acceleration by forcing the spin tune to be constant and equal to 1/2, independent of the beam energy. A compact helical dipole snake, which creates only a small orbit excursion inside the snake, could be used in EPIC. Frequent polarization reversals, or spin-flips, of a stored polarized high-energy beam could greatly reduce the systematic errors of spin asymmetry measurements in a scattering asymmetry experiment. Such polarization reversals can be done by ramping an rf-dipole or an rf-solenoid magnet's frequency through an rf-induced depolarizing resonance. The strength of an rf-solenoid decreases with the beam's energy due to the Lorenz contraction of the solenoid's longitudinal $\int B \cdot dl$, while the strength of an rf-dipole remains almost unchanged. Thus, it is more practical to use an rf-dipole spin-flipper in EPIC's electron and ion rings.

INTRODUCTION

In a circular accelerator ring, each proton's spin precesses around the vertical magnetic fields of the ring's bending dipoles, with a frequency, called the spin precession frequency f_s, that is related to the proton's circulation frequency f_c by:

$$f_s = f_c \nu_s, \qquad (1)$$

where ν_s is the spin tune, which is the number of spin precessions during each turn around the ring.

If there are only vertical magnetic fields, then the vertical beam polarization remains unchanged; however, whenever there is a periodic horizontal magnetic field whose tune is equal to the spin tune, a depolarizing resonance occurs, which can destroy the polarization. The ν_s is proportional to the proton's energy via:

$$\nu_s = G\gamma, \qquad (2)$$

where G=1.792847 is the proton's anomalous magnetic moment, and γ is the Lorentz energy factor; thus the protons will encounter many depolarizing resonances as they are accelerated to a high energy.

An elegant method to overcome these depolarizing resonances, proposed by Derbenev and Kondratenko in 1976 [1], involves using a spin rotator called the Siberian snake that makes ν_s equal to 1/2 independent of the beam energy. This energy-independent ν_s eliminates most of the depolarizing resonances [2,3].

Once a polarized proton beam is accelerated to a high energy and stored, it is important to be able to frequently reverse its polarization direction to reduce the systematic errors in various polarized scattering asymmetry experiments. Studies at the IUCF Cooler Ring [4–6] showed that these polarization reversals, or spin-flips, can be done using an rf magnet, which creates an rf depolarizing resonance at frequency:

$$f_s = f_{rf} + nf_c, \qquad (3)$$

where f_{rf} is the rf magnet's frequency, f_c is the circulation frequency, and n is an integer. After crossing such a resonance by varying the rf magnet's frequency from a value below the resonance to a value above the resonance, the final beam polarization P_f is related to the initial beam polarization P_i via the Froissart-Stora formula [7]:

$$P_f = P_i[2e^{-\frac{(\pi \epsilon f_c)^2}{(\Delta f / \Delta t)}} - 1], \qquad (4)$$

where ϵ is the resonance strength, and $\Delta f/\Delta t$ is the resonance crossing rate, while Δf is the frequency's range during its linear variation time Δt. If the resonance is sufficiently strong and/or the crossing rate is sufficiently slow, the final polarization is reversed with respect to the initial polarization, while its absolute value remains almost unchanged; this is called a spin flip.

I SNAKE DESIGN

A full Siberian snake could be made of either solenoids, whose strength is energy-dependent due to the Lorenz contraction of their longitudinal $\int B \cdot dl$, or dipoles, whose transverse field integrals do not change with energy in the particles' rest frame and whose strength is thus nearly energy-independent. Both designs have their advantages and disadvantages. A properly adjusted solenoidal snake does not cause any beam distortion, while inside a dipole snake there is a large orbit excursion, especially at low energies. However, at higher energies, the longitudinal $\int Bdl$ required for a full snake, becomes very large. Also, a solenoidal snake's current must be ramped as the beam is accelerated; this is difficult with a superconducting solenoid. An important benefits of a dipole snake is that it requires a nearly fixed transverse $\int B \cdot dl$, which needs essentially no adjustment while accelerating the beam from low energies. Such a snake could be made of conventional warm dipole magnets and requires no cryogenics to operate.

As was shown in the LISS snake design [8], a good choice for a compact dipole snake to work with polarized protons is a so-called helical snake, which is shown in

Figure 1. The helical snake's main body is an 8.6-meter-long helical dipole, which produces a 180° spin rotation; two pairs of small tilted regular dipoles D1 and D2 on either end of the helix restore the beam's orbit without affecting the spin direction.

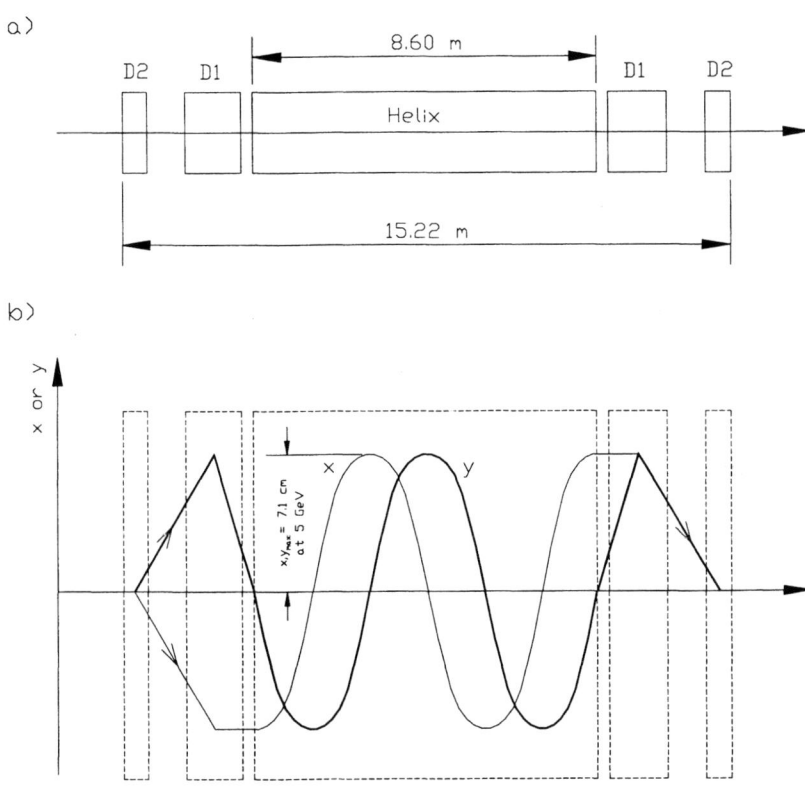

FIGURE 1. Helical Siberian snake layout (a) and beam orbit inside the snake (b).

Notice the rather small beam orbit displacement within the snake, which does not exceed 7.1 cm for protons at EPIC's 5 GeV injection energy, and decreases rapidly at higher energies. This allows one to make the snake magnets' apertures quite small for lower weight and lower power consumption. A conventional Steffen-Lee snake, made of a series of standard dipoles, would create much larger orbit excursions, and thus needs apertures exceeding 40 cm [9] at EPIC's 5 GeV injection energy.

II SPIN-FLIPPER

Of the two possible designs of an rf-magnet spin-flipper: an rf-solenoid or an rf-dipole, the rf-dipole has the very important advantage of its energy-independent spin resonance strength, while the amount of spin rotation by an rf-solenoid decreases with the particles' energy. This rf-dipole could be similar to the dipole spin-flipper, shown in Figure 2, which was designed by Michigan Spin Physics Center for the MIT-Bates South Hall electron ring. It is a ferrite-loaded single-turn magnet, whose inductance would form a tuneable parallel LC-resonant circuit with a variable capacitor (not shown). The LC-circuit would be tuned to have its resonance at the same frequency as the rf depolarizing resonance frequency in order to enhance the rf-dipole's magnetic field and thus increase its spin-flipping resonance strength. Ramping the frequency of the rf-dipole through the rf depolarizing resonance could be used to spin-flip the stored beam.

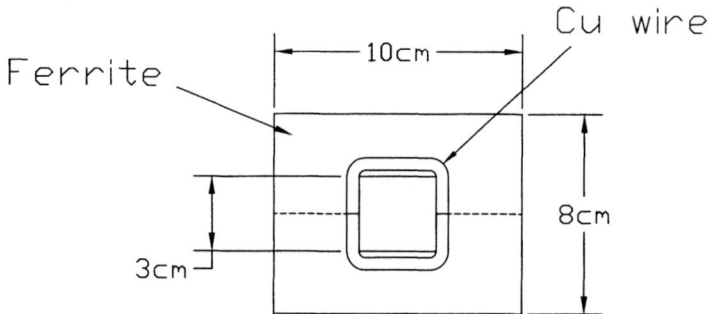

FIGURE 2. Ferrite rf-dipole spin-flipper (beam view).

However, such an rf-dipole could have a strong effect on the stored beam's betatron motion by driving coherent beam oscillations at the rf-dipole's frequency. These oscillations could be so strong that they would eventually kick the beam out of the ring's vacuum chamber aperture. Transverse beam cooling (electron cooling of the protons or synchrotron radiation cooling in the electron ring), and slow, or adiabatic, turn-on of the rf-dipole, may help reduce these oscillations and prevent the beam loss.

Another effect of such coherent beam oscillations is that they would effectively generate an additional rf $\int B \cdot dl$ in the ring's quadrupole magnets, which could interfere with the rf-dipole's $\int B \cdot dl$ and could either increase the effective spin-depolarizing resonance strength, or decrease it. Both effects should be taken into account in designing the actual spin-flipper.

III ACKNOWLEDGMENTS

I would like to thank A.D. Krisch and P. Schwandt for useful discussion and their help, and V.A. Anferov, Ya.S. Derbenev, T. Kageya, V.S. Morozov, D.W. Sivers, B. von Przewoski and V.K. Wong for their help. This research was supported by grants from the U.S. Department of Energy and National Science Foundation.

REFERENCES

1. Ya. S. Derbenev and A. M. Kondratenko, Sov. Phys. Dokl. **20**, 562 (1976).
2. A.D. Krisch *et al.*, Phys. Rev. Lett **63**, 1137 (1989).
3. J.E. Goodwin *et al.*, Phys. Rev. Lett. **64**, 2779 (1990).
4. D.D. Caussyn *et al.*, Phys. Rev. Lett. **73**, 2857 (1994)
5. B.B Blinov *et al.* Phys. Rev. Lett. **81**, 2906 (1998).
6. V.A. Anferov *et al.*, Phys. Rev. ST Accel. Beams **3**, 041001 (2000).
7. M. Froissart and R. Stora, Nucl. Instrum. and Methods **7**, 297 (1960).
8. T.L. Gerig *et al.*, *Design of a Helical Siberian Snake for LISS*, (1996).
9. V.A. Anferov *et al.*, *Acceleration of Polarized Protons to 120 GeV and 1 TeV at Fermilab*, UM HE 95-09 (1995).

Second Workshop on Physics with an Electron Polarized Ion Collider
MIT
September 14-16, 2000

Little Kresge Auditorium, MIT

AGENDA

Thursday September 14, 2000 - Chair: P. Paul (BNL)

09:00	Welcoming Remarks – R. Milner (MIT)
09:05	Overview of EPIC - J. Cameron (IUCF)
09:30	Overview of QCD - F. Wilczek (MIT)
10:15	*Coffee*
10:45	Spin Structure of the Nucleon: Experimental overview - A. Miller (TRIUMF)
11:30	Overview of off forward parton disributions - A. Schaefer (Regensburg)
12:15	*Lunch (not provided)*
14:00-18:00	Parallel Sessions
18:30-20:30	Reception at MIT Faculty Club (corner of Main and Wadsworth Streets)

Friday, September 15, 2000   Chair: A. Krisch (Michigan)

09:00	Spin Structure of the Nucleon: Theoretical overview - R. Jaffe (MIT)
09:45	Current fragmentation in semi-inclusive leptoproduction - P. Mulders (NIKHEF)
10:30	*Coffee*
11:00	Probing the Generalized Parton Distributions with JLab at 12 GeV - S. Stepanyan (JLab)
11:45	eA physics with a collider - G. Garvey (LANL)
12:30	*Lunch (not provided)*
14:00-18:00	Parallel sessions
18:30	Cocktails and dinner at La Groceria Restaurant (not included) 853 Main Street, Cambridge

Saturday, September 16, 2000   A.M. Chair: V. Hughes (Yale)

09:00	Summary of EPIC physics - E. Kinney (Colorado)
09:40	Summary of eRHIC physics - R. Venugopalan (BNL)
10:20	Summary of machine design- V. Lebedev (JLab)
10:50	*Coffee*
11:20	Overview of European plans - D. von Harrach (Mainz)
12:00	SLAC plans - P. Bosted (U. Mass)
12:30	*Lunch (not provided)*

P.M. Chair: R. McKeown (Caltech)

14:00	JLAB 12 GeV plans - L. Cardman (JLab)
14:40	Open discussion
15:40	Coffee
16:10	Workshop summary - K. Rith (Erlangen)
17:00	End of meeting

The conveners of parallel sessions are as follows:

A. Spin and flavor structure of the nucleon - Antje Bruell (MIT)
B. Semi-exclusive processes - Andreas Schaefer (Regensburg)
C. Heavy quark physics and target fragmentation - Timothy Londergan (IUCF)
D. eA Physics - Abhay Deshpande (Riken-BNL)
E. Machine - Peter Schwandt (IUCF)

Parallel Sessions Schedule

Parallel Session A
Spin and flavor structure of the nucleon - Antje Bruell (MIT)

Thursday, September 14, 2000 **Location: 5-134**

1:30 pm	A. Deshpande (Yale)	Inclusive Scattering: low x extrapolation, Bjorken Sum rule and Delta_G from QDC fits
2:10 pm	U. Stoesslein (Colorado)	Delta_s from semi-inclusive scattering
2:50 pm	X. Jiang (Rutgers)	Establishing evidences of factorization in semi-inclusive reactions
3:10 pm	A. Bruell (MIT)	Gluon distribution and polarization from charm production
3:45 pm	*Break*	
4:15 pm	R. Ent (JLab/Hampton)	(Vector) Meson production and duality
4:50 pm	R. Milner (MIT)	Prospects for Deeply Virtual Compton Scattering

Parallel Session B.
Semi-exclusive processes - Andreas Schaefer (Regensburg)

Thursday, September 14, 2000 **Location: PDR 1**

1:30 pm	M. Burkardt (NMSU)	Off-forward parton distributions and impact parameter dependence of parton structure
2:10 pm	A. Afanasev (JLab)	Inclusive Photo- and Electroproduction of mesons at high transverse momentum
3:00 pm	R. Jakob (Wuppertal)	Large angle Compton scattering

3:30 pm	*Break*	
4:00 pm	R. Holt (Argonne)	Structure of the Goldstone Bosons
4:30 pm	C. Weiss (Bochum)	Model calculations of distribution functions, wave functions and SPD's at a low scale in the large-N_c limit
5:10 pm	M. Radici (INFN-Pavia)	Calculation of fragmentation functions in two-hadron semi-inclusive processes

Friday, September 15, 2000   **Location: 5-217**

2:00 pm	V. Korotkov	TESLA-N
2:50 pm	E. Stein (Regensburg)	Hadronic wave functions and exclusive processes
3:30 pm	C. Hyde-Wright (ODU)	Experimental Aspects of Deeply Virtual Compton Scattering in a electron-nucleus collider
4:10 pm	*Break*	
4:40 pm	M. Boglione (VrijeU.)	Azimuthal and single spin asymmetries
5:20 pm	K. Oganessyan (Frascati)	Single spin asymmetries in semi-inclusive electroproduction: access to transversity

Parallel Session C.
Heavy Quarks/Target Fragmentation - Timothy Londergan

Thursday, September 14, 2000   Location: PDR 2   Chair: A. Szczepaniak (Indiana U.)

1:30 pm	W. Melnitchouk (Jlab/Adelaide)	Tests of Quark-Hadron Duality with a Collider
2:20 pm	A. Kotzinian (CERN)	Lambda Production in the Target Fragmentation Region
3:00 pm	D. von Harrach (Mainz)	Prospects for Open Charm Production
3:30 pm	*Break*	
4:00 pm	M. Stratmann (Regensburg)	QCD Corrections to Polarized Photoproduction of Heavy Flavors

Friday, September 15, 2000   **Location: 5-232**   **Chair: W. Melnitchouk (JLab/CSSM, Adelaide)**

2:00 pm	M. Pichowsky (IndianaU)	Diffractive Physics at Collider Energies
2:50 pm	M. Zomer (Orsay)	Strange and Antistrange Quark Densities in the Nucleon
4:00 pm	*Break*	

Parallel Session D.
eA Physics - Abhay Deshpande

Thursday, September 14, 2000   **Location: West Lounge**

1:30 pm	J. Jalilian-Marian (UAz)	QCD at eRHIC and EPIC
2:30 pm	Y. Kovchegov (UWa)	QCD at High Parton Density
3:30 pm	*Break*	
4:00 pm	L. Mankiewicz (INS)	Nuclear Parton Distributions in Coordinate Space
5:00 pm	A. Schaefer (Regensberg)	Comments on pA and eA physics

Friday, September 15, 2000   **Location: 5-234**

2:00 pm	Y. Kovchegov (UWa)	QCD Instantons and the Soft Pomeron
3:00 pm	M. Sargsian (FIU)	Diffractive Electroproduction of Vector Mesons from Light Nuclei
4:00 pm	*Break*	
4:30 pm	J.C. Peng (LANL)	Physics of p-A collisions and its relation to e-A collisions
5:30 pm	ALL	Discussion
6:00 pm	End of Session	

Parallel Session E.
Machine - Peter Schwandt (IUCF)

Thursday, September 14, 2000   **Location: 5-231**   **Chair: C. Tschalaer (Bates)**

1:30 pm	Y. Shatunov (Novosibirsk)	An electron ring/proton ring collider
2:30 pm	L. Merminga (JLab)	An electron linac/proton ring collider
3:30 pm	*Break*	
4:00 pm	D. Barber (DESY)	Electron-positron polarization at HERA: Past and Future
5:00 pm	T. Zwart (Bates)	Polarized Electrons at Bates: Source to Storage Ring

Friday, September 15, 2000   **Location: 5-231**   **Chair: K. Jacobs (Bates)**

2:00 pm	I. Ben-Zvi (BNL)	The eRHIC collider
3:00 pm	S. Nagaitsev (FNAL)	High-energy Electron Cooling: Current Status and Future Prospects
4:00 pm	*Break*	
4:30 pm	D. Barber (DESY)	Proton and Electron Polarization in Storage Rings: some basic concepts
5:00 pm	B. Blinov (UMich)	Siberian Snakes, Spin Flippers in Storage Rings
5:30 pm	Discussion	

List of Participants

Dr. Andrei Afanasev
NCCU
Physics Deparment
303 Robinson Science Building
Durham, NC 27707
afanas@jlab.org

Dr. Charles Benesh
Wesleyan College
Dept. of Chemistry and Physics
Wesleyan College
4760 Forsyth Road
Macon, GA 31210
cbenesh@wesleyancollege.edu

Prof. Ricardo Alarcon
MIT-Bates Laboratory
P. O. Box 846
Middleton, MA 01949
alarcon@mitbates.mit.edu

Prof. Aron Bernstein
MIT
26-419
77 Massachusetts Avenue
Cambridge, MA 02139
bernstein@mitlns.mit.edu

Dr. Desmond Barber
DESY
Notkestrasse 85
22603 Hamburg
Germany
mpybar@mail.desy.de

Prof. William Bertozzi
MIT
26-437
77 Massachusetts Avenue
Cambridge, MA 02139
bertozzi@mitlns.mit.edu

Prof. Douglas Beck
University of Illinois at Urana-Champaign
1110 West Green Street
Urbana, IL 61801-3080
dhbeck@uiuc.edu

Dr. Andrea Bianconi
Universit' di Brescia
Dipartimento di chimica e fisica
Via Valotti, 9
25123 Brescia, ITALY
bianconi@axpbs1.bs.infn.it

Dr. Ilan Ben-Zvi
Brookhaven National Laboratory
NSLS
Building 725C
Upton, NY 11973-5000
benzvi@bnl.gov

Dr. Boris Blinov
University of Michigan
Spin Physics Center
1239 Kipke Dr., suite 2341
Ann Arbor, MI 48109
bblinov@umich.edu

Dr. Mariaelena Boglione
University of Durham
Dept. of Physics
South Road
Durham DH1 3LE
United Kingdom
boglione@nat.vu.nl

Dr. Antje Bruell
MIT
26-551
77 Massachusetts Avenue
Cambridge, MA 01239
abr@pierre.mit.edu

Dr. Peter Bosted
University of Massachusetts
Physics Department
Amherst, MA 01003
bosted@slac.stanford.edu

Ms. Ginny Bullard
MIT-Bates Laboratory
P. O. Box 846
Middleton, MA 01949
vbullard@mit.edu

Dr. Tancredi Botto
IASA
P. O. Box 17214
Athens 10024
Greece
tancredi@ralph2.mit.edu

Dr. Matthias Burkardt
New Mexico State University
Department of Physics
MSC 3D
Las Cruces, NM 88003
burkardt@nmsu.edu

Dr. Maurice Bouwhuis
DESY
HERMES Group
Notkestrasse 85
D-22607 Hamburg
Germany
maurice.bouwhuis@desy.de

Prof. John Cameron
IUCF
2401 Milo B. Sampson Lane
Bloomington, IN 47405
cameron@iucf.indiana.edu

Ms. Kristin Broughton
MIT-Bates Laboratory
P. O. Box 846
Middleton, MA 01949

Dr. Lawrence Cardman
Jefferson Laboratory
12000 Jefferson Avenue
Newport News, VA 23606
cardman@jlab.org

Dr. Kees de Jager
Jefferson Laboratory
12000 Jefferson Avenue
Newport News, VA 23606
kees@jlab.org

Dr. Manouchehr Farkhondeh
MIT-Bates Laboratory
P. O. Box 846
Middleton, MA 01949
manouch@mit.edu

Dr. Yaroslav Derbenev
University of Michigan
Physics Department
500 E. University
Ann Arbor, MI 48109-1120
derbenev@umich.edu

Dr. Wilbur Franklin
MIT
Building 26-402
77 Massachusetts Avenue
Cambridge, MA 02139
wafrankl@mitlns.mit.edu

Dr. Abhay Deshpande
RIKEN BNL Research Center
Building 510, Physics Department
P. O. Box 5000
Upton, NY 11973-5000
abhay@bnl.gov

Prof. Haiyan Gao
MIT
26-413
77 Massachusetts Avenue
Cambridge, MA 02139
haiyan@mitlns.mit.edu

Prof. T.W. Donnelly
MIT
6-300
Center for Theoretical Physics
77 Massachusetts Avenue
Cambridge, MA 01239
donnelly@mitlns.mit.edu

Dr. Gerald Garvey
Los Alamos National Laboratory
MS H*846
Los Alamos, NM 87501
garvey@lanl.gov

Dr. Rolf Ent
Jeffferson Laboratory
12000 Jefferson Avenue
Newport News, VA 23606
ent@jlab.org

Prof. Shalev Gilad
MIT
26-449
77 Massachusetts Avenue
Cambridge, MA 02139
gilad@mitlns.mit.edu

Dr. Douglas Hasell
MIT
26-415
77 Massachusetts Avenue
Cambridge, MA 02139
hasell@mitlns.mit.edu

Prof. Robert Jaffe
MIT
6-311
77 Massachusetts Avenue
Cambridge, MA 02139
jaffe@mitlns.mit.edu

Dr. Roy Holt
Argonne National Laboratory
Physics Division
Argonne, IL 40439
holt@anl.gov

Dr. Rainer Jakob
University of Wuppertal
Theoretical Physics Department
Gauss-Strasse 20
42097 Wuppertal
Germany
rainer@theorie.uni-wuppertal.de

Dr. Vernon Hughes
Yale University
Physics Department
260 Whitney Avenue, P. O. Box 208121
New Haven, CT 06520-8121
hughes@hepmail.physics.yale.edu

Dr. Jamal Jalilian-Marian
University of Arizona
Physics Department
1118 E. 4th Street
Tucson, AZ 85721
jmarian@physics.arizona.edu

Dr. Charles Hyde-Wright
Old Dominion University
Department of Physics
4600 Elkhorn Avenue
Norfolk, VA 23529
chyde@odu.edu

Dr. Xiaodong Jiang
Jefferson Laboratory
MS 12H
12000 Jefferson Avenue
Newport News, VA 23606
jiang@jlab.org

Dr. Kenneth Jacobs
MIT-Bates Laboratory
P. O. Box 846
Middleton, MA 01949
kjacobs@mit.edu

Dr. Edward Kinney
University of Colorado
Campus Box 390
Boulder, CO 80309-0390
edward.kinney@colorado.edu

Dr. Hauke Kolster
MIT
26-551
77 Massachusetts Avenue
Cambridge, MA 02139
hauke@pierre.mit.edu

Dr. Krishna Kumar
University of Massachusetts
Department of Physics
Amherst, MA 01003
kkumar@hysics.umass.edu

Dr. Vladislav Korotkov
Desy Zeuthen
Platanenallee 6
D-15738 Zeuthen
Germany
korotkov@ifh.de

Dr. Valeri Lebedev
Jefferson Laboratory
12000 Jefferson Avenue
Newport News, VA 23606
lebedev@jlab.org

Dr. Aram Kotzinian
CERN
CH-1211
Geneva 23
Switzerland
aram.kotzinian@cern.ch

Dr. Tong-Uk Lee
MIT
26-402
77 Massachusetts Avenue
Cambridge, MA 02139
tong@mitlns.mit.edu

Dr. Yuri Kovchegov
University of Washington
Theoretical Nuclear Physics Group
Dept. of Physics
Seattle, WA 98195-1560
yuri@bnl.gov

Dr. Nilanga Liyanage
Jefferson Laboratory
MS 12H
12000 Jefferson Avenue
Newport News, VA 23606
nilanga@jlab.org

Dr. Alan Krisch
University of Michigan
Randall Lab of Physics
Ann Arbor, MI 48109-1120
krisch@umich.edu

Dr. Earle Lomon
MIT/CTP
6-302
77 Massachusetts Avenue
Cambridge, MA 02139
lomon@mitlns.mit.edu

Dr. Timothy Londergan
Indiana University
Dept. of Physics
Swain Hall
Bloomington, IN 47405
tlonderg@indiana.edu

Dr. Wolfgang Lorenzon
University of Michigan
Physics Department
2477 Randall Lab
Ann Arbor, MI 48109
lorenzon@umich.edu

Ms. Dianne Mac Carthy
MIT-Bates Laboratory
P. O. Box 846
Middleton, MA 01949
dianne@batespop.mit.edu

Ms. Anne Mac Innis
MIT-Bates Laboratory
P. O. Box 846
Middleton, MA 01949
macinnis@mit.edu

Dr. David Mack
TJNAF
12000 Jefferson Avenue
Newport News, VA 23606
mack@jlab.org

Prof. June Matthews
MIT
26-505
77 Massachusetts Avenue
Cambridge, MA 02139
matthews@mitlns.mit.edu

Prof. Robert McKeown
106-38 Kellogg
Caltech
Pasadena, CA 91125
bmck@krl.caltech.edu

Dr. Wally Melnitchouk
Jefferson Laboratory
MS 12H2
12000 Jefferson Avenue
Newport News, VA 23606
wmelnitc@jlab.org

Dr. Lia Merminga
Jefferson Laboratory
12000 Jefferson Avenue
MS 58B
Newport News, VA 23606
merminga@jlab.org

Dr. C. Andrew Miller
TRIUMF
4004 Wesbrook Mall
UBC Campus
Vancouver, BC
Canada V6T 2A3
miller@triumf.ca

Prof. Richard Milner
MIT-Bates Laboratory
P. O. Box 846
Middleton, MA 01949
milner@mitlns.mit.edu

Prof. Rory Miskimen
UMass, Amherst
Department of Physics
Lederle GRT 417C
Amherst, MA 01003
miskimen@physics.umass.edu

Dr. Piet Mulders
Vrije Universiteit Amsterdam
Dept. of Theoretical Physics, FEW
De Boelelaan 1081
1081 HV Amsterdam, The Netherlands
mulders@nat.vu.nl

Dr. Sergei Nagaitsev
Fermilab
P.O.Box 500
Batavia, IL 60510-0500
nsergei@fnal.gov

Prof. John Negele
6-308
MIT
77 Massachusetts Avenue
Cambridge, MA 02139
negele@mitlns.mit.edu

Dr. Wolf-Dieter Nowak
DESY Zeuthen
Platanenallee 6
D-15738 Zeuthen
Germany
nowakw@ifh.de

Dr. Karo Oganessyan
INFN-LNF
via Enrico Fermi 40
C.P. 13, I-00044
Frascati (Roma) Italy
kogan@hermes.desy.de

Dr. Peter Paul
Brookhaven National Laboratory
40 Brookhaven Avenue
Upton, NY 11973-5000
ppaul@bnl.gov

Dr. David Peaslee
Physics Department
University of Maryland
College Park, MD 20742
peaslee@umdhep.umd.edu

Dr. Jen-Chieh Peng
Los Alamos National Lab
Mail Stop H846
Los Alamos, NM 87545
peng@lanl.gov

Prof. Gerald Peterson
Nuclear Physics
Deparment of Physics
University of Massachusetts
Amherst, MA 01003
peterson@physics.umass.edu

Dr. Oscar Rondon
University of Virginia
Physics Department
P. O. Box 400714
Charlottesville, VA 22904-4714
or@virginia.edu

Dr. Michael Pichowsky
Indiana University
Nuclear Theory Center
2401 Milo B. Sampson Lane
Bloomington, IN 47405
pichowsk@indiana.edu

Dr. Misak Sargsian
Department of Physics
Florida International University
Miami, FL 33199
sargsian@fiu.edu

Dr. Marco Radici
INFN-Sezione di Pavia
via Bassi 6
I27100 Pavia
Italy
radici@pv.infn.it

Dr. Andreas Schafer
Universitaet Regensburg
Inst. fuer Theoretische Physik
D-93040 Regensburg
Germany
andreas.schaefer@physik.uni-regensburg.de

Dean Robert Redwine
MIT
4-110
77 Massachusetts Avenue
Cambridge, MA 02139
redwine@mit.edu

Dr. Peter Schwandt
Indiana University
Physics Department
Bloomington, IN 47405
schwandt@iucf.indiana.edu

Dr. Klaus Rith
Physikalisches Institut
Abteilung II
Erwin-Rommel-Strabe 1
D-91058 Erlangen
Germany
klaus.rith@desy.de

Dr. Yuri Shatunov
Budker Institute of Nuclear Physics
11 Lavrentiev Str.
630090, Novosibirsk
Russia
shatunov@inp.nsk.su

Dr. Simon Sirca
MIT
26-402
77 Massachusetts Avenue
Cambridge, MA 02139
sirca@mitlns.mit.edu

Dr. Stepan Stepanyan
TJNAF
12000 Jefferson Avenue
Newport News, VA 23606
stepanya@jlab.org

Dr. Edward Six
MIT-Bates Laboratory
P. O. Box 846
Middleton, MA 01949
ed.six@asu.edu

Dr. Uta Stoesslein
Eschenbachstrasse 4
12437 Berlin
Germany
uta.stoesslein@ifh.de

Dr. Timothy Smith
MIT-Bates Laboratory
P. O. Box 846
Middleton, MA 01949
tim_smith@mit.edu

Dr. Marco Stratmann
Inst. for Theoretical Physics
University of Regensburg
D-93040 Regensburg
Germany
marco.stratmann@physik.uni-regensburg.de

Dr. Erhard Steffens
Physikalisches Institut

Erwin-Rommel-Str. 1
D-91058 Erlangen
Germany

Dr. Adam Szczepaniak
Physics Department
Indiana University
Bloomington, IN 47405
aszczepa@indiana.edu

Dr. Eckart Stein
University of Regensburg
Institut fuer Theoretische Physik
Universitatsstrasse
93047 Regensburg
Germany
eckart.stein@physik.uni-regensburg.de

Dr. Brad Tippens
U.S. Department of Energy
19901 Germantown Road
Germantown, MD 20874-1290
brad.tippens@hq.doe.gov

Dr. Chris Tschalaer
MIT-Bates Laboratory
P. O. Box 846
Middleton, MA 01949
chris@bates.mit.edu

Dr. Evgeni Tsentalovich
MIT-Bates Laboratory
21 Manning Avenue
Middleton, MA 01949
evgeni@bates.mit.edu

Dr. William Turchinetz
MIT-Bates Laboratory
P. O. Box 846
Middleton, MA 01949
billt@bates.mit.edu

Dr. Raju Venugopalan
Nuclear Theory
Brookhaven National Laboratory
Building 510A
Upton, NY 11973
raju@quark.phy.bnl.gov

Dr. Steven Vigdor
Department of Physics
Swain Hall West 117
Indiana University
Bloomington, IN 47405
vigdor@iucf.indiana.edu

Dr. Dietrich von Harrach
Institut fuer Kernphysik
55099 Mainz
Germany
dvh@kph.uni-mainz.de

Mr. Bernard Wadsworth
MIT
26-561
77 Massachusetts Avenue
Cambridge, MA 02139
wadswort@mit.edu

Dr. Fuhua Wang
MIT-Bates Laboratory
P. O. Box 846
Middleton, MA 01949
fwang@mit.edu

Dr. Christian Weiss
Institut f. Theoretische Physik II
D-44780 Bochum
Germany
weiss@tp2.ruhr-uni-bochum.de

Dr. Frank Wilczek
MIT
6-305
77 Massachusetts Avenue
Cambridge, MA 02139
wilczek@mit.edu

Dr. Claude Williamson
MIT
26-431
77 Mass. Avenue
Cambridge, MA 02139
cfw@mitlns.mit.edu

Dr. Scott Wissink
Cyclotron Facility
2401 Milo B. Sampson Lane
Bloomington, IN 47408
wissink@iucf.indiana.edu

Dr. Victor Wong
University of Michigan
1071 Beal Avenue
Ann Arbor, MI 48109-4919
vkw@umich.edu

Dr. Zilu Zhou
MIT
26-452
77 Massachusetts Avenue
Cambridge, MA 02139
zzhou@mitlns.mit.edu

Mr. Vitaliy Ziskin
MIT
26-547
77 Mass. Avenue
Cambridge, MA 02139
vziskin@mit.edu

Dr. Abbi Zolfaghari
MIT-Bates Laboratory
21 Manning Road
Middleton, MA 01949
abbi@mit.edu

Dr. Fabian Zomer
LAL
Universite Paris Sud
BP34
91898 Orsay Cedex
France
zomer@lal.in2p3.fr

Dr. Townsend Zwart
MIT-Bates Laboratory
21 Manning Avenue
Middleton, MA 01949
zwart@aesir.mit.edu

Author Index

A

Anselmino, M., 255

B

Barber, D. P., 338, 350
Barone, V., 286
Ben-Zvi, I., 204
Bianchi, N., 260
Bianconi, A., 240
Blinov, B. B., 355
Boer, D., 240
Boffi, S., 240
Boglione, M., 255
Booth, E., 343
Burkardt, M., 199

C

Cameron, J. M., 1
Casagrande, F., 343

D

De Sanctis, E., 260
Dow, K., 343

E

Ent, R., 182

F

Farkhondeh, M., 343
Frankfurt, L. L., 307
Franklin, W., 343

G

Garvey, G. T., 104

H

Hasell, D., 187
Hoffstätter, G. H., 350
Holt, R. J., 234

I

Ihloff, E., 343

J

Jacobs, K., 343
Jaffe, R. L., 54
Jakob, R., 229, 240
Jalilian-Marian, J., 293
Jiang, X., 176

K

Kinney, E. R., 171
Koop, I. A., 319
Korotkov, V., 245
Kotzinian, A., 272
Kovchegov, Y. V., 297, 302
Krafft, G. A., 204

L

Lebedev, V. A., 142, 204
Londergan, J. T., 1
Lorestelev, M. S., 319

M

Matthews, J., 343
Melnitchouk, W., 267

Merminga, L., 204
Miller, A., 34
Milner, R., 187, 319, 343
Milner, R. G., 1
Mulders, P. J., 75
Murgia, F., 255

N

Nesterenko, I. N., 319
Nowak, W.-D., 260

O

Oganessyan, K. A., 260
Otboev, A. V., 319

P

Parkhomchuk, V. V., 319
Pascaud, C., 286
Peng, J.-C., 312
Perevedentsev, E. A., 319
Pichowsky, M. A., 282

R

Radici, M., 240
Reimer, P. E., 234
Reva, V. B., 319

S

Sahmovsky, V. G., 319
Sargsian, M. M., 307
Satilov, D. N., 319

Schäfer, A., 49
Shatunov, Y. M., 319
Skrinsky, A. N., 319
Smith, T., 343
Stein, E., 250
Stepanyan, S., 89
Stösslein, U., 171
Stratmann, M., 277
Strikman, M. I., 307

T

Takase, K., 187
Tschalaer, C., 319, 343
Tsentalovich, E., 343
Turchinetz, W., 343

V

Venugopalan, R., 121
Vogt, M., 350
von Harrach, D., 155

W

Wang, F., 343
Wilczek, F., 13

Z

Zomer, F., 286
Zwart, T., 343